THE FOREST SERVICE

Published in cooperation with the
College of Forestry, Wildlife and Range Sciences,
University of Idaho

SECOND EDITION
REVISED AND UPDATED

THE FOREST SERVICE

Michael Frome

Foreword by Carl Reidel
Afterword by R. Max Peterson

Westview Press / Boulder, Colorado

Westview Library of Federal Departments, Agencies, and Systems

Copyright © 1984 by Westview Press, Inc.

Published in 1984 in the United States of America by
Westview Press, Inc.
5500 Central Avenue
Boulder, Colorado 80301
Frederick A. Praeger, President and Publisher

Library of Congress Cataloging in Publication Data
Frome, Michael.
  The Forest Service.
  (Westview library of federal departments, agencies, and systems)
  Bibliography: p.
  Includes index.
  1. United States. Forest Service. I. Title. II. Series.
SD565.F7 1983   634.9′0973   83-10504
ISBN 0-86531-177-3 (hard)

Printed and bound in the United States of America

5   4   3   2   1

To the Memory of Four Friends

Guy M. Brandborg, of Hamilton, Montana, who sparked the battle over management of the Bitterroot National Forest, but who never lost the sense of loyalty and of belonging to his old outfit.

Clinton Leon Davis, of Unadilla, Georgia, a country boy with imagination and energy who became director of information of the Forest Service. He enriched the lives of people he touched, this author included.

John P. Saylor, of Johnstown, Pennsylvania, who rowed upstream in Congress for almost a quarter-century and left his mark on history, alongside Theodore Roosevelt and Gifford Pinchot, as Republican conservation trail blazers.

Dick Smith, of Santa Barbara, California, a newspaperman who made the backcountry of the Los Padres National Forest his special beat. He showed resource experts the beauty in yet a different kind of wilderness.

Each of these friends was an individualist out of his own mold. In common, however, they shared a belief in the American system and in their own capacity to make it work for the public good.

Because I believe their efforts—and inspiration—should not be forgotten, I dedicate this book to their memory.

CONTENTS

Foreword, Carl Reidel ix
Preface xiii
Acknowledgments xvii

1   A Resource Agency in the Hectic Eighties 1

2   From the Frontier to the Age of Scarcity 12

3   Organization Form and Function 32

4   Opening for Opportunity and Constituency 63

5   The National Forests: Purpose and Pressures 71

6   The Planning Process and Problems 87

7   Timber Cutting: Benefits and Costs 102

8   Rangeland Use and Abuse 123

9   Re-Creation in the Forests 139

10 Forest Wildlife: Conservation and Quantification 155

11 "To Secure an Enduring Resource of Wilderness" 174

12 Searching for Energy and Conserving It 198

13 Research and Putting It into Action 219

14 Cooperative State and Private Forestry 231

15 Integrating Pest Management 248

16 Global Horizons 264

17 Working with Other Agencies 269

18   Forestry Before Congress and the Courts 283

19   Forestry Education of Yesterday and Tomorrow 298

20   The Individual's Role and Goal in Public Forestry.......... 307

Afterword, R. Max Peterson.................................... 313
Notes 318
Glossary and Abbreviations 346
General Bibliography... 348
Index 349
About the Book and Author..................................... 365
Other Titles in This Series 367

Photograph section appears following p. 197.

Carl Reidel

FOREWORD

> The most interesting feature of this book is that it appeared after Mike Frome was fired from the staff of *American Forests* for his outspoken criticism of the Forest Service. The deep concern and devotion he displays for the Forest Service must have come as a complete surprise to many foresters who had expected this book to be Frome's revenge.
>
> —Arnold Bolle
> July 1974

Although many foresters may have been surprised when they read the first edition of *The Forest Service*, it only confirmed what those of us who knew Mike Frome well had always known. This hard-hitting critic of forestry cares deeply about the future of natural resource conservation in America, and especially about the men and women of our profession. I didn't always agree with the opinions expressed in Mike's *American Forests* column. But I can agree with George Olson, as quoted in the preface of this edition, in saying that "I feel we share a common bond in our interest regarding proper stewardship of the national forests."

This new edition of *The Forest Service* is a testimony to Mike Frome's skill as a creative writer and dedicated journalist. It reflects his years of thorough research and wide experience with the people and issues about which he writes. But unlike some historians and economists who have sought to discover the genius of the Forest Service, Frome brings a refreshing sense of intimacy to the story. Here is a lively and convincing

Carl Reidel is Daniel Clarke Sanders Professor of Environmental Studies, professor of forest policy, and director of the Environmental Program at the University of Vermont; past president of The American Forestry Association; and general editor for the University Press of New England series on the futures of New England. He is editor of *Prospects of New England* (University Press of New England, 1982).

saga that offers new insights into the spirit of the Forest Service by letting the people involved speak for themselves. It is Frome's extensive use of quotes from employees, friends, and critics, supported with solid facts and insightful anecdotes, that gives the story life. And perhaps most importantly, it reflects his own extensive personal experiences with people and places throughout the nation. Few, if any, have traveled more extensively in the national forests or talked with more people at all levels of the agency and throughout the forestry profession. Mike Frome has been there and his book proves it.

One cannot read this book without being amazed at the reach of the Forest Service into all aspects of natural resource conservation, worldwide as well as in the United States. Chapters on timber management, wilderness, recreation, wildlife, energy resources, water, cooperative programs, education, and research reveal the enormous scope of this premier conservation agency. Frome sketches the romantic early years of the agency, the dedication of its first leaders, and the agency's unending struggles with special interest groups seeking to shape government policies to their own benefit.

Throughout the story, Frome invokes the ghost of Gifford Pinchot, the agency's founder and philosophical leader. Writing at the former Pinchot estate in Pennsylvania, now the Pinchot Institute for Conservation Studies, Frome admits his admiration for the first Chief: "Pinchot . . . was a fighter, writer, leader, social crusader for industrial democracy, and an individualist. He saw forestry not as a technical end in itself, but as a democratic force."

Frome's admiration for Pinchot and his agency does not blind him to the agency's shortcomings, however. Throughout the book he gives Forest Service critics equal time, recounting the bitter infighting and failures that mark the history of the American conservation movement. He laments the influence that commodity interests, especially the timber and mining industries, have had in shaping agency programs. The chapter "Forestry Before Congress and the Courts" provides remarkable insights into the politics of conservation. He shatters the civics-class view of the legislative and legal system with a vivid portrayal of how industrial and environmental special interests vie for control of national conservation programs and dollars. Frome warns that even during the so-called "environmental" 1960s and 1970s, short-term exploitation was "confirmed repeatedly in the budgets allocated by Congress for the Forest Service," with timber sale administration almost fully funded, while reforestation, recreation, wildlife, and water programs were consistently underfunded. Worse, natural resource budgets continued to decline as a percentage of the total federal budget at the same time that planning and environmental impact analysis workloads increased dramatically.

His warnings about the power of special interests to influence Forest Service policies through litigation and legislation are remarkably timely in the face of Ronald Reagan's 1983 budget recommendations. As these lines are written, major changes in budget priorities and levels will, unless rejected by Congress, redirect Forest Service programs toward timber and mineral production and dramatically and seriously undermine long-range multiple-use planning efforts.

Frome's criticism of the Forest Service is clear but balanced throughout the book. His evenhandedness is especially evident in the chapter on wilderness resources, a topic about which he has strong personal feelings. Although he presents the emotional positions of those on both sides of the issue in the quotes cited, his presentation is fair and journalistically objective. Some may even find Frome overly supportive of the Forest Service, a critic turned advocate. If so, I expect it is because he has gained new insight into the nature of this agency and the people who make it run out on the national forests and ranger districts. If there is an underlying theme pervading this book it is that the Forest Service has an enormous capacity to change and adapt, even to admit its mistakes and learn from its errors.

I expect that Mike Frome would be the first to protest, even if he agreed with that conclusion, that the future will demand even more responsive leadership than the past. Another Pinchot? Probably not, as Frome admits in the last chapter. But if this book has a clear message for the Forest Service, it is that the key to the future lies in the best traditions of the agency's conservation leadership, not mere management or analytical methods. We must lead, not just administer.

The message of *The Forest Service* is not, however, only for those who direct the U.S. Forest Service; it is a message for foresters everywhere. Frome challenges us to examine our professional goals and the fundamental tenets of resource management and conservation. He asks us to evaluate the purposes and direction of our research, to restructure our educational program, and to rethink our traditional alliances and loyalties. He calls on us to cherish the best of our past and to have faith in our future. One can ask no more from a loyal critic than that.

On the several occasions that Mike Frome has been a guest lecturer at the University of Vermont, he has always challenged me and my students to recommit ourselves as advocates for conservation. He has not asked us to abandon either our scholarship or professionalism. To the contrary, he called on us to work harder as scholars, teachers, practitioners, and citizens. He never asked us to be a "Mike Frome." I am convinced that his concern and admiration for foresters is rooted in a realization that this profession is the best prepared, in tradition and education, to be the flag-bearer for conservation. We have, perhaps, put more emphasis in recent times on perfecting our science and management technique to the neglect of the arts of leadership and communications. In an era when the fundamental values of conservation are being challenged and decades of progress in natural resource management are being reversed, the environmental conservation movement clearly needs the professional leadership that foresters can and must provide.

Our place in the history of conservation lies in the future. Our past is only prologue to what yet must be accomplished.

PREFACE

> If there is any duty which more than another we owe to our children and our children's children to perform at once, it is to save the forests of this country, for they constitute the first and most important element in the conservation of natural resources of the country.
>
> —Theodore Roosevelt

What this book is about, or what I hope it deals with more extensively and effectively in this new and expanded edition than in the original of a decade ago, is people. The fate of our forests, after all, rests with people. Laws don't make things work, nor do rules and regulations; they have their place, but it takes the human hand, heart, and mind to give them meaning, whether for better or worse.

A lot of people have been helpful to me in the course of preparing this edition. The work was done in two installments. The first was a year spent in writing and revision at the Pinchot Institute for Conservation Studies, housed at the former Pinchot family estate at Milford, Pennsylvania. The second installment began at the University of Idaho in September 1982, following a summer of travel in national parks and national forests of the West. As a visiting associate professor, I was given access to the facilities of this university while working on research notes and structure. Through both phases, I felt close to the people of the Forest Service, of past, present and future.

It strikes me today that the agency as an institution is what its people, from top to bottom, make of it. More important, they are what they make of themselves. I'm concerned about the forester, or forest conservationist, as an individual, about his or her motivation and philosophy. Working within a large structure can be difficult and at times discouraging, but those who are rooted with sound outlook will persevere, I hope, without perishing.

Gifford Pinchot was the man who showed how. I derived considerable and continuing inspiration from him while at the Pinchot Institute. Pinchot, founder and first Chief of the Forest Service, ally and close comrade of Theodore Roosevelt, was a fighter, writer, leader, social crusader for industrial democracy, and an individualist. He saw forestry not as a technical end in itself, but as a democratic force. In his challenging book, *The Fight for Conservation*, he wrote as follows (p. 79): "The central thing for which Conservation stands is to make this country the best possible place to live in, both for us and for our descendants. It stands against the waste of the natural resources which cannot be renewed, such as coal and iron; it stands for the perpetuation of the resources which can be renewed, such as the food-producing soils and the forests; and most of all it stands for an equal opportunity for every American citizen to get his fair share of benefit from these resources, both now and hereafter."

And again, on a loftier plane (p. 88): "Seen from the point of view of human welfare and human progress, questions which begin as purely economic often end as moral issues. Conservation is a moral issue because it involves the rights and duties of our people—their rights to prosperity and happiness, and their duties to themselves, to their descendants, and to the whole future and progress of this Nation."

The conservationist's cause lies in the protection and perpetuation of life: the cause of life everywhere in all its forms, the interwoven life-environment of land, water, and sky, embracing material, tangible substances reducible to scientific formulae and economic assessment, yet also embracing intangible aesthetic, ethical, spiritual, and even cosmic values. I think of the pre-Confucian Chinese who described a society wherein "not only a man's family is his family but all men are his family and all the earth's children his children." I think, too, of the reverence for life implicit in the ancient Buddhist doctrine of rebirth. "The forest," according to Gautama Buddha, "is a peculiar organism of unlimited kindness and benevolence that makes no demands for its sustenance and extends generously the products of its life activity, it affords protection to all beings, offering shade even to the axeman who destroys it."

Henry David Thoreau expressed a somewhat similar thought. "I would," he wrote, "that our farmers when they cut down a forest felt some of that awe which the old Romans did when they came to thin, or let in the light to a consecrated grove, that is, would believe that it is sacred to some god."

A forester of my acquaintance told me of an experience on his own property. He had examined an old dead tree, found it unsightly and useless, and prepared to remove it. His young son pleaded with him not to. They walked together for a close look and found it rather like an ecosystem of its own, with insects, birds, and mammals scattered over the limbs, inside the trunk and among the leaves on the ground.

And the forester changed his mind, deciding the tree wasn't so unsightly and useless after all.

All that is green and trees does not quite constitute the forest. Forests are not random collections of animal and plant species. All elements of the life community must be present for a plant or animal to be something other than what is in a zoo. This holistic approach equates humanism with ecology and practical management.

Such ideas are wholly compatible with those of Pinchot and are complementary to them. "Forestry made good its position in the United States before the conservation movement was born," he wrote. He believed one led directly to the other. The principles of conservation and those governing the Forest Service were indivisible. Pinchot preached conservative use, restraint, respect for natural laws as the essence of "national efficiency." His warnings against waste and overconsumption are as valid today as when he wrote them early in this century. Simply stated, "A nation whose natural resources are destroyed must inevitably pay the penalty of poverty, degradation, and decay" (*The Fight for Conservation*, p. 123). Thus conservation begins with biological systems and their protection for the long run, rather than with extraction and exploitation. The practice of conservation is right only when it preserves the integrity, stability, and beauty of the community, wrote Aldo Leopold in *A Sand County Almanac*. "Conservation is a state of harmony between men and the land" (p. 243).

Leopold achieved lasting distinction in his own right, yet he was a product of Pinchot's Forest Service. Pinchot stimulated organizational esprit and individual imagination, a get-up-and-go attitude hard to match in government during any period in history. In its early years the Forest Service fought on many fronts for the protection of natural resources. It was born in the West, where it pioneered in bringing order out of chaos on the public domain, in fighting fires, establishing wilderness as a national concept, then extending the National Forest System to the East, spreading the gospel of forestry, initiating forestry research and education, serving through individuals like Leopold as a foundation for wildlife management, and, later, working with the young in the Civilian Conservation Corps and other relief programs of the Depression years.

As it budded and blossomed, the Forest Service, unusual in government for its fearlessness, stimulated a sense of belonging and spirit of loyalty—derived essentially from Pinchot's crusade, conducted in concert with President Theodore Roosevelt—to protect natural resources from being devoured by powerful economic interests. "It is the honorable distinction of the Forest Service that it has been more constantly, more violently, and more bitterly attacked by the representatives of the special interests than any other Government bureau," wrote Pinchot. "These attacks have increased in violence and bitterness just in proportion as the Service has offered effective opposition to predatory wealth" (*The Fight for Conservation*, p. 116).

For many years the agency adhered to Pinchot's principle of "preservation through use," with use based on respect for natural systems. Whether it continues to do so today is subject to some question. Pressures of all kinds are real. Beyond that, however, as times change, organizations and institutions may or may not keep pace with them. In government bureaus across the board, thirst for battle and martyrdom is rare. Technological training and expertise in all professions tend to become a shelter.

The personality of the chief, or director, of any public agency certainly influences attitudes and actions in the ranks, public support, and political effectiveness. Pinchot was an archetype. J. Edgar Hoover was impregnable and untouchable during his years in command of the Federal Bureau of Investigation. Appropriate leadership style and function are discussed later. Suffice it to say for the present, survivorship of a director and his agency is a continuing challenge, an interplay of program, personality, and politics.

I hope that I may present a fair picture of both the strengths and weaknesses of the Forest Service, the controversies surrounding it, and the challenges of the 1980s. I have my own biases, of course. Through the years I have shared diverse experiences with personnel at all levels, from Washington to wilderness; we have at one time been the closest of collaborators and at another, critics of each other.

"While you and I differ in our opinions on some matters," wrote George Olson, supervisor of national forests in North Carolina, in a letter I received in July 1981, "I feel we share a common bond in our interest regarding proper stewardship of the national forests. You are embarked on an important task. I will be pleased to offer any help that I can."

Such warm encouragement and cooperation have been mine from virtually everyone in the Forest Service.

Michael Frome

ACKNOWLEDGMENTS

This book represents about 25 years of intimate experience with the Forest Service and its diverse activities. In 1959, after I had already been writing for a time about national parks and national forests, Clinton Leon (or Clint) Davis, Director of Information and Education of the Forest Service in Washington, invited me to join him on a pack trip in the West. We traveled with the Trail Riders of the Wilderness, a group sponsored by the American Forestry Association, in the Bridger Wilderness of western Wyoming.

That experience inspired me to explore the National Forest System. During the next two years I traveled in every section of the country except Alaska and reported my findings in *Whose Woods These Are: The Story of the National Forests*, which Doubleday published in 1962. James B. Craig, editor of *American Forests*, who had been with us in the Bridger Wilderness, invited me to contribute a monthly column which I continued until 1971. In that same period Orville L. Freeman, Secretary of Agriculture under Presidents Kennedy and Johnson, asked me to collaborate with him on a book, *The National Forests of America*, which Putnam published in 1968. I note herewith that I have never encountered a public official more considerate or appreciative than Secretary Freeman.

The point of this brief retrospection is to indicate the scope of my continuing and lasting absorption with forests, public and private, and with foresters. I have been the guest of Weyerhaeuser in the Pacific Northwest, Blandin in the Lake states, International Paper in Maine, and Champion (now Champion International) in the southern Appalachians. I have participated in forestry, recreation, and wildlife programs at various universities in this country. On one trip to Europe I visited the École Nationale Forestière, the school Pinchot attended, at Nancy, France. I have toured and consulted with officials in city, county, and state forests and in forests of various dimensions in other countries of the world.

In the course of my travels and adventures, many individuals at all levels of the Forest Service (and at most levels of society) have been of help to me. Many are identified in the pages of the book or in notes following the main text.

I am deeply indebted to Chief R. Max Peterson for his personal interest and openness with me. "You will be doing both the public and us a long-time service if you write as objectively as you can, which I am sure you plan to do," Chief Peterson wrote to me in October 1980. The Chief deserves the last word, which I have invited him to express after reviewing my manuscript. Particular appreciation is due to J. Lamar Beasley, Deputy Chief, and George Castillo, of the public affairs office, who have helped me to be factual and fully informed, without ever once trying to influence my editorial judgment.

Others to whom I owe particular appreciation for aid in preparation of this book include the following. In the Washington office: Robert E. Buckman, Roy Feuchter, Norman E. Gould, Bruce Hronek, E. Delmar Jaquish, Dale A. Jones, David E. Ketcham, Robert M. Lake, Allen J. Schacht, Wallace T. Shiverdecker, and Jetie B. Wilde, Jr. Among the regional foresters: Charles T. Coston, Craig W. Rupp, Zane G. Smith, Jr., Deputy Regional Forester Everett L. Towle, and also Ray Karr, Information Chief of Region One. At the Pinchot Institute: Thomas G. Ellis, Curt Johnson, Allan R. Knox, and Ralph Segman. Among people in the field: Ralph H. Cisco, George D. Davis, H. Reid Jackson, Tom Kovalicky, John J. Lavin, John Marker, George A. Olson, Gray F. Reynolds, Robert Tyrrel, and Lynn Mitchell.

Earl Sandvig, a Forest Service alumnus who has never lost his commitment to resource protection, provided special help on range and a variety of other issues. Other alumni who gave valuable counsel and data include William W. Huber, Dr. Leon Minckler, Nolan O'Neal, Hank Rate, and Dr. Carl Reidel, as well as Dr. M. Rupert Cutler, former Assistant Secretary of Agriculture. And in the private sector: Thomas J. Barlow, Edward C. Fritz, Jean Hocker, Philip Hocker, Peter Kirby, Robert Maynard, Clifton R. Merritt, and Gordon Robinson.

I owe a special note of recognition to Dr. John Gray, Director of the Pinchot Institute, and his associates, who made my stay at Grey Towers a memorable experience. Dr. Gray read the script in three or four sections as I was writing it and provided extremely constructive criticism and ideas, based on his own extensive experience. President John F. Kennedy, in dedicating the site in September 1963, after it had been conveyed to the Forest Service by the Pinchot family, declared it part of "a journey to preserve the past and protect the future." And so I hope this special place (where Pinchot headquartered many conservation battles and wrote his autobiography, *Breaking New Ground*) will always be.

At the University of Idaho I received significant counsel and encouragement from Dean John W. Ehrenreich of the College of Forestry,

Wildlife and Range Sciences; Associate Dean James R. Fazio, head of the Department of Wildland Recreation Management; Dr. William McLaughlin, forest planner and adviser *extraordinaire*, and Dr. George Savage, editor. I also received assistance from Dr. Edwin Krumpe, Dr. Sam Ham, and a special cadre of intimates, the "gourmets of the Palouse." I owe a special word of gratitude for research assistance to Lori Kuykendall, an undergraduate student majoring in wildland recreation management at the University of Idaho. She, in turn, was helped in special instances at the University Library by Dennis Baird, Winifred Dixon, Gail Eckwright, and Donna K. Smith. Susan Wagner and Carmen Savage are thanked for their typing skills.

Dr. Ernest Griffith, distinguished consulting editor of this series, provided invaluable guidance (as he did for the original edition), subsequently supplemented by some members of the editorial staff of Westview Press, notably Kathleen Jones, Alice Levine, Barbara Ellington, Jeanne Remington, and Beverly Plank, all thoroughly constructive and patient.

Finally, I would also like to acknowledge with appreciation and pride the research grant provided to me by colleagues in journalism through the medium of the National Press Foundation as part of its program to advance and enrich the profession.

Michael Frome

1

A RESOURCE AGENCY
IN THE HECTIC EIGHTIES

> The Forest Service . . . has both a responsibility and an opportunity to be a leader in assuring that the Nation maintains a natural resource conservation posture that will meet the requirements of our people in perpetuity.
> —National Forest Management Act, 1976

The Forest Service is more than a federal agency. It is an institution of American life. Consider that in large portions of the West the histories of community development and forest protection are intertwined—forest ranger stations were often among the earliest, if not the first, buildings erected in many towns and counties.

The heritage of this organization derives from Gifford Pinchot's creed that no generation can be allowed needlessly to damage or reduce the future general wealth and welfare by the way it uses or misuses any natural resource. Conservation by his definition means the wise use of the earth for the lasting good of men.

In the national forests alone, this principle has safeguarded—without question until recent years at least—many millions of acres. Through protection of high-elevation watersheds, life has been made more livable and feasible in scores of downstream communities. In addition, the national forests are open to all, with a minimum of restraint in keeping with conservation of the resource.

The lands managed directly by the Forest Service are scattered across the continent, spanning all kinds of climate and topography, from the icy glaciers of southeast Alaska to the rain forests of Puerto Rico. Yet they represent only one phase of the agency's activity. Forests cover about 30 percent of the total land area of the United States, and Forest Service responsibility and authority reaches it all.

The agency's mission embraces the last vestiges (and potential for rebirth) of America's urban forests. It extends beyond the nation's border to the world's tropical forests, in which soils, terrain, temperature, patterns of rainfall, and distribution of nutrients are in precarious balance as a consequence of massive deforestation. In addition, the research arm of the agency conducts studies into diverse aspects of forestry, rangeland management, and utilization of forest resources.

Though forests serve many purposes, the area of forestland in the United States has decreased steadily and continues to decrease. The very process of loss accentuates its value. Forests, no matter how small, serve the nation and the community. They provide a haven for birds, fish, and mammals, a steady and stable streamflow to the cities (with protection from floods), a relief from urbanism, as well as many wood-based products. But millions of rural acres are being lost, converted to highways, shopping centers, subdivisions, and cities. Or they are converted to cropland, "plowed up and planted to the fences," then treated with fertilizers and chemicals.

While forests are shrinking, the demands upon them are unrestrainedly rising, demands for industrial wood products, fuelwood (particularly since the onset of the petroleum-based energy shortage and price increase in 1973), waste disposal, wildlife, recreation, range, water, oil and gas, and mineral resources. Because it is impossible for the nation's forests to meet these demands, the Forest Service has been thrust into a cockpit of conflict and controversy.

Conflict: Finite Resources Versus Infinite Growth

Forests are called "renewable resources," but forest soils are slowly renewable at best. When soils are lost faster than formed, future productivity is jeopardized. John R. McGuire, former Chief of the Forest Service, dealt with one aspect of this question in the course of a Horace M. Albright Lecture at the University of California at Berkeley in February 1981, but his views can be equated with the broadest possible concern.

> Those on the side of an increase in the pace of timber harvesting [on the national forests] argue that the current rate produces too low a return on the timber investment and delays the replacement of slow growing, overmature trees by a faster growing, more juvenile inventory. They also point to benefits from stabilizing employment and dampening the rising trend of stumpage prices with larger federal timber sale offerings. They typically urge substitution of an economic model for the biological model on which the concept of sustained yield policy is based.[1]

Consideration of soil and water is vital in the biological model, though easily overlooked in the economic model. Soil erosion was dramatically thrust into national conscience in the 1930s, when dust storms threatened to scatter the nation's agricultural wealth to the winds. But loss of soil may be the most serious threat to civilization in the 1980s. There is no substitute for soil, as there is for oil.

National forests were established at least in part to protect soils and to maintain favorable streamflow conditions. The groundwater resource is a vast strategic reserve, like the nation's coal deposits. Forest ecosystems have a great capacity for absorbing rainfall, releasing stored water gradually through streams and springs. And forests have the ability to remove from groundwater supplies such pollutants as acids, heavy metals, and radioactive substances—thus yielding higher quality water to springs and streams for ultimate human use and consumption.

As William F. Hyde points out in *Timber Supply, Land Allocation, and Economic Efficiency:* "Not only does the Forest Service manage a large number of acres with the largest single inventory of old growth, but many of these acres are of low potential timber productivity and located at higher elevations; that is, they are the marginal timberlands that should be carefully scrutinized for allocation among alternative uses."[2]

This resource professional viewpoint notwithstanding, the nation's forests and the Forest Service are seen in different lights by different groups. Economists deplore what they consider a huge inventory of standing timber being allowed to rot or burn on public lands before it can become lumber, plywood, and newsprint, and, equally so, the nonproductivity, in commercial timber terms, of millions of acres held privately in small holdings. "The Neglected Promise of Our Forests" is the title of an article that appeared in *Fortune* magazine in 1979, asserting that forests are producing only one-fourth of their potential. The author, Lee Smith, writes: "And if all available land were planted and nurtured intensively with fertilizers and insecticides, some timber economists calculate that the forests could yield us an astounding 60 billion cubic feet annually [as compared with the current rate of 14 billion cubic feet]—or more—without harming the future yield of the forests."[3]

Production must be boosted in this manner, according to the author, for two main reasons. First, the limitation on supply has sent prices skyrocketing, outrunning even the general inflation rate. And second, the U.S. trade deficit in forest products tripled in the 1970s, though the country could be a major timber exporter. The key to the future, he concludes, is to increase our harvest of timber everywhere. "The yield of our underutilized national forests could probably be doubled if Congress and penny-wise budget makers would give the Forest Service enough money and manpower to match private industry's pacesetting performance in intensive cultivation."[4]

Money and manpower doubtless serve certain objectives, but whether they can squeeze more out of the "biological model" referred to by Mr. McGuire over a long period is questionable. Finite resources cannot support an infinite population and infinite economic growth. There is an ultimate biological limit. Resource scarcity is a permanent reality, to which competing influences and society in general must accommodate their demands.

Improved utilization is one approach. Reports from western Oregon and Washington show that residue volumes following logging on national forests are two and one-half times greater than on industrial forestland, and that about 60 percent of this residue is in relatively large logs that could be removed at reasonable cost with existing equipment. On the other hand, old ideas are being challenged by the necessities of the 1980s. Should, for example, ever increasing quantities of forest products be produced for an ever increasing population with an insatiable appetite for tissues, paper bags, and newsprint? Or is it better to advocate a reduction in consumption in order to save the remaining natural forests for their noncommodity amenities? The indication that we could eliminate about 80 percent of the country's trash disposal problems by recycling paper and paper packages points to a course of action that is not yet being pursued.

Criticism: Do Foresters See More than Timber?

Foresters and Forest Service personnel are not of one mind. Change comes slowly in a large organization that deals with formulas and directives, and foresters have criticized themselves for "not seeing the forest for the timber"; yet Aldo Leopold once classified his fellow professionals in two distinct groups:

Group A is quite content to grow trees like cabbages, with cellulose as the basic forest commodity. It feels no inhibition regarding violent manipulation of nature, since its ideology is agronomic.

Group B sees forestry as fundamentally different from agronomy. This group employs natural species and manages a natural environment rather than creating an artificial one. It worries on biotic as well as economic grounds. It worries about diverse forest functions, including wildlife, recreation, watershed, and wilderness.

Both groups are represented in the Forest Service. It is not a monolithic institution, as it may have appeared to be in times past, especially from the mid-1960s through most of the 1970s. That was the era when the policy and practice of clearcutting prevailed, about which more will be written in due course. Foresters inside and outside the agency examined this system and themselves, identifying weaknesses, to effect a change in direction.

William E. Towell, then executive vice-president of the American Forestry Association, declared at a 1967 Symposium on "Undergraduate Forestry Education" at Roanoke, Virginia: "Some foresters have failed to give proper consideration to their environmental responsibilities. They have clearcut steep slopes opening the way to floods and erosion. They have wiped out fish and game populations in whole watersheds in attempts to control forest pests by spraying. They have destroyed roadside vistas and the beauty of the landscape through careless and thoughtless harvest methods. Most foresters seem to be particularly weak in aesthetic appreciation."

A 1970 memorandum dispatched by Neil Rahm, Regional Forester of the Northern Region, headquartered at Missoula, Montana, declared: "Some of our own people are feeling and expressing doubts. The doubts are whether we can perform as well as we tell people we can. Do we have the expertise on the ground to perform an acceptable job? I think the answer is no. Our measuring stick for the quality of timber sale impact is whether or not a national forest has met the Sell and Cut goals."

Edward P. Cliff, then Chief Forester, acknowledged this problem in a memo to all Forest Service employees dated September 18, 1970: "Many employees have recently expressed concern on the direction in which the Forest Service seems to be heading. I share this concern. Our programs are out of balance to meet public needs for the environmental 1970s and we are receiving mounting criticism from all sides."

This criticism led Senator Lee Metcalf of Montana to invite investigation by a committee of professors at the University of Montana of clearcutting on the Bitterroot National Forest. The committee, headed by Dr. Arnold W. Bolle, Dean of Forestry, issued its report to the U.S. Senate Committee on Interior and Insular Affairs in 1970. The report, A University View of the Forest Service, charged: "Such cutting practices abuse the multiple use principle. And they make a mockery of the sustained yield concept which decrees that all resources—but particularly the key ones, soil and water—must be sustained. Above that, many consider it foolish economics. The short-term gains are offset by long-term losses in both economics and environmental quality."

This criticism, or recognition, intensified in 1972. In April of that year, Justice William O. Douglas declared in his dissent in the Mineral King case (Sierra Club v. Morton), involving a proposal for a large-scale ski development in Sequoia National Forest, California: "The Forest Service—one of the federal agencies behind the scheme to despoil Mineral King—has been notorious for its alignment with lumber companies, although its mandate from Congress directs it to consider the various aspects of multiple use in its supervision of the national forests."

Edward C. Crafts, a former deputy chief of the Forest Service (who became first director of the Bureau of Outdoor Recreation), issued a stern warning to the annual convention of the Society of American Foresters at Hot Springs, Arkansas, in October 1972: "Do not deceive yourself that environmental concern is a passing fancy. It is here to stay. Further, I say in all seriousness that so far we foresters have been flunking the course."

John McGuire, who succeeded Mr. Cliff as Chief, conceded in an interview published in American Forests in October 1972, that "roads have been cut where they shouldn't have been permitted. Erosions have followed that make it impossible to get a forest of quality or even any forest, in that area again."

Thus criticism and self-examination continued through the 1970s. In May 1976 Craig Rupp, Regional Forester of the Rocky Mountain Region, headquartered at Denver, stated: "Nationally, I feel the Forest Service is still looked upon as a timber management agency, and this in itself causes us problems. I see indications at all levels that we are still unconsciously selling this approach. This is no longer applicable or acceptable in the Rocky Mountain Region. Here we feel our key resource values are wildlife, recreation, and scenery, and that timber management is justified only to the extent that it can help maintain or enhance these key resource values."

Some opinion leaders among the public have been critical of the entire "forestry establishment." Huey Johnson, California's Secretary of Natural Resources, for one, declared at an Urban Forestry Symposium at Berkeley in May 1978: "It is unfortunate that the leadership and social and political awareness displayed by Pinchot and other early leaders in American forestry have not remained characteristics of the profession of forestry. Today, forestry, as a profession and as an industry, has declined in stature and retreated from leadership in dealing with social issues as well as management of natural systems."

This viewpoint, as expressed by Huey Johnson, may be valid, in part or in whole; yet the Forest Service is not as monolithic as it may appear, and neither is forestry. In May 1974, at the height of the clearcutting controversy, L. Keville Larson, a consulting forester of Mobile, Alabama, told the University of Alabama Conference on Land-Use Planning and Management Alternatives for Forest Resource Managers:

> The promotion of plantation management over the past several years has been so intense as to leave the impression that managing existing stands and relying on natural regeneration is inferior, no matter who the owner is or what his objective. As a simple method of mass production of pulpwood, plantation management may be the best available. But to meet the varying ownership objectives of millions of private non-industrial timberland owners, the range of forest products companies and various government bodies, there is no panacea.
>
> To those who have always believed this, it was frightening a few years ago when most of the industry, most of the forestry schools and the United States Forest Service adopted, or appeared to have adopted, clearcutting and planting as the recommended system for southern pine. Hopefully, the pendulum is swinging. The Forest Service has changed its policy and now considers the range of management methods. Through the Southern Forest Institute, lumber and diversified forest products interests have given credit to other methods, and perhaps the forestry schools are giving more attention to natural stand management. Environmentalists are part of the reason for change, but so too are those foresters and owners who have continued to reap the rewards of managing natural stands.

Changes in Command and Directions

In his memorandum to the field of September 18, 1970, Chief Edward Cliff insisted that "our direction must be and is being changed." Rapid change, however, is difficult to thrust on personnel accustomed to gradual movement. More than a decade later the agency is caught between generations: one trained in commodity production, the other in amenity values, and the transition isn't easy for either. The advent of resource specialists—biologists, hydrologists, soil scientists, landscape architects, recreation planners, and archaeologists—implies that foresters no longer have all the answers and tends to threaten their decision-making confidence. On the other hand, the nontimber professionals are mostly young and sometimes frustrated.

Change during the 1960s and 1970s was dictated to a large extent by laws—the 1969 National Environmental Policy Act, 1974 Resources Planning Act (the Forest and Rangeland Renewable Resources Planning Act, or RPA), 1976 National Forest Management Act (NFMA), and others—in some ways adding to the frustration through requirements of paperwork, but also adding to the challenge and opportunity. The latter two laws define a planning process involving forests and rangelands of all ownerships, private and state-administered as well as federal lands administered by the Forest Service and other agencies.

By 1980 everything was in place for the Forest Service to draft careful solutions to management problems for individual national forests. President Carter, in the midst of his uphill campaign for reelection, had instructed the Secretary of Agriculture in a memorandum of June 12, 1979, "to use maximum speed in updating land management plans on selected national forests through departure from the current nondeclining evenflow policy." It was clearly an effort to win political support through increased logging on the national forests—exactly what forest industries have persistently demanded. In the 1980 Report to Congress, the Forest Service identified the sixteen national forests with potential for departures (or an increase in cut), noting that the action must be based on the congressionally mandated planning process. "In response, the Forest Service has reordered its priorities for completing individual National Forest plans so as to concentrate planning efforts on additional National Forests which have a large inventory of old-growth softwood suitable for home construction."[5]

The 1980 election changed many things, including Forest Service prospects. The Reagan administration enunciated goals to balance the budget, control inflation, reduce unnecessary regulation, and lower interest rates. The Forest Service subsequently was advised that it must show increased receipts from the national forests by selling more timber, reducing road costs, and accelerating leasing of mineral, oil, and gas resources. In March 1981, Vice President Bush directed the agency to review a long list of environmental regulations established under the NFMA (including protection of streamsides, limitation on size of clearcuts, and mandating of plans for individual forests and regions by 1985). The Vice President, acting on behalf of President Reagan, said the review would help "to streamline and speed up the process."

Inside the Department of Agriculture, the new assistant secretary with direct authority over the Forest Service, John Crowell, said his highest priority would be "to increase the harvest levels on the national forests." It was a considerable reversal of the policy pursued by his predecessor, Dr. M. Rupert Cutler, a wildlife biologist and well-known environmentalist (who became senior vice-president of the National Audubon Society after leaving government). In a speech at the University of Michigan in April 1978, Cutler had declared:

> Our scientific training helps us approach decisions objectively. But, because these decisions are largely social or economic in their implications, they often transcend the narrow technical content of our profession.
>
> Resource managers are being called upon to make decisions which affect people—decisions that have definite social impacts. We must recognize this, involve the public in our decision-making processes, be truly open-minded and responsive and seek the breadth of vision these decisions demand.
>
> We have learned, though, that decisions made solely on the basis of short-term economic benefits are sometimes very costly in their ecological consequences.

Mr. Crowell brought a different viewpoint from a different background. He formerly had been general counsel of the Louisiana Pacific Company, a major purchaser of timber from the national forests (whose Alaska subsidiary depends entirely on the Tongass National Forest for its supplies). He was very actively involved in industry efforts to increase cutting. Shortly after taking office he appointed as his deputy Douglas MacCleery, who had formerly served on the staff of the National Forest Products Association.

In a message to the Forest Service on August 18, 1981, Mr. Crowell declared the Administration's design for the Forest Service: to hold down costs while increasing receipts to the Treasury from timber sales on national forests. He suggested that national forests have the capacity to yield 30 to 36 billion board feet of timber annually; he called this a theoretical figure based on biological capacity, and proposed increasing the harvest to between 20 and 24 billion board feet, which he considered practical and attainable under multiple-use constraints. The current Forest Service plan, developed through the 1980 RPA process, called for offering for sale 16.4 billion board feet by the year 2030 (an increase from a level of 11.9 billion board feet in 1982).

During the early years of the 1980s the timber market was badly off, due in large measure to the effect of high interest rates on housing construction. Nevertheless, Crowell insisted that national forests prepare to meet demands of an oncoming boom. He focused attention on the Pacific Northwest, stronghold of the last great virgin forests. In a speech before the Western Forestry Center, at Portland on January 19, 1982, Crowell said the national forests had not been playing their rightful role:

> By regenerating nonstocked areas, by harvesting and regenerating mature stands, and by applying intensive management measures such as spacing control and use of genetically improved trees, we can greatly increase current growth and can offer the prospect of continued high harvest levels for the indefinite future.

This was absolutely necessary, he said, since "privately and industrially owned lands were being harvested at more rapid rates than could be sustained indefinitely from private lands alone." But to call it overcutting would be a simplification, since industrial owners actually were bringing old-growth timber "under good forest management for the benefit of future generations." Moreover, Crowell contended:

> Between 1952 and 1977 the three states [Oregon, Washington, and Idaho] lost 3.3 million acres of commercial timberland, with the major portion of it set aside in parks or wilderness areas. In addition, reclassification of land, from primary timber production to less intensive utilization for timber as a tradeoff for scenic and environmental values, has further restricted output from public lands.

Crowell pledged recognition of wildlife and other "amenity" values. Yet the Forest Service budget for 1983, as outlined by the Administration for consideration by Congress in 1982, placed heavy stress on timber and mineral development, both commodity values. And in the name of simplifying things, efforts were made to restructure Forest Service regulations implementing the National Forest Management Act. The regulations had been adopted following lengthy review and open hearings, during which Crowell and MacCleery led the industry opposition. Under President Reagan, they, and the industry, plainly were determined to have their inning.

These developments are detailed in subsequent chapters. Here they illustrate the surge of conflict and challenge bearing down on public administrators and land managers. The Reagan administration instituted still further policies that would seriously affect the Forest Service and its programs:

- On February 25, 1982, the President issued an executive order establishing a Property Review Board to dispose of unneeded federal land. The order directed that each land-management agency establish an annual target amount of real property holdings to be designated as excess and available for disposition. While there is no legal authority for large-scale disposal (except for limited sites, mostly beyond national forest boundaries), the Forest Service was assigned the unhappy task of identifying lands that could be sold and the legislative changes necessary in order to raise, in concert with the Bureau of Land Management, annual revenues of somewhere between $500 million and $2 billion, starting in fiscal 1984. Further references to privatization, or "asset management," are found in Chapter 5.

- James G. Watt, Reagan's appointee as Secretary of the Interior, demonstrated early and continuing determination to open wilderness areas in the national forests for commodity development. The Interior Secretary has long held jurisdiction over mineral resources on public lands of the West. Watt clearly became the Administration spokesman on resource issues and embarked on programs to appease and please the "Sagebrush Rebels," a commodity-oriented western power bloc.

- The 1980 election changed prospects in Congress as well as in the executive branch. In the session following the election, California Senator S. I. Hayakawa, chairman of the Forestry Subcommittee of the Committee on Agriculture, introduced legislation—supported by Senator Jesse Helms of North Carolina, chairman of the full committee, and by the Administration—setting rigid deadlines for consideration of wilderness designation on the national forests. Between 1971 and 1973 the Forest Service had conducted the first Roadless Area Review and Evaluation (RARE I) to determine suitability, or nonsuitability, of roadless and undeveloped areas that had not been considered under the Wilderness Act of 1964. In 1977 another nationwide review, RARE II, was instituted; two years later 15 million acres were recommended for classification as wilderness, 11 million acres for "further planning," and 36 million acres for nonwilderness use that became subject to development. The drive to accelerate "release" of contested roadless areas from the planning process represents another expression of the conflict between economics on one hand versus "amenities" on the other hand.

In summary, it is impossible to review the role of the Forest Service without relating in one way or another to critical issues confronting America in the 1980s. These include chemical contamination of the environment; the fate of rare and endangered species of wildlife; material wants for wood, water, and energy sources that relate to a productive economy and expanded employment opportunity; and soil stability and productivity over the long run.

The forest ranger is no longer a backwoodsman, nor even a small-town person. Forests have been engulfed by patterns and problems of urban America. National forests (the Los Padres, Angeles, San Bernardino and Cleveland) compose the backyard of Southern California, the last large greenbelt to protect its cities from soil erosion, flood, and mud slides. To one extent or another the urban interface is pressing against the forest in virtually every section of the country. Today's way of making things work must be different from yesterday's, and tomorrow's even different from today's. Whether the Forest Service, through its leadership and field personnel, is fully equipped to shift gears in order to keep pace, or to get ahead of the race, represents a considerable question. The answer lies not only inside the agency but outside of it as well. In our ever more complicated structure of life, the nation's forests are as essential to social welfare and progress as the armed forces are to defense. Somewhere in recognition of these factors rests the key to the eighties and beyond.

2

FROM THE FRONTIER TO THE AGE OF SCARCITY

> In the administration of the forest reserves it must be clearly borne in mind that all land is to be devoted to its most productive use for the permanent good of the whole people and not for the temporary benefit of individuals or companies.
>
> —James Wilson, Secretary of Agriculture, in a letter to Gifford Pinchot, Chief Forester, February 1, 1905 (the date on which the Transfer Act was approved by President Theodore Roosevelt).

Gifford Pinchot, the pioneering first Chief of the Forest Service, was a key figure in the movement of social reformers and trust busters—along with Jane Addams, Lincoln Steffens, Ida Tarbell, and Theodore Roosevelt. At one period he had given serious thought to a career in social work. Forestry, his chosen profession, he conceived as a means of blocking big business from steamrolling the government and the country. He and his collaborators, notably Henry Solon Graves, Overton Price, Herbert A. Smith, and Raphael Zon, were as much missionaries as technicians. They were part of the thread of history, while raising forestry and conservation from an experiment to a national movement and public issue.

It was logical that Pinchot should be a close friend of Theodore Roosevelt. By the time Roosevelt became president they already had long been active in the Boone and Crockett Club, the elite organization of big-game hunters that played a leading role in wildlife protection. The two men's common interests were reflected in TR's first State of the Union Message, when he identified "the forest and water problems as perhaps the most vital internal questions of the United States."[1]

12

How Public Concern Led to Federal Forestry

The sequence of history traces from this declaration nearly a half-century back to the post–Civil War days. The railroads were rewarded for their efforts in opening the western empire with large gifts of land from the public domain. Cattle syndicates, formed with eastern and European money, followed in their wake, disgorging hundreds of thousands of livestock to devour and destroy the grassland wilderness. Millions of acres designed to aid settlers passed into the hands of lumber companies, mining outfits, and assorted speculators. Immense private principalities emerged.

Few voices were raised against the tide of land theft and resource exploitation. But one significant voice heard was that of Carl Schurz, Secretary of the Interior, who, as early as 1879, warned of an oncoming "timber famine." Ten years later, on October 15, 1889, in a significant address before the Pennsylvania and American Forestry Associations in Philadelphia, Schurz called for a reversal of public opinion "looking with indifference on this wanton, barbarous, disgraceful vandalism; a spendthrift people recklessly wasting its heritage; a Government careless of its future."[2]

As rich native forests fell under intense attack, concern over high-quality timber led several states to adopt tree-growing bounties and tax exemptions. In 1872, a nationwide Arbor Day program began in Nebraska to stimulate planting. The following year, Congress adopted the Timber Culture Act to encourage homesteaders to devote one-fourth of their new land to forest plantations. The desire for forest protection found expression among ethical, intellectual, and scientific leaders, including Ralph Waldo Emerson and Henry David Thoreau; Dr. Wolcott Gibbs, the chemist-physicist president of the American Association for the Advancement of Science; and Interior Secretary Schurz, who understood the value of forest management from observing it in his native Germany.

Organized public effort in behalf of forestry began in earnest in 1871, when the nation was shocked by the worst fire in its history, at Peshtigo, Wisconsin. Fifteen hundred persons lost their lives and nearly 1.3 million acres burned. Two years later, the American Association for the Advancement of Science (AAAS), already disturbed by the wave of fire and destruction, heard a paper delivered by Franklin B. Hough entitled "On the Duty of Government in the Protection of Forests." The Academy then urged Congress and the states to recognize the need of "cultivation of timber and preservation of forests and to recommend proper legislation for securing these objectives." This effort led to organization of the American Forestry Association in 1875 to advance the cause of forestry and timber culture. Courses were instituted at a number of agricultural colleges, mostly to provide instruction in the general study of trees and tree planting.

As a result of the upsurge of public concern, Congress in 1876 considered a bill providing for the "preservation of the forests of the national domain adjacent to the sources of the navigable waters and other streams of the United States." Though it failed to pass, it represents the first attempt to establish national forests and, significantly, it was based on the concept of their value for regulation of streamflow.

That year, however, another bill did pass, calling for a study of and a report on the forest condition and best means for preservation and renewal. Because trees were viewed more as a crop than a resource, the task of preparing the report was given to the Department of Agriculture. Such was the seed of forest administration in that branch of the national government.

The bill provided for appointment of a special agent, a "man of approved attainments," to conduct the investigation. Franklin Hough, the vigorous exponent of governmental action in forestry (who had served as a physician during the Civil War and then had become a tireless writer on nature, history, and statistics), was chosen by the Commissioner of Agriculture for this assignment. His report, delivered in 1878, is regarded as a major compilation supporting the beneficial effects of forest cover on climate, streamflow, and soil, and pointing out the negative effects of forest removal. Hough was succeeded in 1883 by Nathaniel H. Egleston, later termed by Gifford Pinchot as "one of those failures in life whom the spoils system is always catapulting into responsible positions."

In 1886, Bernhard E. Fernow took command of the newly upgraded Division of Forestry. Trained as a forester in his native Germany, Fernow had already worked as secretary of the American Forestry Association. During his twelve-year tenure in Washington, he did pioneering work in silviculture and wood utilization. After leaving the government, he extended his influence into education by organizing in 1898 the New York State College of Forestry at Cornell University. However, he became embroiled in serious controversy as a result of heavy logging of a 30,000-acre experimental state forest in the Adirondacks. The lumbering operations were halted through court action, Fernow was dismissed, and the state college of forestry was shifted from Cornell to Syracuse University. Nevertheless, Raphael Zon, a fellow immigrant who benefited from Fernow's tutelage, paid tribute to him as not just a teacher of forestry, but "a leader of life."

Looking Back to the Pioneers

Just as Pinchot and Roosevelt might view the post–Civil War period, Hough and his contemporaries could cast an eye on two centuries of forest ethics and use in the settlement and growth of America.

When the first colonists arrived, trees seemed to extend as an endless sea across the continent—about half the country we know today was covered with timber. Though serving many useful purposes, trees barred the way of farms, homes, cities. In New Hampshire, the Crown encouraged, even forced, destruction of virgin forests. Grantees were required to clear and cultivate a minimum acreage within five years and to pay a tax of a few ears of corn as evidence that agriculture was progressing to the point of sustaining increasing numbers of subjects loyal to the King. In colonial Maine, the best white pines were sent to England as masts; others were sawn into timber to build ships or to build forts as protection from the French and Indians. Timber was so plentiful that sawmill men discarded slabs thick enough to be considered first grade today. It made no difference, according to prevailing theory, how many trees were felled or burned, since there would always be more.

There were a few early restraints. The British government adopted a policy of reserving for the use of its navy a future supply of naval stores, tall pines for ship masts and pitch pines for tars. William Penn, in his 1681 ordinance for disposal of lands, required that for every five acres cleared, one acre must be left in forest. The Palatinate Germans (or Pennsylvania Dutch) are credited as the only settlers practicing economy in the use of wood, despite its apparent abundance.

Virtually all demonstrations of concern in the country's early years were based on fear of scarcity—principally on fear of not having live oak timber for warships. The Federal Timber Purchases Act of 1799 appropriated $200,000 for the purchase of two tracts of forest land on the Georgia coast. The Federal Timber Reservation Act of 1817 established in Florida the Santa Rosa live oak timber reserve, the first such reservation of public land. (Santa Rosa, a peninsula jutting into the Bay of Pensacola, in 1828 became the first forest experiment station, intended for work in transplanting and cultivating live oak.) The Timber Trespass Act of 1831, while relating specifically to live oak and red cedar, became the basis of the present law for prevention of timber trespass on government land, though the government was powerless to enforce it; moreover, the advent of iron ships ended concern with shipbuilding timbers. Plainly, as long as forests and timber seemed limitless, concern and caution had no place in the scheme of things.

Maine was the first chief lumber source, supplying markets of the Atlantic seaboard for two centuries, then by 1850 showing the effects of unrestrained exploitation. New York took the lead for a decade, followed by Pennsylvania, with its choice hardwoods and pines. Timber demands increased as the country grew and pushed westward. During the 1850s, prairie schooners and canalboats were made of wood and the new railroads were laid on wooden ties. Following the Civil War, new industries, new cities, new homes emerged, all clamoring for wood.

Thus logging came to the Lake states, starting in 1870 and lasting on a large scale for about thirty years. Men by the thousands were employed in logging camps and sawmills, while logs by the millions were swept up in the famous river drives. The Lower Peninsula of Michigan was harvested first, after which lumbermen moved into the wilderness of the Upper Peninsula and west into Wisconsin and Minnesota. The period was marked by wasteful exploitation followed by devastating fire. When the lumber boom town of Peshtigo was engulfed in October 1871, the tornado of flame not only burned every one of its buildings, but caused the rivers and streams to be choked with floating dead fish and the surface of Green Bay to be covered with birds that had burned in flight and fallen. One of the worst fires in history struck northern Minnesota in 1894, wiping out Hinckley and surrounding towns near Duluth and burning at least 400 persons to death. Other fires wiped out settlements across Michigan. With little or no protection, fire often swept through the debris, or slash, left by lumberjacks. The flames were the final toll of progress through logging.

Time of the Land-Grabbers

Another characteristic of this period was land acquisition on a large scale. In South Dakota, the Homestake Mining Company dominated the Black Hills, if not the entire state. In Montana, Marcus Daly, copper king of Anaconda and Butte, controlled the mineral wealth by enacting laws to ruin his opposition; then he swallowed the Bitterroot Valley for his cattle, sawmills, and stable of race horses. In California, Henry Miller, in the heyday of his reign, could ride by horse and buggy the entire length of the state and into neighboring Oregon and Nevada, stopping each night on his own rangeland. Yet the railroads were the giants of them all: Southern Pacific owned over 10 million acres and held California in the hollow of its hand, while Northern Pacific received from the federal government, as "encouragement" to finance construction, almost 40 million acres—an area greater than Pennsylvania, Rhode Island, New Jersey, and the District of Columbia combined.

In the process of settlement, land and water were first claimed or awarded under laws of the colonies, states, or territories. This explains why today there is relatively little federal land in the East, most of that having been acquired through purchase. The public domain, however, was property of the nation and therefore subject to legislative control and disposition by Congress alone. Defined as encompassing areas "acquired by treaty, capture, cession by States, conquest or purchase," lands of the public domain included the Louisiana Purchase, Red River Basin, cessions from Spain and Mexico, the Oregon Compromise, and the Texas, Gadsden, and Alaska purchases. Jurisdiction over the public domain was lodged in the General Land Office, an agency of the Department of the Interior, whose principal mission was to oversee land disposal.

The Homestead Act was intended to be the greatest instrument for distributing land among the people—although, in fact, the railroads were granted more land than all homesteaders combined. When the Act became effective on January 1, 1863, the same date as the Emancipation Proclamation, the world thought it a highly fitting complement to that declaration of liberty, and it was hailed abroad as the greatest democratic measure of all time. Any citizen could earn 160 acres of the public domain if he would live on the land five years, make his home and cultivate the ground, and pay fees of about $16; or he could gain title after only fourteen months by paying a minimum of $1.25 per acre. When he complied with the requirements, the land "went to patent" and became his property.

Other land laws followed the Homestead Act: the Mineral Land Act, 1866; the Timber Culture Act, 1873; the Desert Land Act, 1877; and the Timber and Stone Act, 1878. These were all designed, ostensibly, to encourage, assist, and reward ordinary Americans who would open frontier lands and settle the West. Presently, however, land laws were short-cut and subverted, leading to fraud, land thievery, and land speculation, through which something like one-half of the nation's forests passed into private ownership.

The Timber and Stone Act provided lumbermen their best means to obtain choice land as they moved from the Lake states to the West. This law provided for the purchase of 160 acres of nonmineral land at $2.50 an acre, presumably so that miners and settlers could obtain timber and building materials from adjacent wild lands for use in construction on their sites. Each applicant was required to pledge that he would not pass on the title and would use the materials on the land itself. That stipulation proved merely incidental. The first deception of record was perpetrated by the California Redwood Company in Humboldt County, California, in 1882–1883. The company recruited seamen by the score, mostly at "Coffee Jack's" boarding house in Eureka. About twenty-five at a time, they proceeded to the Land Office and filed for locations the company had already chosen for them. Next stop was a notary public, where the new but temporary landowners sold their claims for $50 each, then returned to their ships, the boarding house, or their favorite saloons. In this manner the heart of the Redwood Empire passed into private hands, most of it to be cut down and scattered across America in the form of railroad ties, bridge timbers, silos, posts, shakes, shingles, sidings, ceilings, doors, furniture, and caskets, with a small fragment to be repurchased by the federal government at a cost of hundreds of millions of dollars for inclusion in a small Redwood National Park.

From then on, lumber companies transported "entrymen" by the trainload. Often they were teachers, delighted to accept a free trip to the scenic redwoods. They swore faithfully that their new land was for personal use, then blithely transferred title to the lumber company that had granted the excursion. Complicity of land officials was not uncommon: in some cases a claim to a quarter-section of 160 acres, or even a quarter-quarter of 40, was stretched for miles, which is how the term "rubber forty" entered the lexicon.[3]

The Wave of Scientific Activism

An influential voice in defending the public interest was that of
Major John Wesley Powell, Civil War veteran, intrepid explorer, and
geologist, whose classic work, *Report on the Lands of the Arid Region of
the United States*, published in 1879, constitutes a primer for land use.
In this "blueprint for a dryland democracy," he urged protecting streams
in public ownership, safeguarding timberlands from fire, and encouraging
cooperative labor and capital for development of irrigation where it
could be justified. Powell was part of a wave of scientific activism,
sparked in large measure by the Smithsonian Institution during the
tenure as Secretary of Joseph Henry and his successor, Spencer Fullerton
Baird.[4]

The inspiration for this movement was derived from *Man and Nature*,
published in 1864 by George Perkins Marsh, a versatile wizard who
served as a Vermont congressman and U.S. diplomat in the Mediterranean.
In this classic work of the nineteenth century, Marsh painstakingly
explained the relationship of soil, water, and vegetative cover to civilization; he illustrated with impressive data how cutting of forests, fire,
and overgrazing contributed to the decline of agriculture, water supplies,
cities, and the prosperity of nations, citing once productive lands of
China, Mediterranean Europe, and North Africa that had turned into
desert. Applying these lessons to the United States, Marsh warned that
it was "of the utmost importance that the public, and especially land
owners, be roused to a sense of the dangers to which the indiscriminate
clearing of the woods may expose not only future generations but the
very soil itself."

Franklin Hough, strongly influenced by Marsh, had sought and
received counsel and encouragement from both Henry and Baird in his
effort to generate interest in forest protection. Such men were involved
in many things of magnitude. Baird, for instance, was the first permanent
secretary of the American Association for the Advancement of Science.
He was not only Secretary of the Smithsonian, from 1878 until his
death in 1887, but also the first commissioner of the United States
Commission of Fish and Fisheries (through which he established the
nation's first marine laboratory at Woods Hole, Massachusetts) and head
of the science department of *Harper's Magazine*. Powell himself served
at the same time as director of the United States Geological Survey, at
the Department of the Interior, and of the Bureau of Ethnology, at the
Smithsonian.[5]

This early scientific movement was instrumental in 1872 in setting
aside Yellowstone National Park as a public trust, withdrawn from any
possible private claim, and in the establishment of Yosemite National
Park in 1890. It was the start of a new course in history, leading to a
network of national parks, national forests, and national wildlife refuges.

The advent of the parks, combined with a growing fear that timber
resources were being consumed at a disastrous rate, laid the foundation

for the Forest Reserve Act of 1891, authorizing the President to withdraw portions of the public domain and designate them as "forest reserves." Dr. Fernow was a prime mover of the law, which marked the beginning of the National Forest System. President Benjamin Harrison proceeded to set aside the Yellowstone Timber Reserve (now the Shoshone and Bridger-Teton National Forests) in western Wyoming, followed by the White River Plateau Timberland Reserve (now White River National Forest) in Colorado. Before his term expired, President Harrison set aside reserves totaling 13 million acres.

The Act of 1891 provided no plan of operation; the reserves were merely closed areas. In 1896, the Secretary of the Interior requested the president of the National Academy of Sciences to appoint a commission to consider and report on questions relating to protection and use of the reserves. The resulting National Forest Commission urged expansion of the reserve, and President Grover Cleveland withdrew 20 million acres—no small feat in the face of severe western opposition. It also recommended specific measures for protection and administration of the reserves, resulting in the Organic Act of 1897 by which Congress set forth a system of administration and qualified the objectives of the reserves as being "for the purpose of securing favorable conditions of waterflows and to furnish a continuous supply of timber for the use and necessity of citizens of the United States."

Pinchot later wrote: "Except for the Act of 1891, the Pettigrew Amendment to the Sundry Civil Service Act of June 4, 1897"—otherwise known as the Organic Act—"was and still is the most important Federal forest legislation ever enacted. It did two essential things: it opened the Forest Reserves to use; and it cleared the road to sound administration, including the practice of Forestry."[6]

Gifford Pinchot was the youngest member of the National Forest Commission. On graduating from Yale in 1889, he had set sail for Europe to study at the National School of Forestry in France, from whence his forebears had come. He became the first native-born American to complete graduate training in forestry, for there was still, at that time, no such course given at an American school. On his return in 1892, he was engaged to develop a forest plan for George Washington Vanderbilt's estate, Biltmore, at Asheville, North Carolina, where he introduced selective logging, removal of defective trees, planned reproduction, and trespass control. Although single-minded and often uncompromising, he inspired others and emerged as the most influential figure in his profession and, quite possibly, in the entire conservation movement in America.

In 1898, Pinchot succeeded Bernhard Fernow as chief of the Division of Forestry in the Department of Agriculture. He began with eleven employees, the nucleus of the Forest Service of today. Pinchot and his disciples spurred expansion of forest conservation and, in fact, brought the word "conservation" into popular usage. They had no federal lands to manage, but through Circular 21, of October 15, 1898, entitled "Assistance to Farmers, Lumbermen, and Other Owners of Forest Lands," they were able to offer free assistance to any farmer, lumberman, or state or local government to study woodlots and provide working forestry plans. Owners of large tracts were required to pay the expenses of men in the field. By the end of the first year, 123 requests had come from 23 states.

In that same year, the first four-year professional course on the college level was started at Cornell University by Dr. Fernow, while Carl A. Schenck, a German *Forstmeister* who had succeeded Pinchot at the Vanderbilt estate, opened a one-year practical curriculum at the Biltmore Forestry School.[7] Two years later, in 1900, Pinchot, with funds contributed by his family, helped to establish the graduate school of forestry at Yale, his alma mater, and dispatched Henry Solon Graves, his close associate (and later successor as Chief of the Forest Service), to serve as its dean. Also in 1900, Pinchot and a handful of colleagues organized the Society of American Foresters to promote professional ideas and ideals in forest sciences. Although forestry as a profession was still virtually unknown, the society's meetings at Pinchot's home in Washington brought young foresters in contact with leaders of the government, including members of the Cabinet and even the President.

Though Pinchot was the head of a minor federal agency, who should normally have communicated with the White House through superiors, he always went directly to the President. He and James A. Garfield, Secretary of the Interior (and son of the twentieth President), occupied unique positions in Roosevelt's administration. "No two men have been as closely identified with so many of the policies for which this administration has stood," TR wrote at the end of his term. And in a letter to Pinchot: "There has been a peculiar intimacy between you and Jim and me because all three of us have worked for the same causes, have dreamed the same dreams, have felt a substantial identity of purpose as regards many of what we three deemed the most vital problems of today."[8]

The Roosevelt years have been called the "First Wave" of conservation in this century, with Pinchot playing an intimate role in shaping policies and programs on a broad front. Roosevelt was enlisted in Pinchot's campaign to create additional reserves and during his term in office set aside 132 million acres of forest and parkland. This stands as a major achievement in itself, even more so considering that the total includes 15 million acres Roosevelt dramatically proclaimed in March 1908, just before signing a bill sent to him by Congress to prohibit any further such action.

Equally important, Roosevelt supported the proposal to bring all forest work under Pinchot's wing in the Agriculture Department, which insisted that trees, being a renewable agricultural crop, should logically be its concern. The Interior Department, which held the public land, had no foresters and was tainted with patronage and scandal.

Applying "Preservation Through Use"

The Transfer Act of 1905 effected the switch that Roosevelt had recommended and opened a new era in government forestry. The small bureau that Pinchot headed now bloomed as the U.S. Forest Service. Use rather than mere custody was now the doctrine to govern the forest reserves (soon to be designated "national forests"). In a celebrated letter to the Chief Forester, likely drafted or instigated by Pinchot himself (printed in full in *Breaking New Ground*, pp. 261–262), Secretary of Agriculture James Wilson declared that "conservative use in no way conflicts with permanent value." The Secretary directed that use of the forest reserves "must be brought about in a thoroughly prompt and businesslike manner," adding as follows:

> The vital importance of forest reserves to the western states will be largely increased in the near future by the continued steady advance in settlement and development. The permanence of the resources of the reserves is therefore indispensable to continued prosperity, and the policy of this Department for their protection and use will invariably be guided by this fact, always bearing in mind that the conservative use of these resources in no way conflicts with their permanent value.

Thus, sale and cutting of timber were instituted and regulated, and fees were charged. "A reasonable charge may be made for any permit, right, or privilege, so long as that charge is not inconsistent with the purposes for which the reserves were created," declared the manual, or "Use Book." The 1897 law had already stipulated that any sale of timber valued at more than $2,000 must be appraised, though in many cases of actual practice, no charge had been made. The western lands had been open to everyone, and the Forest Service moved cautiously in imposing restrictions and charges. Fire control rules, on the other hand, were made and generally enforced.

The new public forest enterprise was not universally hailed. Resentment against restriction and regulation imposed by the Washington bureaucracy was vented in western communities and by their political representatives. As part of his strategy to deal with such hostility, Pinchot sought to build a democratic grass-roots field force combining young forestry graduates from eastern colleges and woods-wise men from western ranches and logging camps.

Managing timber reserves was not the only challenge. Overgrazing by uncontrolled numbers of cattle and sheep had led to deterioration of millions of acres. Despite bitter opposition, in 1905 a grazing permit system was estabished, opening a new field of adventure with issues and conflicts still not wholly resolved to this day. Range reconnaissance in 1910 on the Coconino National Forest in Arizona was another stride forward in the effort to balance use with the capacity of the range to rejuvenate itself.

Research, which had started with the inventory and assessment work of Franklin Hough, was significantly expanded during the Roosevelt-Pinchot era. In 1907, experiments were begun with range seeding, deferred-rotation grazing, and sheep grazing within coyote-proof fencing. The following year, the first forest experiment station was established at Fort Valley, Arizona. And, in 1910, the Forest Products Laboratory was established at Madison, Wisconsin, in cooperation with the University of Wisconsin; the laboratory is now recognized as the world's outstanding institution of its kind.

Nevertheless, the natural resources inventory of the nation was being depleted at an alarming rate. "Cut and get out" was the unwritten law of the lumber industry. As a direct result of Forest Service activities, a movement was launched for conservation of all resources. The first step was the appointment by President Roosevelt of the Inland Waterways Commission in 1907. The Commission's reports stressed the need for river-basin planning, but also warned against monopoly control by private interests of forests, waters, lands, and minerals, charging that an excessive share of natural resources "has been diverted to the enrichment of the few rather than preserved for the equitable benefit of the many." Then came the White House Conference of Governors in May 1908, attended by a thousand persons, including such diverse luminaries as William Jennings Bryan, Andrew Carnegie, James J. Hill, and John Mitchell, and reported on front pages throughout the country. "The Conference set forth in impressive fashion," Pinchot wrote later in *Breaking New Ground,* "and it was the first national meeting in any country to set forth the idea that the protection, preservation and wise use of the natural resources is not a series of separate and independent tasks, but one single problem."[9]

With Pinchot the emphasis always was on wise use, or "preservation through use," rather than on use through preservation. He had objected strenuously to the New York constitutional amendment forbidding timber cutting in the Adirondack Forest Preserve as a bar to scientific forest management. And ultimately he parted company in bitterness with John Muir, the leading apostle of preservation. Born in Scotland and raised in Wisconsin, Muir began his great adventures with a thousand-mile walk across the Southeast, followed by a wide range of travels over the West, including Alaska and across the Bering Sea to Siberia. His special kingdom was Yosemite, where he lived for six years and where President Theodore Roosevelt came to camp with him. "It was like lying in a great solemn cathedral," Roosevelt wrote, "far vaster than any built by the hand of man."

Despite his admiration for Muir (who was in great demand as a writer, and who lived more outdoors than indoors), Roosevelt was most influenced by Pinchot. The two idealogues—Muir and Pinchot—had been friends and fellow members of the National Forest Commission in 1896, but they parted company over the disposition of Hetch Hetchy Valley on the Tuolumne River in Yosemite. As early as 1882 city engineers of San Francisco, 150 miles distant, had scouted the possibility of damming Hetch Hetchy's narrow lower end to make a reservoir for water storage, then of using the fall of the impounded water to generate hydroelectric power. Establishment of Yosemite National Park in 1890 appeared to head that off. But San Francisco politicians pressed the issue for years, from one national administration to another. Muir foresaw a dangerous precedent in the destruction of a beautiful valley, warning, "It behooves us all faithfully to do our part in seeing that our wild mountain parks are passed on unspoiled to those who come after us, for they are national properties to which every man has a right and interest."

Yosemite's proponents looked to Theodore Roosevelt, when he took office in the White House, to defend the integrity of the park; Roosevelt, however, looked for guidance to Pinchot, who recommended utilitarianism and convinced the President to support conversion of Hetch Hetchy on the grounds that the greatest good for the greatest number would be to furnish good water to hundreds of thousands rather than to save a scenic valley for a few hundred. In 1913 the battle was resolved in Congress: the dam was built, flooding a valley that many have said compared in beauty with Yosemite itself, and leaving bitter feelings between the forces of preservation and those of "wise use."

Pinchot was embroiled in another classic encounter after Roosevelt left office in 1909. William Howard Taft had been handpicked by TR to succeed him in hopes of continuing his policies, but Taft forthwith reversed course in the conservation field by dismissing James Garfield as Secretary of the Interior and replacing him with Richard A. Ballinger, a former commissioner of the General Land Office. The old collaboration was now turned into conflict and Pinchot faced a Waterloo of principle and purpose. Taft had promised to retain the Chief Forester, though considering him a radical and crank who refused to honor the protocols and disciplines of government. The showdown came when Pinchot objected to the issuance to powerful economic interests of mineral claims entitling them to exploit coal-rich lands in Alaska.

Pinchot fought the issue in his own style, openly and boldly. In defiance of ritual, he charged that no other question before the nation could be so important, or so difficult to straddle, as the great question between special interest and equal opportunity, between the privileges of the few and the rights of the many, between government by men for human welfare and government by money for profit. He insisted that it was a showdown between those who stood for the Roosevelt policies and those who stood against them. For his efforts Pinchot was dismissed from office, yet not without causing a Congressional investigation, at which a brilliant Boston corporation lawyer, Louis D. Brandeis, washed the Interior Department's linen in public.

For a time after Pinchot's dismissal, the Forest Service was in low repute and its budget was slashed. Henry S. Graves returned from Yale to serve as Chief Forester (1910–1920) and defended the national forests against being shifted back to the Interior Department.

Meantime, there was much pressure to establish national forests in New England and southern Appalachia, where wooded slopes and valleys had been cut over, burned, farmed out, blighted with erosion, and left idle and unprotected, in tax default, contributing only flood waters to downstream valleys. National forests of the West served to protect the flow of navigable streams, but those forests had been carved out of the public domain. There was no effective counterpart in the East until the Weeks Law of 1911 (named for Representative John W. Weeks of Massachusetts), authorizing the purchase of private lands for watershed protection or for the production of timber. Under this law, almost all of the National Forest System east of the Great Basin was established and the National Forest Reservation Committee, consisting at first of the Secretaries of Agriculture, War, and the Interior, was created to approve all purchases. The Pisgah National Forest in western North Carolina was not the first national forest in the East, but it was the first composed of land not originally in the public domain. A substantial tract was made available to the government in 1916 by Mrs. George W. Vanderbilt following her husband's death.

In 1915, forest research took a major leap forward when Earle H. Clapp was appointed by Graves to head a new and separate Branch of Research, according that function what the Chief Forester called "the fullest possible recognition." Clapp would for years be an influential figure (including service as Chief without the full portfolio), spurring the movement that led to passage of the McSweeney-McNary Act of 1928, which increased funds for a broad program of forest research and authorized a nationwide survey of forest resources.

As the agency matured, new ideas emerged in the ranks. Arthur Carhart, the first landscape architect hired by the Forest Service, conceived the idea of protecting a portion of the White River Forest, Colorado, in a roadless, natural condition, and subsequently of saving the marvelous lakes country of the Superior National Forest in Minnesota as a "water-trail wilderness." Carhart resigned from the Forest Service in 1922, frustrated by lack of support from his superiors, but he left a seed to take root. Aldo Leopold, with whom Carhart corresponded, in 1924 successfully won designation of an area of the Gila National Forest in New Mexico for wilderness preservation; from that year forward portions of national forests have been set aside in one way or another for such protection.

Following World War I, the Forest Service embarked on a movement to develop a broad national policy with particular emphasis on public regulation of cutting on private lands, sparking a running controversy with industry. Pinchot, even out of office, and his supporters pressed for expanded public ownership and regulation, through strict inspection and supervision, of private holdings.

Foresters, even within the agency, were divided on the issue. Pinchot's position was rejected by the Society of American Foresters, most of whose members were more conservative than he, and who became more industry-oriented with passing years. Robert Marshall insisted that the forests were essential to national welfare and so merited ownership and regulation. In *The People's Forests*, published in 1933, he reasoned as follows:

> As sources of greatly needed raw material they [the forests] play a vital part in raising the physical standards of American life. As conservers of soil and water they are absolutely necessary if we are not willing to have our country become as denuded and flood-swept as the Chinese hillsides and valleys. As environment for the highest type of recreational and esthetic enjoyment, they are essential to the happiness of millions of human beings. Economic, physical and social considerations all demand that we maintain a bountiful forest resource.[10]

Chief Forester William B. Greeley (1920–1928), highly influential both during and after service with the government (when he joined the industry as Secretary-Manager of the West Coast Lumbermen's Association), preferred to follow the route of federal-state-private cooperation. This approach was embodied in the Clarke-McNary Act of 1924, which Marshall later derided as "private ownership with public subsidy" because it provided federal money for protection of private lands from fire, as well as for free planting and free advice, including help in seeking tax relief.

A National Plan and New Deal Programs

Regulation and expanded public ownership once again became policies of the Forest Service during the tenure of Chiefs Robert Y. Stuart (1928–1933) and Ferdinand A. Silcox (1933–1939). They argued that the best, most productive, most accessible three-fourths of the nation's forest lands were privately owned, that they furnished most of the timber used, and that timber was continually subject to destructive cutting practices. Through the 1930s and early 1940s the Forest Service pressed for a nationwide action program based on: (1) public ownership and management of more forestland by communities, states, and the federal government; (2) continuation and extension of public cooperation with private owners of forestland; and (3) public regulation of woods practices on privately owned forestland. Bills to attain this program were introduced in Congress in 1941, 1946, and 1949. Pressures of the industry, fearful of restrictions, blocked them all.

Kenneth Crawford, a well known Washington correspondent of the New Deal era, recorded his impression of the regulation issue in his book *The Pressure Boys: The Inside Story of Lobbying in the United States* (New York, 1939). Crawford wrote (pp. 198–199):

When F. A. Silcox, the present Chief Forester, took office in 1933, he found himself up against a system known in the service as "Greeleyism." This was Forester Greeley's policy of "co-operation" between the Forest Service and the lumber industry, a policy described by Col. George P. Ahern, Pinchot associate, as "the lumbermen leading the Forest Service by the hand." Silcox wanted public regulation of private timber lands (similar in principle to government regulation of privately owned coal mines), which had been advocated by Pinchot before the Greeley regime, but, recognizing the power of the combination he was up against, he tried to edge along by labeling his plan as "co-operation and regulation."

Silcox never even got the Forest Service back of him, to say nothing of the lumber industry. The system of decentralization prevailing in the Forest Service makes the biggest and most distant regional offices, at San Francisco and Portland, practically independent. These offices were filled with men who got their training under Greeley and remained intimate with him after he left the Forest Service in 1928.

The crusade for regulation reached a high-water mark with publication in 1933 of *A National Plan for American Forestry* (Senate Document No. 12, 73d Congress, 1st Session), known as the Copeland Report, after Senator Royal S. Copeland, a member of the Committee on Agriculture and Forestry. More than 1,600 pages long, the Copeland Report contained by far the most detailed statistics ever gathered on all phases of American forest conditions, including timber, research, economics, watershed, range, wildlife, and recreation. It advanced progressive interpretations of federal responsibility, proposing a large extension of public ownership and more intensive management of all timberlands. One significant premise was an estimate that 308 million out of 615 million acres of forest and brush exercise major influence on watershed conditions. The report alleged that private initiative had failed, even when aided by public subsidy, since fire damage was eleven times greater on private lands than on federal lands; only 0.85 percent of private forests were managed to insure continual growth of timber, and men who wanted to work in the woods were obliged to migrate with the industry, often to work under unsafe conditions.

The Copeland Report proposals were not accepted, at least not as stated, but the New Deal era saw enhancement of the Forest Service role—almost comparable to that of Theodore Roosevelt's day. The agency was an integral part of the Franklin D. Roosevelt social crusade, even the catalyst in combining programs of land restoration and self-sustaining public work. Its effectiveness was due in no small measure to the continuing influence of Gifford Pinchot, then serving as Republican governor of Pennsylvania, and to the President's personal interest in natural resources. Mr. Roosevelt called on the Forest Service for major special studies, for Great Plains shelter-belt planting, for leadership in salvaging millions of trees blown down by the New England hurricane of 1938, and made funds available for purchase of lands within the boundaries of national forests authorized under the Weeks Law.

One of the most imaginative New Deal programs was the Civilian Conservation Corps (CCC), which began April 10, 1933, when the first camp (appropriately named Camp Roosevelt) was opened in the George Washington National Forest, Virginia. During the nine years of the CCC program, more than two million young men participated. They were known as "Roosevelt's Tree Army," for, of all the forest planting in the history of the nation, more than half was done by the CCC. Most camps worked on projects in national, state, or private forests, under direction of the Forest Service.

Despite improved management of some industrial land during his tenure, Acting Chief Forester Earle H. Clapp (1939–1943) strongly advocated regulation of private cutting and substantially increased the area of public forests. Despite consummate ability and experience, Clapp was blocked by opponents from ever becoming Chief. On one hand was the industry, resisting regulation, and on the other hand the self-styled "old curmudgeon," Secretary of the Interior Harold L. Ickes, fighting to bring the national forests "home." Ever since the 1905 Transfer Act, every Interior Secretary had aspired and conspired to reclaim control of the national forests from the Department of Agriculture, and none more aggressively than Ickes. In the course of a long struggle, Ickes very nearly succeeded, but the Forest Service and friends outside (notably including Pinchot) maneuvered to hold him off.

With World War II, new responsibilities arose. The Forest Service operated the Timber Production War Project, mainly to stimulate the output of wood with minimum waste, and undertook an emergency effort for the production of rubber-bearing plants. The war also led to opening to logging formerly remote old-growth virgin forests, particularly in the Pacific Northwest. The industry pressed for harvesting Douglas fir in Olympic National Park, but the war was won without it.

In the Mainstream of Modern Life

In 1945, the Forest Service made a reappraisal of the country's forest condition, involving for the first time a field survey of timber-cutting practices. The report showed that sawtimber volume had declined 43 percent in thirty-six years, that it was being drained one and a half times as fast as it was being replaced by growth, and that there had been a marked deterioration in timber quality, as well as quantity, with cutting practices "poor" to "destructive" on 64 percent of all private land. On this basis Chief Lyle Watts (1943–1952) again raised the issue of public regulation, but his efforts resulted only in the Cooperative Forest Management Act of 1950, giving cooperative management aids to private landowners and processors of forest products.

The issue persisted. In 1952 the Forest Service announced plans for a national inventory of timber resources—that is, an examination of timber growth measured against consumption. The industry was ap-

prehensive that it would show a deteriorating forest condition, leading
to proposals for regulation and expansion of federal land holdings.
Following a period of controversy, the 700-page report *Timber Resources
for America's Future* (Forest Resource Report No. 14, January 1958, USDA
Forest Service), known as the Timber Resources Review, made its
appearance. The timberland base, the report stated, had shrunk under
urban and agricultural expansion. "There is no 'timber famine' in the
offing," declared Chief Richard A. McArdle in introducing the report,
"although shortages of varying kinds and degrees may be expected."
Actually, the study showed annual sawtimber growth 9 percent higher
than a decade earlier, but more desirable trees losing to those of poor
quality. Edward C. Crafts, who directed the Timber Resources Review,
wrote an extensive retrospect in an article titled "The Saga of a Law"
(*American Forests*, June 1970). This report was published during the
decade-long tenure of Chief Richard McArdle (1952–1962), a period
when forestry was thrust into the mainstream of problems created by
the booms in population, technology, consumption, mobility, and leisure
activities. In 1957, a five-year program called "Operation Outdoors"
was launched to improve and expand recreational facilities in the national
forests overtaxed by the postwar surge of hiking, camping, canoeing,
fishing, and other outdoor pursuits. But this was plainly not enough;
the act of balancing conflicting resource uses against rising demands
became more difficult. The agency sought a specific Congressional
directive, which resulted in the Multiple-Use–Sustained-Yield Act of
1960, declaring that national forests shall be administered for outdoor
recreation, range, timber, watershed, wildlife, and fish—based on "the
most judicious use of the land for some or all of these resources." The
Multiple-Use Act was intended not only to give legislative sanction to
long-standing programs, but to step back from priority on timber. Chief
McArdle saw the Act as critical to "continued management of the national
forests for 'the greatest good of the greatest number in the long run,'"
though controversy would overrun best intentions and legislative drafts-
manship.

    Also in 1960, about four million acres of "land utilization projects"
were given permanent status as national grasslands. These were submarginal farm lands, primarily on the Great Plains, which the government
had acquired during the Dust Bowl days of the 1930s. They had been
restored by the Soil Conservation Service in a notable achievement of
land husbandry and now were transferred for administration as part
of the National Forest System.

    With the onrush of the 1960s, Congress gave the Forest Service,
under Chief Edward P. Cliff (1962–1972), a new set of directives not
always to its liking. The agency was being told increasingly what to
do and not to do, while its flexibility and professional expertise were
circumscribed. First came the Wilderness Act of 1964, drafted by wilderness enthusiasts who felt the Service had inadequately protected the

wildlands in its trust. (And when President Lyndon B. Johnson signed the Act in the Rose Garden, Secretary of the Interior Stewart L. Udall was at his elbow, while Secretary of Agriculture Orville L. Freeman was not among those present.) Then followed the Rare and Endangered Species Act of 1966; National Trails System Act and National Wild and Scenic Rivers Act, both of 1968; and the National Environmental Policy Act, signed by President Richard M. Nixon in 1970, though it was passed by Congress in 1969.

Although recent Chiefs of the Forest Service have carefully avoided the issue of regulation, opting instead for cooperative state and private programs, federal laws do indeed establish regulatory programs or set national standards that affect private lands. These include the Clean Air Act, Federal Water Pollution Control Act, Coastal Zone Management Act, and Federal Insecticide, Fungicide, and Rodenticide Act.

In addition, between 1969 and 1982 more than half a dozen states enacted their own laws to regulate timber management practices, in some cases including rules governing chemical and fertilizer application, slash disposal, and precommercial thinning. The California Forest Practices Act of 1973 is the most far-reaching: it requires preparation of a plan by a professional registered forester, complete with stipulations to assure protection of streams and unique areas.

Protests and lawsuits erupted widely through a tumultuous period. A project to kill underbrush in a portion of the Tonto National Forest by spraying toxic chemicals via helicopter brought charges from residents of Globe, Arizona, that women's reproductive organs were being damaged and birth defects were caused in goats. A Congressional committee in 1970 found the charges serious enough to warrant investigation. Five families pressed legal action. The case attracted nationwide interest, since the basic chemicals (2,4-D and 2,4,5-T) were the same as those used in Agent Orange in Vietnam (and the basis of a major lawsuit brought in behalf of former servicemen). In March 1981, Dow Chemical Company and the federal government reached a settlement with the Globe families just as their case was due for trial. A significant payment of damages was involved, but the exact amount was not disclosed on order of the court. The Forest Service has always insisted, however, that a level of dioxin equivalent to that found in Agent Orange has never been allowed in herbicides that it uses. (Further details on the Globe case are found in Chapter 15.)

Lawsuits also arose over management of Douglas-fir forests of the Northwest, spruce forests of Southeast Alaska, lodgepole and other pine forests of the Rocky Mountains, and mixed forests of Appalachia. Such suits charged that timber management plans unlawfully impaired recreation, ignored soils and watershed, failed to protect endangered species, or failed to consider other mandated forest uses.

The Bitterroot National Forest in western Montana became a particular storm center even without litigation. A task force appointed by the regional forester issued a detailed report in April 1970, covering its investigation. It noted serious mistakes and recommended that:

- Any lingering thought that production goals hold priority over quality of environment must be erased.
- Multiple-use planning must be developed into a definite, specific, and current decision-making process that it is not today.
- Quality control must be emphasized and reemphasized until it becomes the byword of management.
- The public must be involved more deeply than ever in developing goals and criteria for management.[11]

Another step in this direction was taken the following year when a Senate subcommittee, headed by Senator Frank Church of Idaho, conducted lengthy hearings on clearcutting practices on the national forests. The subcommittee's report, *Clearcutting Practices on National Timberlands*, appeared in March 1972. The subcommittee, with input from the Forest Service, enunciated a set of guidelines allowing flexibility but counseling moderation and caution that agency leaders found acceptable.[12] By contrast, a presidential executive order, drafted by the Council on Environmental Quality, to restrict clearcutting was blocked in January 1972 at a White House conference, where participants included representatives of industry, Secretary of Agriculture Earl Butz, and Forest Service officials. The sharp struggle over the executive order was another episode in the running classic contest over the public lands, in this case with industry the clear victor.

The failure of the Multiple-Use–Sustained-Yield Act of 1960 to achieve balance in use, or even general acceptance of the principle, was evident throughout the 1960s and 1970s. "The poor man can't buy a house. Withdrawals for single-purpose are a luxury the nation no longer can afford," industry groups charged in referring to recreation, endangered species, and wilderness programs they had tried unsuccessfully to defeat; and the Forest Service was represented as a conservative obstructionist in the path of economic and social progress.[13]

The industry mounted a national campaign, resulting in the introduction of the National Timber Supply Act of 1969, which would have directed the Secretary of Agriculture to increase immediately annual allowable cuts on some national forests. It would also have defined logging as the "optimum" use on 97 million acres of the National Forest System, designated specifically as "commercial timber lands."[14]

The bill was defeated in the House, by a vote of 150-229, on February 26, 1971. This vote followed receipt by Congress of a Forest Service letter warning that efforts to manage national forests solely for optimum timber productivity would lead to "an immediate and basic conflict with the long-standing principles of multiple use and sustained yield."

Although this bill was defeated, its rationale reappeared continually during the tenure of Chief John R. McGuire (1972–1979) and well into that of his successor, R. Max Peterson. In September 1973, for example, President Nixon reported on the three-year findings of his Advisory Panel on Timber and the Environment (headed by former Interior Secretary Fred R. Seaton). The panel reported an abundance of resources.[15]

Or, as President Nixon declared:

> It found the nation faces no scarcity of forest land, or of standing timber; no scarcity of forest wildlife, recreational opportunities, or existing and potential wilderness areas.
>
> It also found the national forests overstocked with mature and overmature timber. As a result, the annual growth per acre in our national forests is less than half of that found on other commercial forest lands. According to the Panel, the main forestry issue facing us in the next several decades is the rate at which this old-growth timber in the national forests is converted to new, well-managed stands of trees.
>
> The considered judgment of the Panel was that timber growth on all forests of the nation, despite problems of different ownerships and diverse physical characteristics, could be doubled by the year 2020 with an increase in management output.

This approach proved hardly satisfactory. Litigation and nagging controversies led to passage of the Resources Planning Act of 1974 and National Forest Management Act of 1976. On the face of it, they appear to address the issues of clearcutting, balanced multiple use, even-flow "departures," biotic diversity, and pesticide use. The planning is designed to be national and long range. The RPA Assessment portion embraces private and public lands, the Program portion addressing Forest Service activities in particular. NFMA was drafted to deal with the legal impasse over timber harvesting. It was written, however, to prevent environmental abuses of logging and to ensure equal consideration of all the renewable resources. Like the Multiple-Use Act of 1960, these laws identify problems rather than solve them; they define procedures rather than goals, providing a complicated course for action in the 1980s.

3

ORGANIZATION FORM
AND FUNCTION

> Those of us in positions of administrative responsibility need
> to be reminded frequently that we are making resource decisions
> for future generations. What we do today—good, bad or other-
> wise—will be reflected in how well the national forests meet
> some of the nation's needs 100 years from now: for water,
> wilderness, timber, recreation, forage, life-style, wildlife, air
> quality, and other factors we may not be aware of today.
> —Charles T. Coston, Regional Forester,
> Northern Region, 1981.

The Tradition of Loyalty and Esprit

The Forest Service is a widely respected institution. Despite abundant
criticism from some quarters, the country generally thinks well of it.
In 1981 a study titled "A Comparative Analysis of Successful Orga-
nizations" made by the Productivity Resource Center of the Office of
Personnel Management and Pennsylvania State University rated the
Forest Service among the ten most successful organizations in the country.
Selections were based on three criteria:

1. That the organization produced a well-respected product, whatever the nature of the product.
2. That it appeared a good place to work.
3. That it was "sound" and "healthy" and had been so over a
   sustained period of time.

The report made the following comment, a useful assessment for
anyone trying to fathom what makes the agency tick:

> The United States Forest Service appears to offer an interesting mix of
> innovativeness and conservatism. The Forest Service is an organization with

a strong sense of tradition whose mission has remained essentially unchanged since its inception. The people who make their careers in the Forest Service tend to come predominantly from rural areas of the United States, and have been educated in the conservative tradition of the land grant colleges. Members have a strong sense of the land ethic, which is itself a conservative philosophy. An integral part of that philosophy, and of the Forest Service tradition, is a firm sense of individuality.[1]

The Forest Service may be overinstitutionalized, with individuality no longer a prominent characteristic, but this may have its blessings, too. For a moment compare the stability of leadership with that of the National Park Service, an agency of the Department of the Interior. Between 1973 and 1980 the Park Service was subjected to a continual change in directors—five in a period of seven years. Two of them were park superintendents without experience in the Washington office, and one was a member of the White House staff without any park experience. Personnel in the Forest Service advance "through the chairs," doing their stints in field, regional, and Washington offices, while preparing for higher responsibilities. At any given moment at least four or five officials of the agency are well schooled in public administration and qualified for appointment as Chief. Democratic and Republican administrations have carefully avoided firing the incumbent Chief (though other heads may roll) and traditionally, to this time, have selected a career employee to succeed him.

In *Breaking New Ground*, Gifford Pinchot wrote: "The Service had a clear understanding of where it was going, it was determined to get there, and it was never afraid to fight for what was right. Every man and woman in the Service believed in it and its work, and took great pride in belonging to it."

Out of the record of doing, as Pinchot wrote, "what to many seemed impossible," esprit and devotion to the agency rose as a binding force between individuals and institution. Cohesiveness and loyalty are illustrated in an incident that occurred when President Dwight D. Eisenhower visited Missoula, Montana, in company with Richard McArdle, then Chief of the Forest Service, to dedicate a new smokejumpers' headquarters on September 22, 1954. At the ceremony the President declared:

> I am not at all surprised that it is such a good outfit. Within the last week I have had a little proof of the qualities of Mr. McArdle himself. It has not been my good fortune to know him, but only two nights ago in Fraser, Colorado, I was visited by a cook, a cook in the Forest Service. He said, "I read in the paper you are going to Missoula. There you will see my boss, Mr. McArdle. Give him my greetings and best wishes."
>
> I was long in the Army, I have seen some of the finest battle units that have ever been produced, and whenever you find one where the cook and the private in the ranks want to be remembered to the General, when someone sees him, then you know it is a good outfit.

The Forest Service blossomed out of the decentralization launched in 1908, when six district offices (later renamed regional offices) were established to bring decision making close to the national forests. This decentralization has been described as the first successful effort by a federal bureau to keep in close touch with current local conditions and problems to ensure a sympathetic, understanding approach and, at the same time, to establish nationwide policies and standards.

In the early days it was virtually every man for himself. Under the forest reserves, the field force consisted of grades from forest inspector down through forest supervisor, forest ranger, and forest guard. Many were old Indian scouts, rodeo artists, or Spanish-American War veterans who loved adventure and the wild lands; others were patronage appointees, a collection of saloon keepers, waiters, doctors, and blacksmiths. Interviews with oldtimers reveal ways that are long gone. "One of the first constructive steps was to send me a rake," according to the recollection of a pioneer ranger, "with instructions to clear up the floor of my district—which was only 250,000 acres. When I asked for directions, my superior said, 'Go and range.' When I asked where, he said, 'You know better than I do. You claim to be a woodsman and I don't.'"

Pinchot changed things, instituting professional standards and conservation-through-use principles. Supervisors were required to keep at least one horse at their own expense and to devote their entire time to the Service. Woe unto him who "moonlighted." But the emphasis on decentralization and grass-roots organization developed and deepened over time.

Ferdinand A. Silcox, after being appointed regional forester in Montana in the late 1920s, wrote: "Efficient functional organization is predicated on the idea of predetermined plans with accomplishment checked by competent technical specialists. Fundamentally the ranger district is the basic unit of our organization. I have therefore taken it as the starting place in the application of the principles of an administrative plan which provides for directive control, competent inspection of accomplishment, determining the ability of each unit to accomplish in accordance with specific standards the quantity of work assigned to it, and checking on the efficiency of personnel."

Silcox was a leader with distinct social conscience, a personality in his own right as well as a Pinchot disciple. "Ferdinand Augustus Silcox," an article by E. I. Kotok and R. F. Hammatt in *Public Administration Review* (Summer 1940, pp. 240–252), reveals that this Forest Service Chief was offered, and declined, remunerative positions in the commercial world and appointment as Undersecretary of the Interior. "The death of Mr. Silcox is a blow to the whole movement for conservation of human and natural resources," Henry A. Wallace, Secretary of Agriculture, declared in 1939. "His belief in truly democratic processes and institutions was deep-seated and passionate, and he had a penchant for getting more done by inspiring people than most people do by driving them."

Thus his idea in the Forest Service was to keep things simple, open, and close to the people. When Silcox became Chief in 1933, as related in the article by Kotok and Hammatt, he warned colleagues against the bureaucratic danger of becoming satisfied with their own decisions and permeated with a holier-than-thou attitude. The Forest Service, he said, must keep wide open the channels by which citizens could see for themselves and judge decisions, actions, processes, and their effects. "Then, and not until then," he stressed, "can you and I and all of us honestly say we are conducting a federal agency on a truly democratic basis, with people and communities having a real and actual voice—not merely a gesture—in vital questions of policy and practice that affect them."

Individuality in the Ranks

Because a democratic atmosphere prevailed, personnel produced fresh ideas. Inspired by his mentor, Gifford Pinchot, Benton MacKaye, who served with the Forest Service from 1905 to 1918, became part of a group of pioneer thinkers in social and land reform. MacKaye had a fruitful career as forester, planner, humanitarian, and author. In *The New Exploration*, published in 1928, he urged control of the world's birthrate, warning, "This subject lies at the bottom of every social question known to man." In the same period MacKaye outlined the concept of an Appalachian footpath from Maine to Georgia, which in time became the Appalachian Trail. For him it was more than a recreational resource, more than wilderness conservation. He viewed it as an instrument of regional planning, the backbone of a whole system of wild reservations and parks, linked by feeder trails into a reservoir for maintaining primeval and rural environments at their highest levels. His approach was not based simply on "back to nature" or "saving everything," but on making each metropolis, whether large or small, a place of individuality and unity, of valid regional culture based on a natural setting of its own.

Guy M. Brandborg offers another example. He joined the Forest Service at the age of 21 in 1914, when the fledgling outfit was loaded with Pinchot's disciples. Brandborg rose through the ranks, imbued with the idea that all wealth derives from the earth. For twenty years he served as supervisor of the Bitterroot National Forest, in western Montana, a domain of high mountain lakes and streams, ponderosa and lodgepole pine, and a thousand kinds of wildflowers. He provided vital data to Bernard DeVoto, who came West in the late 1940s to prepare his series of articles for *Harper's Magazine* in defense of national forests. Twenty years later, while retired in Hamilton in the heart of the Bitterroot Valley, Brandborg was visited by correspondents from prominent national media outlets like the *New York Times, Washington Post,* and CBS, who came to him for guidance in covering the clearcutting controversy on the Bitterroot; and he collaborated closely with local journalists, notably

Dale Burk of the *Daily Missoulian*. Though Brandborg fought the clearcutting policy and practice, he was never vindictive or personal. (Never
get involved in personalities, warned Pinchot, it's hard enough to
accomplish things by sticking to principle.) Many men in the ranks
cheered him, and, following his death in 1977, Brandy Peak, one of
his favorite lookouts above the valley, was named in his honor.

The late Clinton Leon Davis reflects another aspect of individuality
in the ranks. When he retired in 1968, after thirty years with the Forest
Service, he left a record of achievement in public relations that would
be difficult to match anywhere in or out of government. A native of
South Georgia, Davis had come to the agency after working as outdoors
editor of an Atlanta newspaper and as an information specialist for the
Georgia Fish and Game Commission. He was instrumental over the
years in setting up the well-known Smoky Bear fire prevention campaign;
the Cradle of Forestry, outside Asheville, North Carolina; the Pinchot
Institute for Conservation Studies, at Milford, Pennsylvania; and a
Service-wide program of visitor information, or interpretation.

Indeed, as recently as 1960, the agency's high level of efficient field
coordination impressed a Yale University professor of political science,
Herbert Kaufman. In his book of that year, *The Forest Ranger: A Study
in Administrative Behavior*, Kaufman contended that the agency had
proved that public service was not necessarily inefficient, wasteful, or
extravagant. "The Forest Service, despite its success in injecting its own
outlooks into its men," he wrote, "has avoided many of the hazards of
success; it has preserved a good deal of its own flexibility."[2]

The Forest Service may be considered a department in all but name.
It employs more than 60,000 people, including, in 1980, approximately
21,500 permanent full-time, 25,000 temporary full-time, and 15,000 part-
time intermittent employees.[3] This is more than each of the departments
of State, Commerce, Labor, Education, and Housing and Urban Development. Except for the food stamp program, the Forest Service budget
is the largest in Agriculture, larger than the budgets of the National
Park Service, Fish and Wildlife Service, and Bureau of Land Management
combined.

The Forest Service and the Department of Agriculture

The Forest Service for years has been the largest bureau in the
Department of Agriculture. To some critics it appears at times like a
child acting independently of its mother. The inimitable Harold L. Ickes
repeatedly charged the Service with being a law unto itself, a tight
little organization with an extensive lobbying network—an example of
bureaucracy running wild. The Forest Service is, in fact, the only agency
in the Department authorized by law to publish its own annual report.
At one point in recent years the Department asserted itself with a
directive instructing that instead of being the U.S. Forest Service of the

Department of Agriculture, the agency henceforth would become the Forest Service of the U.S. Department of Agriculture.

The Department has been directed by strong Secretaries at times, such as industry-oriented Earl L. Butz under Presidents Nixon and Ford, environmentally oriented Orville L. Freeman under Presidents Kennedy and Johnson, and Bob Bergland under President Jimmy Carter, who issued orders to the Forest Service or overruled it when they chose to. Immediate administrative control, however, is exercised by an assistant secretary (who also has jurisdiction over the Soil Conservation Service).

When M. Rupert Cutler was assistant secretary during the Carter administration, he sought to broaden multiple-use concepts more fully to recognize wildlife values. His successor in the Reagan administration, John B. Crowell, from the outset of his tenure emphasized increased logging on the national forests and private woodlots; yet both men acted with a degree of caution and respect in dealing with the Forest Service and its established systems.

The Forest Service tends to lose control of its destiny when dealing with its budget—the initiative, or at least the control, belongs to the Department, the Office of Management and Budget, and Congress. The Church Senate Committee in its 1972 report on clearcutting noted that: "From 1954–70, the Forest Service received 66 percent of budget increases for timber sale administration, but only 20 percent of its requested increases for recreation and wildlife, 17 percent for reforestation, and 15 percent for soil and water management."

OMB staff personnel throughout the 1960s and 1970s have followed their own consistent path, regardless of change in administration. Commodity programs are favored and funded. Timber sales on every national forest are analyzed, the heavy timber-producing units receiving the most support. Timber management research may appear to benefit disproportionately over wildlife or wilderness research, but that, too, is specified by OMB. Even after Congress acts, OMB has impounded, or "placed in reserve," appropriated funds, usually those earmarked for noncommodity programs. Disputes have occurred between Congress and OMB over this process.

The Chief and Staff

The Chief (once called Chief Forester), headquartered in Washington, D.C., is atop the chain of command. So far, every man appointed Chief (excluding Gifford Pinchot and Henry Graves) has had long experience at different echelons in the field and some previous service in Washington as well. The same is true of his immediate right-hand men (all being men, so far): the associate chief, number two in command, plus five deputy chiefs, including one each for the National Forest System; State and Private; Research; Programs (Planning) and Legislation; and Administration.

38

FIGURE 3.1

Organization Chart
U.S. Department of Agriculture
Forest Service

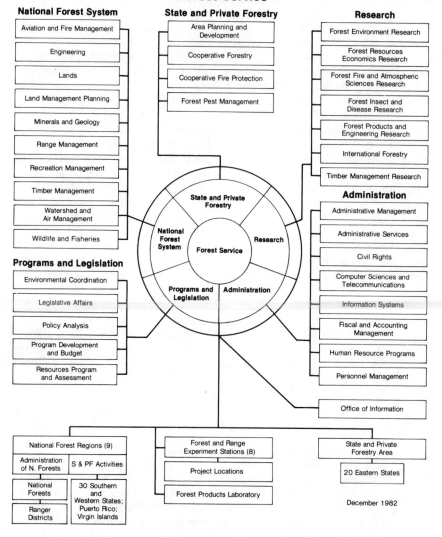

National Forest System
- Aviation and Fire Management
- Engineering
- Lands
- Land Management Planning
- Minerals and Geology
- Range Management
- Recreation Management
- Timber Management
- Watershed and Air Management
- Wildlife and Fisheries

Programs and Legislation
- Environmental Coordination
- Legislative Affairs
- Policy Analysis
- Program Development and Budget
- Resources Program and Assessment

State and Private Forestry
- Area Planning and Development
- Cooperative Forestry
- Cooperative Fire Protection
- Forest Pest Management

Research
- Forest Environment Research
- Forest Resources Economics Research
- Forest Fire and Atmospheric Sciences Research
- Forest Insect and Disease Research
- Forest Products and Engineering Research
- International Forestry
- Timber Management Research

Administration
- Administrative Management
- Administrative Services
- Civil Rights
- Computer Sciences and Telecommunications
- Information Systems
- Fiscal and Accounting Management
- Human Resource Programs
- Personnel Management

Office of Information

National Forest Regions (9)
Administration of N. Forests | S & PF Activities
National Forests | 30 Southern and Western States; Puerto Rico; Virgin Islands
Ranger Districts

Forest and Range Experiment Stations (8)
Project Locations
Forest Products Laboratory

State and Private Forestry Area
20 Eastern States

December 1982

"Chief and Staff" are a composite of leadership, sharing responsibility and expertise at a time when no single individual could possibly have all the answers at his fingertips. Nonetheless, it's the Chief who sets the pace and tone. "All of the leaders of the Forest Service, except Pinchot and possibly Silcox, have been 'bread and butter men,'" in the view of Earl D. Sandvig, provided in a personal interview, who spent many years in the agency. "That is, they depended upon the job for a living. They haven't had private fortunes like Pinchot; if they didn't like policies handed to them, they couldn't tell whomever they wished 'to go to hell.'"

That is one way to look at it. Although the Pinchot style of aggressive, public pronouncement has been scarce in recent years, there are defenders of the low key, laid-back style of leadership. They reason that the fearless, frontal Chief plays a dangerous game, inviting dismissal, politicization of the agency, and loss of freedom to make professional decisions. It is important to the Chief to have the confidence and support of troops in the field. He can achieve this, in part, by protecting their management and decision making from external forces. When the political power structure changes, the Chief shifts with it—proving himself the best soldier of them all—and maintaining for his agency the ability to choose its own leaders and ultimately to render its own decisions.

Among resource agencies the Forest Service has written an enviable record in this regard. The Chief is cautious to avoid defying leaders of industry, for they can get his scalp one way or another. He acknowledges the role of citizen groups, but is careful not to go overboard or to become overly intimate with them, since such behavior is likely to stir up the industry. The best he can do in this era is to define alternatives, and their consequences, in a professional manner and let the public make its choice through the political process.

Chiefs are rooted in their agency. The Forest Service is one large family, of which the Chief is the head. Chief McArdle attended large picnics prepared by field personnel and their families. He would enter a ranger's office and know the secretary by name. These gestures generated good feeling and support, but the Chiefs generally have been democratic without being maudlin about it. And stuffed shirts don't make it to the top.

Washington Headquarters and the Regions

Washington headquarters are in "South Agriculture," a vintage building extremely limited in space for the number of personnel of various agencies working in it. The Forest Service also occupies offices in high-rise private buildings leased by the General Services Administration at Rosslyn, on the Virginia side of the Potomac River. Compared with employment in the field, Washington holds increasingly less appeal. Even with a two-grade promotion, moving to Washington means a reduction in real income. Housing is expensive, particularly with high interest rates. To find the same quality of life they knew elsewhere in America, personnel must locate in Virginia or Maryland suburbs and struggle with morning and afternoon traffic jams.

Under the Senior Executive Service, high-ranking officials of the Forest Service in the early 1980s theoretically were eligible to receive annual salaries of up to $60,000, or more. A pay freeze in effect until 1982, however, set a limit of $50,112 for all employees, further reducing the attractiveness of Washington, since many receive nearly that much and enjoy the quality of life in the field. The pay ceiling more recently has been increased to $67,200.

Divisions of the Washington office (see Figure 3.2) coordinate specific activities and programs, working with the Chief and deputy chiefs, the Department, other federal agencies, and with field offices. The rest of the job, under the principle of decentralization, is delegated to the field. In 1973 the Office of Management and Budget ordered realignment of Forest Service regional offices in order to comply with a federal region concept then considered standard. The idea was to close three regional offices (in Ogden, Utah; Missoula, Montana; and Albuquerque, New Mexico) and two forest and range experiment stations. Fifty witnesses appeared before the Senate subcommittee on forestry and almost all spoke against it.[4]

These witnesses included Floyd Iverson, a retired regional forester, speaking in behalf of the Governor of Utah. The Intermountain Region, which Iverson had headed, covers 18 national forests and one national grassland, almost 32 million acres, in Utah, southern Idaho, Nevada, western Wyoming, portions of Colorado, and eastern California. The regional office has been located in Ogden, Utah, since 1906. Iverson opposed dissolution, citing positive results of the past decentralization policies, inevitable increased management problems due to a greater span of control, inevitable loss of public support, and decline in morale of employees. He insisted that research functions must be free of administrative control, must be organized around physiological and ecological regions, and be responsive to new problems in local areas.

Organizational Structure

The Forest Service structure remained unscathed. It is organized into nine regions, identified by numbers 1 through 10, except for 7. The former Region 7 was combined with Region 9 in 1966 to form an "Eastern Region," stretching from Minnesota to Maine. The regional forester is the "line officer" responsible for administration of all activities in a region, except for research and, in the East, state and private forestry. Staff directors furnish specialized assistance and advice. The timber staff generally is the largest and most influential. Others cover range and wildlife, fire control, lands, recreation, watersheds, fiscal

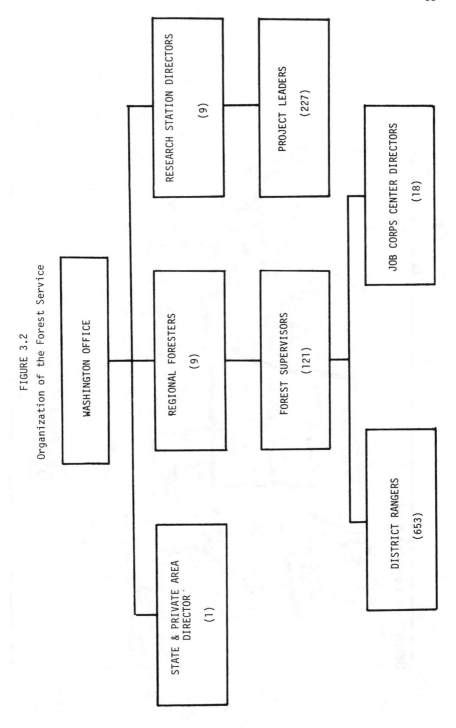

FIGURE 3.2
Organization of the Forest Service

WASHINGTON OFFICE

RESEARCH STATION DIRECTORS
(9)

PROJECT LEADERS
(227)

STATE & PRIVATE AREA DIRECTOR
(1)

REGIONAL FORESTERS
(9)

FOREST SUPERVISORS
(121)

JOB CORPS CENTER DIRECTORS
(18)

DISTRICT RANGERS
(653)

42

FIGURE 3.3

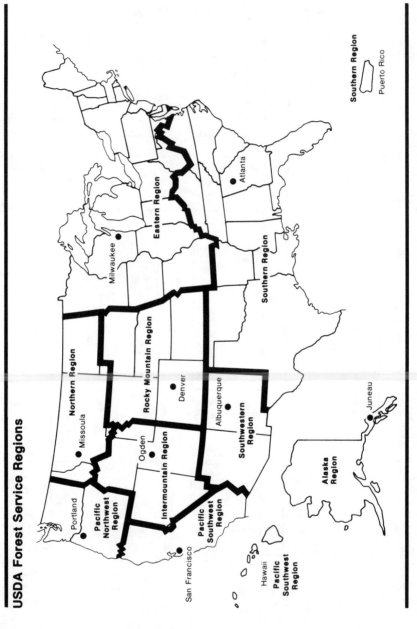

USDA Forest Service Regions

control, information and education, personnel management, and engineering.

The regional forester exercises considerable authority over management covering many millions of acres; responsibilities within his jurisdiction are almost identical to those of the Chief, and he has the authority to commit his agency to positions of national significance (as when the regional forester in the Northern Region ruled in 1981 against seismic exploration in the Bob Marshall Wilderness of Montana). He may be a conservative, timber-oriented forester, or he may be inclined to stress recreation, wilderness, and wildlife—though it's unlikely that a nontimber type would be assigned to a heavy timber region like the Northwest.

The 155 national forests are headed by a total of about 120 forest supervisors, some of them administering combined units. For example, in the Rocky Mountain Region (Region 2), three national forests—Grand Mesa, Uncompahgre, and Gunnison—are under the jurisdiction of a single supervisor (headquartered at Delta, Colorado). The supervisor has "line authority" to protect, develop, and utilize the resources in his care. Being a supervisor is one of the most desirable assignments in the Forest Service—close to the resource and to policy. The supervisor's headquarters usually are located where he and his family can enjoy life in a town or small city, though his pay compares favorably with that of ranking officials in the regional or Washington office.

Because district rangers conduct most national forest business, the supervisor devotes most of his time to reviewing and approving programs and to furnishing guidance to ensure compliance with policies and procedures. In essence, the forest headquarters is the planning unit while the ranger district is the executing unit. The supervisor's office embraces various specialists—in timber, engineering, recreation, wildlife, hydrology and soils, planning, landscape architecture, fire control, occasionally archaeology—to assist in resource management plans and to form interdisciplinary teams. Interdisciplinary planning is considered the new hope for balance in uses.

The 600-plus ranger districts are the basic line units for administration of the national forests. The Chief is at the apex of the pyramid, but the ranger supports it from the base. A ranger may administer more than a million acres in his (or her) district (the average size is 290,000 acres) and supervise a staff of 30 to 150 during the summer (professionals, technicians, and seasonals). Policy decisions and detailed planning from above are based, in one degree or another, on recommendations of the ranger and his staff. He (or she) must work with ranchers and loggers, state and county officials, local business interests, conservation groups, and the media, interpreting national policies to them and representing local viewpoints upward in the organization.

The forest ranger has come a long way since the early days. When the Forest Service began, the ranger on the ground had no automobile, no telephone, few trails, and few boundaries. Chances are he hadn't been to school, didn't know silviculture, and didn't wear a uniform. But as a rule he had community respect, which meant more than official prestige, and this enhanced the efficiency of decentralization.

The late Charles McDonald, known as "Ranger Mac," began his career in 1919 as a summer employee on the old Beartooth National Forest in the northern Rockies of Montana, his native state. He was just out of high school and, after his summer in the field, enrolled as a forestry student at the University of Montana. His entire working career was spent with the Forest Service in Utah, Wyoming, Idaho, and Montana. During his last twenty years he served as district ranger on the Stevensville District of the Bitterroot National Forest under his friend G. M. Brandborg. His field experiences were touched with raw adventure and the potential of a western novel.

"I had to deal with real killers, men who would shoot their neighbors in the back," he recalled in the late 1970s of the time he spent at Jackson Hole, Wyoming, fifty years before. "They were from families of outlaws and fugitives who had moved to that country a hundred years ago."

In southern Idaho, when he tried to deal with problems of an overgrazed range, stockmen defied and threatened him. They challenged him to prove they were grazing beyond the limit of their permits.

> It was rough country but I did it using my four kids, all of whom came into the world at remote ranger stations. I spaced them a half mile apart on horseback and had them go through the district checking for brands, tallying and showing location and comparing them with permitted stock.
>
> The results showed over a thousand illegal cattle. This was already after the ranchers had been warned. So this time I rounded up the cattle and told their owners to come collect them.[5]

Today's ranger tends to be removed from direct contact of this nature. He is pressed to absorb and understand politics; the environmental, ecological, and social consequences of his actions; and a host of new laws and regulations to weigh before he acts. The world, it seems, has beat a path to his back door. The arid states of New Mexico and Arizona, for example, once were lightly populated but now are booming; people are drawn from cool northern states to an area they perceive as a land of sunshine and low energy costs. But the water table is dropping steadily, forcing people to drill deeper for the dwindling, more costly water supply. The national forests are directly involved since they yield more than 60 percent of the water supply of the two states. Western Colorado is booming, too. Vail and Aspen began as ski resorts in the heart of national forests. Today Vail alone attracts a million skiers a year. These and other resorts have become desirable year-round communities, preempting winter wildlife habitat, pressing the forests for recreation and water. And now western Colorado is subject to still another growth cycle as a focal point in the search for energy resources.

Critics complain that decisions can be difficult to obtain and responsibility hard to pin down. "The Forest Service," according to the official manual, "is dedicated to the principle that resource management begins—and belongs—on the ground. It is logical, therefore, that the ranger district constitutes the backbone of the organization." But in recent years rangers and forest supervisors have been subject to transfer more than in the past, creating unfamiliarity and unwillingness to make commitments. The rangers complain of more paperwork, of increasing emphasis on process at the expense of resource program. As a result of such "bureaucratic creep," decisions are difficult to reach; and it is sometimes hard to find the official who will rise and say, "I did it."

Research is conducted through eight regional experiment stations, plus the Forest Products Laboratory at Madison, Wisconsin, the Institute of Tropical Forestry at Rio Piedras, Puerto Rico, and cooperating universities. Regional experiment stations function independently of national forest operations in order to ensure freedom of research. The station director and assistant directors serve as connecting links among projects, experiment stations, universities, and other scientific institutions, and with the Washington office in its nationwide program and policy considerations. Effective application of research is also the responsibility of the station director and staff.

State and Private Forestry (S&PF) cooperative programs are administered through an area office serving twenty northeastern states in which relatively little forest land is federally owned. Elsewhere in the country S&PF activities are conducted through the regional offices. These activities are treated in detail in Chapter 14.

Information and Education, or "I&E," has always been a key activity. Forestry in the Department of Agriculture was started to undertake research and provide information—a public information and education program. The Organic Act of May 16, 1862 establishing the Department declares one function shall be: "to diffuse among the people of the United States useful information on subjects connected with agriculture in the most general and comprehensive sense of the word."

Pinchot may have been the best press agent of his time. (Stephen T. Mather, first director of the National Park Service, also independently wealthy, was another—enlisting magazines, newspapers, and railroads to publicize the parks; he hired Robert Sterling Yard, brilliant writer and editor, later setting up Yard as executive secretary of the National Parks Association, to provide an outside voice of support.) Those early crusaders never ducked confrontation on tough issues; they demonstrated offensive skills by searching for and provoking crisis after crisis and fighting them in the arena of public judgment. As Theodore Roosevelt wrote in his *Autobiography*:

> It is doubtful whether there has ever been elsewhere under the Government such effective publicity [as that of the Forest Service]—purely in the interest of the people—at so low a cost. Before the educational work of the Forest

Service was stopped by the Taft Administration, it was securing the publication of facts about forestry in fifty million copies of newspapers a month at a total expense of $6,000 a year. . . . It was given out freely and published without costs because it was news. Without this publicity the Forest Service could not have survived the attacks made upon it by the representatives of the great special interests in Congress; nor could forestry in America have made the rapid progress it has (p. 67).

The Budget and Organization

Budget is integral to the agency's operation, the balance or lack of balance in its programs, its ability to manage federal lands properly and to promote better use of state and private lands. The Forest Service is directed by law to pursue these major programs:

- Administration of 191 million acres of national forests, national grasslands, and land-use projects.
- Research aimed at all types and ownerships of forests and associated rangelands, and their industrial, environmental, and other uses.
- Cooperative forestry programs conducted in conjunction with state forestry agencies in efforts to protect and improve 1.4 billion acres of forests, rangelands, and associated resources in private and nonfederal public ownerships.
- Human and community development programs aimed at employment and training of the disadvantaged and at benefitting human and community values.

The Forest Service is considered a three-sided operation—National Forest System, State and Private, and Research—but on the principle that where the dollars go so goes the power and power begets dollars, the National Forest System receives the highest appropriation and has the most personnel and influence. The 1981 budget of almost $2.3 billion included more than $1.8 billion for National Forest System Management, approximately $132 million for Forestry Research, and $88 million for Cooperative State and Private.

Despite increases in funding and personnel, the budget has been used to emphasize specific resource values and to de-emphasize others. Economic theories governing the budget process in modern government do not recognize social and environmental benefits because they cannot readily be measured or "quantified." It is difficult, for instance, to account for wildlife habitat protection or improvement in terms of tax benefits, and perhaps it shouldn't be necessary, but such "amenities" cannot compete in the minds of analysts in the Department of Agriculture or the Office of Management and Budget with definable values like timber, in terms of specific investments and returns.

A review of changes between 1981 and 1983 shows the use of the budget system for policy purposes. The 1981 budget, still reflecting the influence of the Carter administration, provided for a national forest timber sale program of 11.9 billion board feet (300 million board feet less than 1980), a volume generally considered sound in terms of economics and the environment. It included funds for advance land management planning where "temporary departures from nondeclining even-flow"—a technical way of expressing a sacrifice of the sustained-yield principle—might be found possible. Reforestation was budgeted to reforest 460,000 acres, a level believed necessary to eliminate feasibly a backlog of reforestation needs by 1985.

Funding for recreation rose 6 percent in 1981, to increase opportunities for heavily populated areas of the East and West Coasts. Funding for wildlife and fish habitat management rose 25 percent above 1980. State and private allocations provided extra funds for integrated pest management, while research was given more funding to improve forest productivity, wood utilization, western range desertification studies, and integrated pest management.

With the change in administration, a new viewpoint took hold. "We will use the budget system as the excuse to make major policy decisions," Secretary of the Interior Watt declared before a meeting of national park concessioners in 1981. He was referring specifically to the Land and Water Conservation Fund, the source of revenue for federal and state land acquisition, but the same idea was applied to resource programs throughout the government. The Council on Environmental Quality took a deep cut, reducing its activity and influence. Solar and other renewable energy programs were drastically curbed. The Environmental Protection Agency also lost much of its funding, despite facing legislatively assigned workloads in coping with hazardous wastes and toxic chemicals. On the other hand, the Reagan budget proposed huge increases for seventy water development projects in the western states (from $179 million to $950 million) and for processing leases of federal lands (from $21 million to $1.8 billion).

Dollar values prevail over nonmarket values, or biosystems. Assistant Secretary Crowell in the early 1980s set a course to "operate the Forest Service like a corporation and show a profit," although a public agency answers to a board of directors quite different from the board of any profit-making corporation.

"Our priorities are quite clear," Crowell declared before the North American Wildlife Conference at Portland, Oregon, on March 29, 1982. He defined priorities for the Forest Service as "more fully to utilize the timber resources of the national forests for this generation, to manage timber so that future generations can enjoy even greater bounty, and to encourage development of the timber, oil, gas, and mineral resources which are so plentiful on the public lands."

These priorities were clearly evident in the 1983 budget proposal of $2.13 billion. Funding emphasized timber and minerals development.

The Congressionally endorsed 1980 RPA program containing Forest Service integration of balanced multiple use appeared to be overlooked. Funding for recreation, fish and wildlife habitat, and watershed programs was largely to support timber and mineral activities, but significant reductions were made in activities to promote and enhance their own special values. Research was reduced by 11 percent, State and Private by 25 percent.

The 1983 timber budget called for an increase in cutting to 12.3 billion board feet (from 11 billion board feet for which appropriations were made in 1982). The budget proposed spending more than $585 million on roads, an increase of $70 million over 1982, for construction in "difficult terrain" with "access problems." On the other hand, trail maintenance was to be reduced, and the Forest Service said that 50,000 miles likely could not be maintained.

Budget allocations related to quantifiable "products" add to the emphasis on commodity production and to the influence of commodity-oriented professionals in the ranks. As of 1982, the Forest Service was staffed by 22,475 professional and technical personnel. Of 11,443 professionals, foresters totalled 5,823; followed by civil engineers, 1,345; range conservationists, 519; fish and wildlife biologists, 523; soil scientists, 309; landscape architects, 252; and hydrologists, 257. More nonforesters, in diverse disciplines, have been engaged in recent years, but since they are newer and younger, they usually are at lower grades and lower pay scales. A few are in key positions and are line officers, but generally their programs are more easily reduced when the budget needs cutting.

"Fish and wildlife are a major consideration," Assistant Secretary Crowell insisted before his audience in Portland. He announced a five-year research program for old-growth forests on the west flank of the Cascades from the Canadian border to northern California. "It will," he said, "identify plants and animals which depend on old growth or find their optimum habitat there, and try to determine their biological requirements. It will describe, inventory and classify old-growth ecosystems, and will evaluate different ways of managing old-growth stands."

However, Russell Sadler, a Portland *Oregonian* columnist, writing on the same day as Mr. Crowell's speech, noted budgetary threats to wildlife research. Beneath a headline "Old-Growth Timber Faces Political Saws," he observed:

> As plans increasing timber cutting on federal timberlands grind on with the methodical pace of a bulldozer, the administration also is reducing and eliminating efforts to conduct research into the consequences, all in the name of reducing the size and scope of the federal government.
>
> The Fish and Wildlife Cooperative Research Unit—a joint effort of the U.S. Fish and Wildlife Service, state wildlife agencies, various universities and the Wildlife Management Institute to study wildlife habitat—are losing their federal funding. They may be eliminated unless the states are willing to pick up the tab.

The Pacific Northwest Forest and Range Experiment Station, headquartered in Portland, faces an 18 percent budget cut that could mean a reduction of 50 of its 300 employees and the closure of its Silvicultural Laboratory in Bend, cuts in programs at the Forestry Sciences Lab in Corvallis and the Wildlife Lab in La Grande mean less research in range and wildlife habitat, the use of fire as a forest management tool, insect and disease control, timber management, forest economics and watershed management.

Some scientists suspect that budget cuts have more to do with politics than fiscal prudence. Last year plant ecologist Jerry Franklin and a group of scientists produced a study "Ecological Characteristics of Old Growth Douglas-Fir Forests." The study adds new weight to biologists' contentions that old-growth forests are a valuable economic asset as wildlife habitat, seed gene pool reservoirs and watersheds that produce the purest drinking water—assets the biologists insist are frequently overlooked in the calculations of timber-oriented forest managers.

Another powerful budget influence relates to the principle that spending begets funding, recognition, and promotion. Performance standards are based on what are called "hard outputs," the quantifiables, such as volumes of timber logged or volumes of visitors in campgrounds. Performance of everyone in the federal government (as in most large organizations) is rated periodically and systematically. In the Forest Service, performance standard is defined as "the expressed measure of the level of achievement established by management for specific performance elements of a position or group of positions."[6]

Performance elements and standards derive from position descriptions, work plans, instructions, project proposals, goals, targets, job analyses, equal opportunity action plans, individual development plans and other sources that assign or fix responsibility for accomplishment. Standards and elements are identified jointly by the employee and supervisor and are documented in writing. They include a critical element spelled out in Chapter 430 of the USDA Personnel Handbook: "A performance element of sufficient importance that performance below the minimum standard established by management requires remedial action and denial of a within-grade increase and may be the basis for removal or reducing the grade level of the employee."

These factors help to explain—they are essential to understanding—motivations and modus operandi and pressures to perform of personnel in a public agency. In the Forest Service there is something still more: the idea of belonging, of being fully accepted by one's peers.

The "Weitzman Report," or "Lessons from the Monongahela Experience," by Sidney Weitzman, an in-service analysis of the controversy over the Monongahela National Forest, was prepared in 1977. It is discussed in Chapter 7, but the following portion is relevant at this point:

There is a strong sense of loyalty and belonging to the Forest Service "family." I have seen such loyalty to an organization and its people only in the armed forces under wartime conditions. Thus, requests made from the Administration or Congress through the Chief's office received highest priority. No one wants to "let the Chief and the organization down" in meeting their assigned targets. The organization, in turn, fostered and rewarded loyalty, ability and performance. Conversely, it withheld rewards from non-performers. In this case, it took the form of awarding unobligated "crash" timber sale money near the end of the fiscal year to those forests or districts that could perform by getting out the cut. This money was usually allocated on the basis of "Can you sell X volume of timber if you receive X dollars?" This practice assured performance, but, coming late in the year, did not encourage full planning or care in sale design and implementation.[7]

Continuing Changes in Form and Function

Many things have changed in both internal and external relations in order to overcome past weaknesses. The National Environmental Policy Act of 1969 signaled a need for consideration of a range of alternative courses, based on evaluation of biophysical, economics, and social effects in a human environment broader by far than traditional choices of supplying products and services demanded by society. The Office of Environmental Coordination in the Washington office represents a departure in itself, directing the integration into normal agency functions of laws governing clean air, clean water, coastal zone management, use of herbicides and pesticides, and other aspects of the environment. This Office coordinates interests and activities of the Forest Service with those of Council on Environmental Quality, Environmental Protection Agency, and President's Domestic Council. It endeavors to ensure that all timber sale planning conforms with the NEPA, although such conformity has remained far from uniform.

Socially Responsive Management (SRM) represents another means to consider social impacts. Initiated in the Rocky Mountain Region (Region 2), it recognizes that impact analyses are crucial to socially responsible and politically supportable resource decisions.[8] At one time or another the Forest Service has engaged social scientists as consultants and interdisciplinary team members in order to alert decision makers to significant social changes a management action could cause and to clarify the issues and concerns of affected people—in other words, to reflect social values connected with resource activities.[9]

In May 1981, the Forest Service proposed to systematize this approach, announcing in the Federal Register a new social analysis policy. "The social scientist or social science coordinator should develop a matrix of the effects of Forest Service alternatives on the social variables selected," according to this announcement. "The matrix should consist of an overview of the projected effects of Forest Service actions for each alternative and should be accompanied by a more detailed narrative summary." Time will tell whether the SRM program will prove a burden of paperwork or the means to be socially responsive to changing times.

One of the most obvious of contemporary social changes is the increased participation of women in virtually all professions. The Forest Service is yielding to this change, but it took a long time. In *Men Who Matched the Mountains*, a history published by the Southwestern Region (Region 3) in 1972, the emphasis is on the roles performed by men. The book does record, however, how a young woman named Anita Kellogg came to work for the Forest Service in 1917. Three years later she took and passed the Civil Service examination to qualify as a ranger. The supervisor of the Santa Fe National Forest offered her a job, which the regional office disapproved because she was an "unattached female." Next year, while working on the Coronado National Forest, she was made a special deputy fiscal agent to pay off the firefighters; however, when she was sent to fill the same position on a Class C fire on the Mogollon Mountains on the Gila National Forest, the supervisor there objected to having her, again because she was an "unattached female."

In a field long (and still) dominated by white males, the Forest Service has made substantial headway in hiring women and minorities. (Human resource programs, designed to benefit the land and society, are the subject of the next chapter.)

The Forest Service has demonstrated its capacity for change. Chief Peterson is a graduate engineer, rather than a forester (as were his ten predecessors). During mid-level training he was sent to the Kennedy School of Government at Harvard University, and others participate in similar programs to broaden their horizons. The influx of biologists, sociologists, historians, archaeologists, and representatives of other disciplines ensures that foresters are not talking only to each other.

The Intergovernmental Personnel Act of 1970 (IPA) is a mechanism specifically designed to broaden experience. It enables the Forest Service to assign personnel to—or to accept personnel from—states, local governments, and institutions of higher education for up to two years (and in some circumstances for an additional two years as well). "Mobility assignments" can be made to: share scarce experience, provide operating experience in a counterpart organization, provide general developmental experience, improve management of programs and make more efficient use of available resources, strengthen intergovernmental understanding, transfer new technology, and encourage use of research findings. From 1972 through 1981 the Forest Service participated in 558 IPA assignment agreements, or about 55 per year.

Another means available to broaden experience and outlook is provided by the program called Volunteers in the National Forests— but only if Forest Service personnel choose to use it as such. Under terms of the Volunteers in the National Forests Act of 1972, individuals are able to donate time, talent, knowledge, and expertise. Though compensation is not authorized, the Act allows unlimited opportunities for doing and learning. During 1980 more than 15,000 volunteers participated (of whom 36 percent were women and 15 percent minorities). It is an aspect of involvement that can work both ways, providing that volunteers are given actual responsibility rather than menial tasks and that permanent staff personnel heed the aspirations and expressions of retired educators, minorities, and young people of diverse backgrounds and training.

Generally, volunteers work at information centers, conduct natural history walks, build and maintain trails, and serve as campground hosts. They also act as wilderness rangers, research assistants, and library aides. They receive legal protection as well as insurance for work-related injuries; in some cases they can be compensated for travel, food, and lodging expenses. The Appalachian Mountain Club has long been a major volunteer influence in the White Mountain National Forest and elsewhere in New England. With a grant from the Richard King Mellon Foundation in 1981, AMC became involved elsewhere in the country as well.[10]

Complexity in modern life seems to have dictated a pattern of something gained and something lost, a shift of decision making away from the grass roots. The Forest Service Manual, which emerged from the slender Use Book devised by Pinchot, by the early 1980s spanned 27 volumes and an estimated 20,000 pages—and there also were more than 100 handbooks from the Washington and regional offices, even including a handbook on how to write directives.

"Demands for more services, for better and faster decisions to resolve conflicting interests, and a trend toward more involvement by a concerned public require us to take advantage of all technology that can help us reach our objectives," wrote Chief Edward Cliff in the foreword to *A Computer Systems Study*, published in 1971. "Computers are tools we have had for years, but we find that the potential of these amazing devices has barely been tapped."

The Forest Service has been a leader in computer use, employing more than 250 computer specialists, programmers, and analysts. Two of the major programs are ADVENT to allocate budgetary resources and FORPLAN for land-management planning, but computer developments make it conceivable that even small ranger districts can make use of one technique or another.

The Fort Collins (Colorado) Computer Center (FCCC) is the central computing source, providing reference works for a wide range of users. For instance, RIM (Recreation Information Management) system is used to collect, store, retrieve, and summarize information on sites, maintenance costs, and recreation use on over 22,700 national forest recreation sites and areas. It provides data summaries for reports, public information, budget planning, recreation use reporting, recreation planning, and land-management planning.

FORPLAN, or Forest Planning Model, begun in 1980, is considered a primary tool for analysis in preparing national forest plans under the

National Forest Management Act.[11] In a letter to the field in March 1979, Associate Chief Douglas R. Leisz outlined FORPLAN in a verbiage old line rangers might find intimidating:

> The Forest Planning Model is a linear programming package for resource allocation and activity scheduling, with linkages to program planning. Resource allocation and activity scheduling determine when, where, and how specific practices or activities (i.e., management prescriptions) will be applied to the land. Allocation is affected by the ability of the schedule to meet periodic targets, and scheduling is affected by assignment of land to management prescriptions and intensities given it through the allocation process. While allocation and scheduling may be separated for operational reasons, they must be modeled as parts of a single process.

There is a tendency among computer advocates to feel that these mechanical devices solve virtually any problem and that decision makers lived in a primitive world without them. According to the *Systems Management Plan*, which the Forest Service issued in 1980:

> Well-managed systems and related information technology create new opportunities for responding effectively to current and future challenges. Also, technology now makes it possible to improve the decisionmaking process through analysis techniques, simulation, modeling, etc., not previously available or feasible. . . . Forest Service experience has shown that without automation it is physically impossible to complete necessary analysis. Data and alternative complexity require the use of advanced analytical techniques including modeling and simulation. Without the support of information technology to aid this effort the Forest Service would be unable to carry out its responsibilities.

Possibly not, but the computer is no more efficient than its operator and the data fed to it. "Garbage in, garbage out," as the saying goes. The end product depends upon interpretation and application by resource professionals. The challenge is to utilize data processing systems as aids in solving management problems, rather than to rely on them to furnish the questions and answers.

Another aspect of change with mixed benefits involves unionism. Since 1970 some employees, mostly at technician levels, have been unionized; in many circumstances they can't contribute time and effort to useful projects even when they want to. (Union activity has been strongest on large forests and in urban areas.)

On this particular point of unionization, in 1962 President John F. Kennedy issued an executive order allowing collective bargaining for federal employees. In December 1980, the Forest Service became the first federal agency to sign a nationwide contract with the National Federation of Federal Employees. Although district rangers and assistant rangers have generally been considered as part of management and not eligible for union membership, it may still seem strange for an old line, tightly knit organization like the Forest Service to be in front of the movement to unionize. A mixed blessing, perhaps, but in some ways a credit to both the agency and its personnel.

The Dangers of Losing Touch

Inherent in any large organization, whether designed for profit or public service, is the tendency to internalize one's interest and focus. In federal career agencies, particularly those recruiting largely from one profession, promotion is largely from within. This has its pluses as well as minuses, since it protects the agencies from the influence of political appointees. Still, candidates for promotion are sometimes more successful it they are graduates of colleges and universities that have traditionally furnished personnel for the agency. (The same holds true in the media, where a newspaper editor is likely to favor a job applicant who has attended his own journalism school.) Such people move up faster. Nonconformists, however, are either weeded out or dispatched to positions without significant responsibility.

In the case of federal foresters, the badge and uniform set them apart. Requirements to become a forester under Civil Service (the 460 series) include completing a four-year college program with a major in forestry or a closely related field. At least 24 semester hours in forestry must include such courses as silviculture, wildlife habitat management, forest ecology, forest soils, and biometrics. Courses in such subjects as sociology, government, range ecology, or law, may be considered desirable but are not required. There is no requirement for the humanities. Transfers, particularly in early career years, tend to give the forester one primary reference group, his own organization; unless he strives hard to prevent it, the individual, despite involvement in local communities and far-flung assignments, becomes internalized.

The problems of inbreeding were accented in a critical article in *Harper's* (April 1962) by a Washington correspondent, Julius Duscha:

> Over the last half century the Forest Service has developed an *esprit de corps* that detractors refer to as a priesthood or a classic example of the faceless organization-man system. Made up almost entirely of forestry school graduates, the Service has an ingrown merit and promotion system that covers even the Chief Forester and permits practically no transfusion of new ideas or new blood except at the very bottom. And the man who comes in with ideas soon submerges them in the interest of regular and choice promotions or else quickly leaves the Service in a revulsion against its monolithic structure.

Somewhat the same idea was expressed differently in *A University View of the Forest Service*, issued in 1970 by faculty members of the University of Montana Forestry School following their review of the clearcutting controversy:

Bureaucratic structures such as the Forest Service not only alienate public support, they also inhibit effective exploitation of key personnel. In order to maximize local community resources and to attract local community support, those persons in the Service most intimately associated with local community interests must be free to act. They require a latitude and flexibility of operation which is denied them within the conventional bureaucratic structure. The person most sensitively located to relate constructively to local people is the district ranger. He represents the Forest Service. He makes administrative decisions within limits imposed by agency policy. By and large, the image he projects is likely to determine the way in which those within his district perceive the total organization. Yet his authority is severely limited and all too frequently his decisions and answers are bureaucratically detemined.

Despite reservations or frustrations that he may feel, his ultimate action is likely to be taken within the context of his supervisor's office and eventually of the regional forester's office. He is therefore denied the flexibility to meet issues and problems on an ad hoc basis. It might also be said that his decisions are always predetermined, at least with respect to major issues and problems.

The idea of management close to the resource and close to people has been variously expressed. Hank Rate joined the Forest Service in 1958 and resigned eight years later with a degree of disillusionment. He joined after obtaining his master's degree in timber management at Colorado State University. In a letter written in February 1982, he recalled that he had signed on with a government agency doing a good job managing a little understood resource:

We had a handful of non-professionals who had started up through the CCC program. We planted trees, cruised and marked timber, built campgrounds and fought fire under their wing. They were some of the finest men ever to serve the U.S. government. Most were amazingly tolerant of university-trained foresters destined to climb past them.

Rate advanced through the ranks of the Gallatin National Forest and became a district ranger. When he decided to quit he wrote a lengthy letter of resignation to the forest supervisor dated August 3, 1965. It included the following:

No matter how hard we have tried to get wildland management computerized and put into formulas, it still remains an art and not a science. An artist must have the feel of his work. Moving the ranger stations into town and loading the rangers with nonfield duties that take priority over field work is making it impossible for the man who is making the decisions to have this "feel." The grass roots character of the Forest Service dies when the ranger loses contact with the ground.

I see an increasing tendency on the part of resource managers to want to regulate things that are not in need of regulation at this time, to get people to feel they must come crawling to us to get permission to do anything on the national forest. This is a joke to most of the people who live near the forest. We are not taxpayers in the sense that local landowners are. We do not share the burdens of ownership, and cannot expect the respect that goes with them. The only way we can command respect is as humble, modest civil servants. We must be firm when firmness is called for, but at all costs avoid the impression that we are big wheels and that people must come crawling to us.

Rate feels that money is the root of evil.

The best thing for the Forest Service would be to cut back funds and put the money that is left out on the ranger districts. Get the man who is making the decisions out on the ground. He needs to cruise the timber, clean the campgrounds, and get the feel of the ground. Young professionals often *never* do these things in their careers.

Blowing the Whistle—and the Cost

Is there room for independent views within the agency? Should nonconformity be tolerated?

Senator Patrick Leahy of Vermont answered these questions in a statement submitted to a hearing before the Subcommittee on the Civil Service of the Committee on Post Office and Civil Service of the House of Representatives on March 4, 1980 (published under the title "Civil Service Reform Oversight, 1980—Whistleblower" by the Government Printing Office, 1980). Senator Leahy declared (pp. 4, 6):

> As the Federal Government grows larger and more complex, the op-portunities for inefficiency, corruption, mismanagement, abuse of power, and other inappropriate activities become more frequent. When these opportunities are exploited, the public finds itself spending more money for less service. The people in top policymaking positions are often too far removed from the programs to be able to spot such cases. The people in the best position to bring examples of these activities to policymakers are the Federal employees involved in program implementation. . . .
>
> The core of the whistleblower issue is the question of the Federal Government's responsibility to the people. Disclosure of waste and abuse by Government officials should be seen as a sincere commitment to make this Government more responsive to people's needs and more worthy of their trust. Taken in this light, these disclosures can be used to strengthen and improve the Government, not to weaken or disrupt it.

Leahy and others have insisted that freedom of expression, whether called "whistleblowing" or by any other term, needs recognition and defense as an essential element of good government—that diversity of opinion and even dissent should be allowed to circulate within and without an agency, like a danger warning.

"One thing is certain," in the view of A. Claude Ferguson, who served the Forest Service for thirty years before his dismissal in 1974, expressed in a personal letter to the author, "You will not get sound decisions and sound resource management until you get good personnel and turn them loose. As long as those who try are persecuted and the uninspired are rewarded, you can expect things to go downhill."

When a revelation is made public, the question asked inside any federal agency tends to be "Who leaked it?" rather than "Is it right or wrong?" The best known of all whistleblowers, A. Ernest Fitzgerald, the Air Force cost analyst who revealed the mammoth Lockhead 5A cost overrun, said, "When I objected to the cost overruns, the two billion dollars was not the problem. *I* was the problem."

As a consequence of experience and extensive testimony by Fitzgerald and others, the Civil Service Reform Act of 1978 included key provisions aimed at ensuring employees the right to make public, without reprisal or fear of reprisal, information concerning acts or failures to act by their employer which they believe harmful to the public interest. The Act established the Merit Systems Protection Board and an independent special counsel to receive information and complaints, with authority to investigate and decide employee appeals. Certainly aggrieved employees are not always in the right, and there are two sides to consider, but the odds have been tilted in favor of authority. "Whistleblowing is both career and institutional suicide," observed Senator James Abourezk, of South Dakota, in support of the protective provision in the 1978 Act. "The moment a whistleblower surfaces, the full force of bureaucratic oppression is brought to bear on him or her."

The Forest Service is quite conscious of the applications of whistleblowing. Between February 1980 and March 1982 the agency processed 104 whistleblower complaints. A total of 27 resulted in some type of corrective action (although in a number of cases investigations were said to show complainants lacked information or understanding of regulations and laws). Chief Peterson has supported whistleblowing as "a mechanism of positive change and self-correction." Before the meeting of the regional foresters and research directors (RF&D) in 1979, he advised: "Be alert to and prepared to deal constructively with whistleblowers. I'd suggest you start by assuming there is a possibility the whistleblower has a point. To do otherwise will seriously undermine our ability to deal with those at our levels."

As a case in point, Claude Ferguson objected to plans to construct, with what he called misappropriated funds, offroad-vehicle trails for motorcycles and four-wheel-drive vehicles on Indiana's Hoosier National Forest. He charged that these would be routed through public recreation areas, across steep grades, and through young pine plantations, in violation of the agency's own standards of safety, erosion control, and wildlife protection. After failing through internal channels to halt construction of the trails, Ferguson provided an affidavit to the Izaak Walton

League which brought legal action in 1974; his affidavit proved instrumental to the League in obtaining a court order temporarily restraining the Forest Service from opening the trail. Three years later, just before the case was due for trial, the Service withdrew its plan.

The League filed a complaint in 1974 charging that his superiors were harassing Ferguson, one of its key potential witnesses. He was then ordered transferred to regional headquarters in Milwaukee; on declining the transfer, he was dismissed within eighteen months of eligibility for retirement. Ferguson was not only honored by national conservation organizations but received contributions from old colleagues to pursue a legal appeal. In 1976 he was granted retirement rights. And two years later a district ranger conceded in a local Indiana radio broadcast that he had, in fact, followed instructions to construct the trails in violation of prescribed regulations and with misappropriated funds.[12]

Monty Montagne, a Forest Service biologist with twenty-five years service, and four other biologists in 1971 discovered the presence of the rare blunt-nosed leopard lizard in Ballinger Canyon, a part of Los Padres National Forest, California, and recommended against offroad-vehicle use until all environmental impacts could be studied. His superiors, however, proceeded to designate much of the canyon as an "open-use" area, emphasizing the "family fun" implicit in motorcycling, though promising that the most critical part would be fenced.

"Unless Ballinger Canyon receives complete protection from any ORV use," persisted Mr. Montagne, "the total gene pool of the blunt-nosed leopard lizard will become extinct in a very few years. The forest supervisor has the authority to designate the forest system lands in Ballinger Canyon as a Biological Natural Area to accomplish this preservation of a very unique and endangered species."

For his troubles Montagne was ordered transferred to the regional office. Pete Sorenson, special assistant to Assistant Secretary Cutler, was dispatched to investigate complaints in Montagne's behalf. In his report Sorenson wrote that he was shocked at the extreme erosion and denudation of vegetation on the site, then adding:

> What is more disconcerting than the damage is the apparent disregard of the provisions of President Carter's Executive Order (directing protection of public lands from ORV impacts).
> Aside from the offroad vehicle problem, I recommend that we should ask the inspector general or general counsel to look into Mr. Montagne's case. Although there may be a need for a wildlife biologist in the regional office, it seems more than coincidental that the wildlife biologist the agency chose to transfer to San Francisco is Mr. Montagne. The Department has a responsibility to protect those staff professionals in the agencies who render professional judgments to line officers.

Forest Service tradition, however, dictates that line officers (ranger, forest supervisor, regional forester), rather than staff officers (including biologists), make decisions. Consequently Montagne was removed despite strong support from the scientific community.

Dr. Robert C. Stebbins, professor of zoology and curator of herpetology at the University of California, declared: "I am appalled that such an action could be taken against a person attempting to do his duty. Offroad vehicles are clearly out of control. They must be properly managed or we will suffer losses of unique biological materials, irreplaceable soil, and visual appeal of many of our wild lands. Mr. Montagne should be commended rather than disciplined for his actions." Agency officials, explaining their position at a later date, stated that Montagne's forced transfer was initiated by the forest supervisor "in order to strengthen the biological program on the Los Padres." Although Montagne was separated from the Forest Service with a "discontinued service retirement," his efforts and those of others spurred appointment by the Chief of a task force to review ORV management in the national forests. (This activity is covered in Chapter 9.)[13]

Arthur Anderson was hired in 1975 as a highway engineering inspector on the Lolo National Forest, in Montana. From that year through 1979 he persistently complained of alleged waste, mismanagement, and disregard for professional engineering standards with reference to timber sales. In October 1978, under terms of the Civil Service Reform Act, he directed charges via the Special Counsel against his superiors, insisting that contractors were not given adequate credit for road building, thus being encouraged to construct substandard roads into marginal timber sales areas.

In March 1979, the Lolo National Forest proposed his discharge from employment for insubordination. Later this was changed to a "directed reassignment" to the staff of the Klamath National Forest in California, "because the management-employee relationship had been rendered inoperable in Region One." Meanwhile, Secretary Bergland investigated his charges and found them justified. In September 1979, he directed the Forest Service to develop "a system for reviewing and considering reports of improper resource management actions, so that specific complaints of this nature can be reviewed at a high enough level to insure uniform policy application with a specific finding on each report received."

One month later the Department's Office of General Counsel reported on an audit of Anderson's charges:

Our review showed that the Forest needed to improve its practices for designing, inspecting, and accepting public works and timber sale contractor-constructed roads. In this regard, we found that many of the complainant's allegations of non-compliance were technically correct.

With respect to road design, Forest personnel had not followed, for the most part, Forest Service manual instructions. We found deficiencies in design relative to intersections of roads, turnarounds at ends of roads, reconstruction of existing roads, criteria for handling dead trees, and temporary spur roads. Based on our review, Forest personnel had placed an unwarranted trust in the ability and willingness of road contractors to follow plans and specifications.[14]

Anderson protested his reassignment to California and was dismissed. "I should remain to insure that corrective actions are carried out," he wrote the Special Counsel in December 1979. Ultimately the Denver Field Office of the Merit Systems Protection Board ordered that his transfer be set aside and that the Forest Service and Agriculture Department "cease immediately from retaliating against Mr. Anderson for his protected activities." The Forest Service appealed but the Board in Washington on January 18, 1982, affirmed the decision of the Denver Office, ordering his reinstatement.[15]

The Pressures of Politicization

Officials at high levels have taken various steps to open and protect channels of communication. Charges of internal waste and mismanagement tend to be better heard and heeded in the 1980s than in the 1970s. Difficulties of one kind or another have arisen in all agencies of government. President Reagan created the impression of the average federal employee being shielded by protective civil service rules, turning bureaucracy into deadwood. Certainly lower ranks do contain deadwood hard to clean out, but there is plenty of deadwood at higher echelons as well. When the prime criterion for advancement is being "a team player," rather than a boat-rocker or one who "stirs the pot," banality prevails over originality.

From the late 1960s through the early 1980s, the federal administrations (of Richard M. Nixon, Jimmy Carter, and Ronald Reagan) downgraded or politicized federal employment, creating an environment demeaning to able professionals, generating cynicism, and devastating morale.

President Reagan frequently derided the Civil Service and pledged himself to clean out "the puzzle palace on the Potomac." Douglas Bruce McHenry, a longtime professional of the National Park Service, was moved to write the President as follows (voicing the feeling of many in the Forest Service and other agencies):

> There is a growing feeling among your career employees that you do not support those of us who support you. We feel that you and your staff are cutting us down in the eyes of the public. Many of us are loyal to your program and goals. We are dedicated to our agencies, their missions and our duty to serve the public need. We try hard to be efficient and save money. We recognize the need for cuts in spending. We have honor and pride in our jobs. In many cases we have distinguished careers.
>
> We do not see ourselves as being the only cause of inflation. This hurts a career employee. Some of my non-goverment colleagues now look upon

me as less professional. I find myself constantly apologizing for the federal government. Many of my fellow employees are leaving or retiring early in frustration.

The short-term result of this is a reduced work force and reduced federal spending. You will not have difficulty in meeting your goals in this regard. We will, of course, need to reduce services to the public.

The longer term effect of the cuts in government could be more damaging. I believe that we are going to see a government made up of employees who are substandard, poorly educated, poorly qualified and poorly motivated. I feel they will represent those people private industry would not accept. This is being brought about because the better qualified, better educated, highly motivated career employees are the ones who are leaving.

Politicization struck close to the Forest Service in March 1982, when Norman A. Berg was dismissed as Administrator of the Soil Conservation Service. He had been appointed in 1979 after 39 years with the agency and was only the sixth chief since the establishment of SCS almost half a century before. His successor, to whom he gave way, was a political ally of Secretary of Agriculture John R. Block; the new SCS Administrator came aboard without experience in government or in soil conservation. Berg departed gracefully, but with a very telling statement quoted in the media: "Once a decision of this kind is made, it usually doesn't stop there. It usually goes deeper into the agency with succeeding administrations and the attrition of truly professional organizations becomes heavy."

The Civil Service Reform Act of 1978 established a Senior Executive Service ostensibly to encourage leadership at higher ranks. The effects, however, have been questionable at best. "Low Morale at FS Laid to Politics"—thus reads a headline in *Portland Oregonian* of April 20, 1979, above an article by Jim Kadera, a staff writer. His lead paragraph: "Politics is attacking the morale of the U.S. Forest Service in a way that all the wildfires and timber-killing insects could not."

Kadera's article was based on an interview with Rexford A. Resler, who had taken early retirement in 1978 from his position as Associate Chief of the Forest Service to become executive vice-president of the American Forestry Association. Resler identified in the interview two key factors causing morale problems:

1. Political accusations that dedicated people have not managed national forests responsibly.
2. Threat of political appointments stemming from passage of the Civil Service Reform Act.

As he noted, the Reform Act allows an administration to place political appointees into 10 percent of management positions of an agency, which means that such appointments could reach down to the level of regional foresters and affect the performance of forest supervisors. Expressing a vewpoint similar to Berg's, Resler declared: "If we get to the point of making long-term decisions on natural resources for short-term political expediency, we're in trouble."

Critics of the Civil Service Reform Act have also felt that by providing for bonuses for outstanding performance of up to $20,000, the Act discourages "boat-rocking," that it promotes conformity, responsiveness to the hierarchy and immediate supervisors. Taking a tough stand, after all, is apt to make somebody unhappy and incur the opposition of politicians and the displeasure of superiors. Avoiding conflict has become a safer course for the federal employee.

It may be a part of the general social trends and the pressures of political interests on all levels of government, but it is hardly a secret that restoration of the old esprit and devotion is a challenge of the 1980s; clarity of direction, fearlessness, and pride are not quite what they used to be.

4

OPENING FOR OPPORTUNITY AND CONSTITUENCY

> We want to develop all our employees and provide equal access for all Americans. Promoting equal opportunity in employment and services is good management and sound social policy. Putting people productively to work and helping them to realize their potential benefits them as individuals, and the nation as well.
>
> —Jetie B. Wilde, Jr., Director, Civil Rights, Forest Service, 1981

Human Resources Programs

In the chaos of depression America, men and women were jobless and adrift. Many were young. The new administration of Franklin D. Roosevelt had to cope with their fate. It also faced scars brought by generations of waste and misuse of the land. The consequence of planless abuse of the forests and fields was erosion; three billion tons of soil washed downstream each year. A like amount was blown away by wind. Deserts of dust replaced grasslands.

Throughout the campaign of 1932, Roosevelt corresponded with Gifford Pinchot, the Republican governor of Pennsylvania, and other conservation leaders, gathering their views on a possible program to combat soil erosion and timber famine through self-sustaining public work. In mid-November, following the election, Henry A. Wallace, Secretary of Agriculture–designate, and Rexford Guy Tugwell, Roosevelt's economic advisor, asked Robert Y. Stuart, Chief of the Forest Service, if he could develop plans to put 25,000 men to work in the national forests.

Such relief work had already begun on a limited scale in the West, where the Forest Service had cooperated with state and country officials in California and Washington. Accordingly, Stuart responded with certainty that the Forest Service could handle 25,000 men usefully; but within a month he was told the number would increase tenfold.

Thus the Civilian Conservation Corps was born. It was part of the American scene for nine years, until 1942, and part of the lives of 2.5 million young Americans who passed through its ranks. They enlisted in what was primarily a work program but found a healthy return to the land and the making of many of them. CCC also enrolled a cadre of professionals, classified as "technical assistants," to furnish leadership; these included men and women who later became key figures in land conservation agencies.

CCC enrollees planted millions of trees and fought fires (in which forty-seven lost their lives). They opened thousands of miles of trail; built 600 lookout towers, scores of water storage basins and well-digging units to ensure a ready supply of water for tanks and pumps. They built bridges, picnic facilities, and campgrounds, many of which were still in use more than forty years later.

CCC, however, had its weaknesses. The black American did not get his proper share of the program. Education was one of the less successful endeavors. "Development" occasionally resulted in overdevelopment. But it proved to be the turning point in the lives of many Americans, their first chance at fresh air, good food, and a feeling that someone cared.

A generation later Congress established a new force to be called the Job Corps. In 1964, when it began, almost 30 percent of young people entering the nation's work force were school dropouts. They had difficulty finding work because they had no skills, and many were unable to hold employment when they got it. Through the Job Corps they were assigned to camps in national forests and other public lands where work would be directed "toward conserving, developing and managing the public natural resources of the nation and developing, managing and protecting public recreation areas."

They lived together, worked together, played together and had their misunderstandings—blacks, whites, Hispanics and Native Americans. Without training, many undoubtedly would end up on the welfare rolls as nonproductive citizens, or as criminals. Critics called it an expensive program, yet it cost less per capita than keeping a person on relief or in a penitentiary. Some Job Corps enrollees owned toothbrushes for the first time and had to learn to use them, and almost 40 percent could neither read nor write. Some failed and quit, but more than a few found inspiration to pursue their education or to find continued employment in the field of natural resources.

Other human resource programs followed, designed to provide jobs, training, experience, and education for young and old, while contributing to the improvement of public facilities. The Youth Conservation Corps was established through legislation in 1970 (patterned after the program of the Student Conservation Association, which already had operated effectively as a volunteer organization in national forests, national parks, and other reserves). YCC provided summer employment and offered a degree of conservation education for 15- to 18-year-olds from all levels of society.

Success with YCC resulted in the establishment of the Young Adult Conservation Corps through passage in 1974 of the Emergency Jobs and Unemployment Assistance Act. YACC provided for year-round jobs for unemployed and out-of-school young people between the ages of 16 and 23.

Jobs were also provided for older citizens through the Senior Community Service Employment Program, administered under a provision of the Older Americans Act of 1965 and providing for part-time employment, work experience, and skills training for the economically disadvantaged 55 years and older living primarily in rural areas.

Such activities are not without weaknesses and deficiencies; they do not always fulfill their promise of serving the economically disadvantaged or minorities and are not always accessible to the handicapped. But considerable construction, rehabilitation, maintenance, and improvement of facilities on national forests and other public lands could not have been accomplished without them.

Equalizing the Opportunities

Equal opportunity in federal employment is required by law, but that doesn't make it so in fact. The most cursory observation reveals wide variation in compliance, or desire to comply, among the various federal agencies. As for performance by the Forest Service, the natural resources field has long been dominated by men, and especially by white men; yet the agency has made appreciable headway, and the same could be said for the forestry profession and forest products industry.

In 1975 a Civil Rights Staff Unit was established in the Washington office, reporting to the Deputy Chief for Administration. This unit was authorized to develop and administer a comprehensive program to assure equal opportunity in all Forest Service activities. Five years later, equality was plainly still a long way off, but progress was measurable. Between 1976 and 1980 among professionals employed by the Forest Service, representation of minorities (black, Hispanic, Native American, or Oriental origin) rose from 218 (2.7 percent) to 532 (4.7 percent), and of women from 137 (1.7 percent) to 888 (7.8 percent). Among administrative and technical personnel, minorities rose from 573 (7.1 percent) to 1,788 (10.6 percent), and women from 1,205 (15 percent) to 3,365 (20 percent).

Women over the years have been involved in forestry mostly through marriage, sharing the outdoor life as wives of rangers or of lookouts in fire towers. But until recent times the idea of an actual professional career in this field was strictly for men only. Paula J. Williams in 1978 produced revealing statistics about the state of things in the Forest

Service. She reported that two years earlier (1976), women occupied less than 2 percent of full-time professional jobs, 15 percent of administrative and technical jobs, and 84 percent of clerical jobs in the agency. Less than one-fourth of those women professionals were foresters, though forestry was the dominant occupational group and the most influential in decision making. It was the only group with Civil Service ratings higher than GS-15—and of foresters rated GS-15 to GS-18, all were white males. The highest grade levels achieved by professionals showed this pattern: minority women, GS-13; nonminority women, GS-14; minority men, GS-15; and nonminority men, GS-18.[1]

Women in the ranks continue to face a variety of obstacles. Patricia Seubert, a graduate student at the University of Vermont, completed a paper in 1982 titled "Women in the U.S. Forest Service: A Study in Attitude." She reported that: "One young woman with whom I spoke recently said the application procedure, coupled with the lack of positive reinforcement from the recruiter, helped make up her mind that the U.S. Forest Service was not seriously interested in hiring women."

The opportunities are there, Seubert concluded, but hard work is the key, along with the recognition that change takes time and must come from within the system. She suggested forming a "Council for Women in the Forestry Profession," to include members at all grade levels, all job descriptions, full-time and part-time employees, to discuss philosophy, needs, objectives, and actions to be taken for women in this career field. "If forestry is to be more than a stop gap for women as some other professions have been," Seubert wrote, "women already in the field must emphasize the ways this career can bring long-term responsibility and success."[2]

But women are making headway in what has long been considered a man's domain. In 1983 *Women in Forestry*, a quarterly publication, made its debut as a joint venture of the College of Forestry, Wildlife and Range Sciences and the Laboratory of Anthropology at the University of Idaho. It was designed to provide ideas and information for, from, and about women in the forestry profession. As for individual advancement, following are three significant examples:

- *Wendy L. Milner* was appointed in 1979 as the first woman district ranger in Forest Service history. She began her career (following graduation as a landscape architect from the University of Oregon) as a landscape architect at regional headquarters (Region 6) in Portland, with assignments following on national forests in Oregon, Colorado, and South Dakota. The Blanco Ranger District of the White River National Forest in Colorado, where she was assigned, covers 346,000 acres, including most of the Flat Tops Wilderness. Her position made her responsible for management decisions covering extensive recreation, wildlife habitat, livestock grazing, and mineral exploration.

- *Susan Odell,* a forestry graduate, became the second woman district ranger and, at age 28, possibly the youngest. The Mariposa District of the Sierra National Forest in California, to which she was assigned, lies adjacent to the east side of Yosemite National Park. Her husband, an automobile and industrial mechanic, was willing to follow his wife's career from Virginia to Oregon and California, finding a position he enjoyed in each location.

- *Geri B. Larson,* a 1962 forestry graduate of the University of California (who joined the Forest Service as a seasonal employee while still in college), in 1978 became deputy supervisor of the Tahoe National Forest in California, and thus the highest ranking woman line officer in the agency. She had formerly worked in public information and land planning in the regional office (Region 5) in San Francisco. Other women have broken barriers in range conservation, firefighting, wildlife biology, and flying.

In 1980, Charles H. Irby became the highest black line officer in the agency when he was appointed deputy supervisor of the San Bernardino National Forest in California. Irby was experienced and widely traveled, having worked for the Peace Corps in different parts of the world and as a Congressional fellow in Washington, where he became versed in forestry legislation and policy. Also in 1980, Charles Cartwright, another black, was appointed district ranger of the Conconully District of the Okanogan National Forest in north central Washington, a long way from his hometown of Petersburg, Virginia. While a freshman at Virginia Polytechnic Institute, counselors had advised against a career in natural resources or agriculture. The push was on for blacks to enter engineering, physics, and other technical areas, because for many blacks agriculture carried the stigma of "labor on the land," or field labor, left over from slave days. But Cartwright found a summer of employment with the Forest Service encouraging and decided to try to make it a career. "It was an uncommon step for a black to take in the 1960s," he would later recall, "especially since forestry schools then were predominantly white." But he persevered and succeeded; having been appointed a district ranger, he undertook "outreach efforts" in the Seattle area to develop Forest Service communication with black community leaders and to share with young people his personal experience.

According to Civil Rights Director Jetie B. Wilde, Jr., in 1981, Hispanic men were being recognized and advanced, particularly in the Southwestern Region (where Sotero Muñiz, a Hispanic, was deputy regional forester), but this was still far from being the case for Hispanic women, and even less so for Native American women. (Two Native American men, Bob Tippeconnic and John Hossack, were already forest supervisors.)

Besides encouraging equal opportunity in employment, the Forest Service civil rights program focused on attracting minority use of national forest recreation areas and on stimulating economic well-being of minority landowners. In March 1980, for instance, seventy black landowners attended a weekend workshop, the first of its kind, cosponsored by the Southeastern Area of State and Private Forestry (with the Virginia Division of Forestry and VPI) at Blacksburg, Virginia. Participants were presented with data about the forestry outlook in Virginia and financial and other opportunities for nonindustrial owners. The workshop, according to Otis Jones, coordinator of the Area's minority contractor program, was "a noteworthy success and a precedent for similar workshops in other Southeastern Area states."

New Careers and New Constituents

When Geri Larson attended the University of California, Berkeley, she was the only woman in the forestry class of 1962. There had been one woman in the Class of 1954, and the next ones were in 1964 or 1965. That has changed considerably. In 1978 Dr. Richard A. Skok, Dean of Forestry at the University of Minnesota, observed that within ten years the number of women forestry students had grown from a negligible proportion to nearly 20 percent of total enrollment, and growth has probably been more pronounced since then.

The same has not quite been the case for blacks, though progress has been made. In 1968 the Forest Resources Program was initiated at Tuskegee Institute, Alabama, to attract and prepare blacks and other minorities for careers in forestry and related natural resource professions. Basically, it consists of a two-year program to introduce students to the field and to make them aware of opportunities to continue at other universities. In 1977 the Tuskegee Forest Resources Council was formed to assist in recruiting, provide financial aid through scholarships and summer employment, and offer career counseling. The council, composed of representatives of five federal agencies (Forest Service, Fish and Wildlife Service, National Park Service, Soil Conservation Service, and Tennessee Valley Authority) and sixteen private firms, sponsored a conference in 1979 on "Blacks and Careers in Forest Resources." In that year fifty-six students had summer jobs in the field; twelve worked for industrial members of the Forest Resources Council, the remainder for the Forest Service.[3]

The Forest Service during the 1970s and 1980s embarked on another adventure, reaching out to black communities in inner cities. E. Delmar Jaquish, who played an important part, reported in 1981 how this began:

The National Environmental Policy Act formalized the public involvement in the Forest Service and other federal agencies. However, except for a few biggies like the herbicide question and some wilderness issues, the ranger and forest supervisor still heard from the same local people with whom they had communicated prior to NEPA. Some indicators, however, began to show us that a large part of the population was not taking part in our public involvement activities. We could not tell whether it was because these citizens did not know the issues or did not care what decisions were being made. Because of these feelings the idea of introducing or attracting nontraditional audiences to Forest Service activities began to emerge.[4]

In the mid-seventies, while serving as Director of Information for the Eastern Region (Region 9), headquartered in Milwaukee, Jaquish tried to initiate communication with black communities in urban centers like Chicago, Boston, and Detroit, but with little success. In 1977, however, the National Forest Products Association contracted with the National Council of Negro Women (NCNW) to conduct two pilot "natural resource awareness" workshops, which Jaquish attended. When NFPA chose not to go further, the Forest Service entered into an arrangement with NCNW.

"The project had several objectives," Jaquish reported. "First, we wanted to raise the awareness of natural resource issues in the minds of the women attending the workshops. Second, we wanted to learn if inner city minorities, faced with frustrations and pressing issues in their daily lives, had much interest in decisions being made often hundreds of miles away. Third, we wanted to gain experience in things to do or not to do in dealing with nontraditional audiences. We planned to pass this information on to natural resource managers and information specialists throughout the Service."

The first workshops were in New Orleans, Boston, Seattle, Detroit, Pittsburgh, and Denver; a group in Chicago became interested and insisted on one there as well. Dr. Dorothy Height, president of the National Council, became personally involved, attending five of the first seven workshops. Participants raised and discussed questions about housing, recreation, clean water, and opportunities for employment of inner-city youth. In 1980 and 1981 the Forest Service encouraged regional offices to expand these activities and contacts. A Montana workshop involved a group of Vietnamese, along with Native Americans and Hispanics. And in May 1981, a successful workshop was conducted at Riverside, California, under the auspices of the Pacific Southwest Region and an NCNW chapter, with a followup work plan agreed upon.[5]

NCNW continued its interest, acquiring funding to begin a basic work skills training program for young people from the inner city of New York. A few of the young people were placed in beginning technical positions with the forest products industry, and a few others on national forest ranger districts in the Northeast, in the hope that these positions would be the starting point in outdoor careers.

If better appreciation of natural resources is advantageous to conservation and wise use, then such programs will prove their worth. An article in the *Philadelphia Inquirer* in August 1981 illustrated the experience of three inner-city black teenagers hired through the Urban Coalition for a summer work program at Bartram's Garden, the oldest botanical garden in America. "The other kids couldn't understand why we were doing it," said one. "They were sitting around doing nothing, bored all day, while we were out learning, doing something worthwhile." And from another: "When I walk down the street I can point to certain trees and give their names. I couldn't do that before. I have a better understanding of nature and how it works. I relate to Mother Nature, and she relates to me."

5

THE NATIONAL FORESTS: PURPOSE AND PRESSURES

'Righting the Balance' Toward Timber Production on the National Forests
This involves firm and definite resolution of the wilderness vs. renewable resource use issue which dominates national forest planning and management. RARE II, the Resources Planning Act Program, and forest planning under the National Forest Management Act are major battlefields of this issue.
—Agenda for Forest Productivity in the 1980s, National Forest Products Association.

Scope of the System

The National Forest System embraces 191 million acres, approximately 8 percent of the country's land surface. Though national forest land is scattered across the continent, the bulk of it is in the West. National forests comprise 38.5 percent of all of Idaho, 25.3 percent of Oregon, 22 percent of Colorado, 21.3 percent of Washington, 20.2 percent of California. The 24 million acres located east of the 100th meridian amount to only 13 percent of the National Forest System. The largest ratio is in Minnesota, where national forests comprise 13.8 percent, followed by Michigan, 13.3 percent, New Hampshire, 13.1 percent, and Arkansas, 13.3 percent. Of heavily forested Maine, only 0.3 percent is national forest land, and of Pennsylvania, 2.8 percent. Because the eastern national forests were acquired by purchase (rather than by withdrawal from the public domain), they form a patchwork of inter-mingled public and private lands—only 51 percent of land within their authorized boundaries is federally owned, while in many areas the Forest Service controls the surface but not subsurface mineral rights.

The National Forest System was enlarged by two significant actions early in the 1980s:

72

FIGURE 5.1

The National Forests

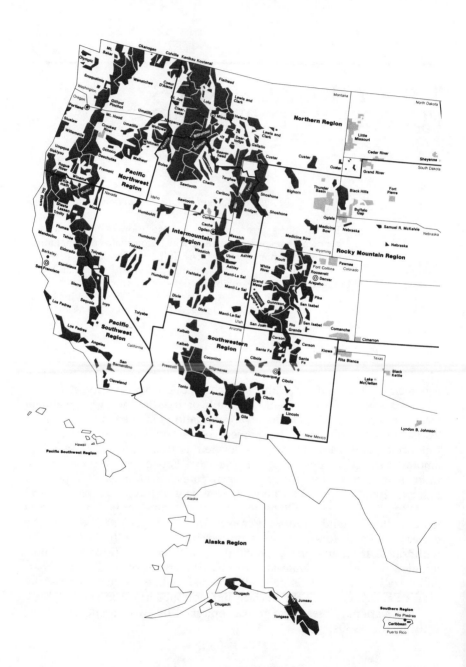

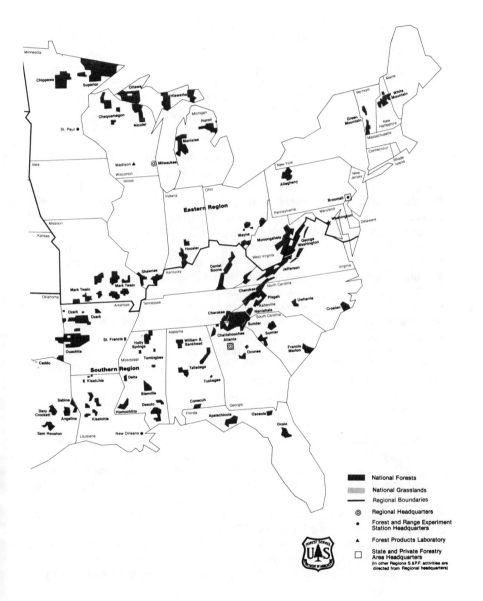

National Forests

National Grasslands

Regional Boundaries

⊚ Regional Headquarters

• Forest and Range Experiment
Station Headquarters

▲ Forest Products Laboratory

☐ State and Private Forestry
Area Headquarters
(In other Regions S.&P.F. activities are
directed from Regional headquarters)

The Alaska National Interest Lands Conservation Act of 1980 added 3.4 million acres, while transferring 600,000 acres from the national forests to national parks, national wildlife refuges, and Alaska native corporations. The Act also designated 5.4 million acres of wilderness of the Tongass National Forest and 2 million acres for wilderness study on the Chugach National Forest. It established two major national monuments, Admiralty Island and Misty Fjords, as part of the National Forest System in Southeast Alaska.

In January 1982, the Pennzoil Company of Houston, Texas, donated 100,000 acres of the half-million-acre Vermejo Ranch in northeastern New Mexico to be incorporated in the Carson National Forest. It was the largest and most valuable acreage ever given to the Forest Service. The Vermejo Ranch, part of an historic Spanish land grant, embraces rugged scenery with numerous alpine lakes and streams in the Sangre de Cristo Mountains along the Colorado border. It shelters 60 species of mammals, 200 species of birds, 33 species of reptiles and amphibians, and 15 species of fish.

National forests are vast and varied. A number of national parks were carved out of forest reserves, yet many spectaculars remain, including the spectacle of space itself. The violent eruption of Mt. St. Helens in 1980 reminded Americans that it lies within Gifford Pinchot National Forest in southern Washington (and that most volcanic peaks of the Cascades—including the Three Sisters, Mt. Jefferson, Mt. Hood, Mt. Adams, Glacier Peak, and Mt. Baker—lie in other national forests). National grasslands of the plains states include the last stronghold of the black-footed ferret, the rarest mammal in America.

Though President Ronald Reagan allied himself with critics of federal land-ownership, he personally chose to establish his ranch on a tract of private land within the boundary of Los Padres National Forest and to benefit from the buffer around it. The vegetative cover of the Los Padres, in fact, protects coastal California cities from soil erosion, flood, and mud slides. The Los Padres is a key sanctuary of endangered species, including the blunt-nosed leopard lizard, peregrine falcon, and California condor. Considering that condors now number thirty or less, without protection of the Sisquoc Sanctuary, in the San Rafael Wilderness, and the Sespe Sanctuary, they would be goners.

National forests could also be called national pasturelands, national water reservoirs, national energy reserves, national fish and game reserves, and national recreation areas. Many of them are underlain with oil, gas, uranium, coal, hard metals, and geothermal resources. They provide habitat for more than half the big game animals of the West, plus spawning grounds of the several species of Pacific salmon, and of the Atlantic salmon as well (in the White Mountain National Forest). Of 236 domestic species on the federal endangered species list, 70 have been identified in national forests, and some only in national forests. And the national forests are documents of human history, holding more than 100,000 identified Indian mounds, ruins of pueblos and cliff dwellings, remnants of old Spanish settlements, vestiges left by fur traders, pioneer settlers, homesteaders, miners, and loggers; many of these are listed on the National Register of Historic Places or are eligible for it.

Besides all these uses, more than a hundred different kinds of special facilities are spread across the national forests. Television transmission towers, power lines, churches, military zones, pipelines, airfields, private recreation and service establishments, and reservoirs are among the recipients of more than 80,000 "special use" permits.

National parks lend themselves to clearer definition and direction. The National Park Service has been assigned by law to preserve parkland unimpaired for future use and enjoyment. Despite the seeming built-in contradiction, park advocates can insist that land must first be preserved in order to be used and enjoyed. It is more difficult to make the same case on behalf of national forests, when they have been declared open to use.

Nevertheless, national forests were established as a protection against land abuse and exploitation by self-serving vested interests. In Alaska, for instance, national forests embrace more than 20 million acres, including most of the coastal woodlands of the temperate Southeast region. Afognak Island, just north of Kodiak, became one of the first withdrawals from the public domain in 1892, through the efforts of conservation activists like George Bird Grinnell, editor, naturalist, and close collaborator of Theodore Roosevelt in the Boone and Crockett Club. Afognak was intended to be a refuge for walrus, seals, sea lions, sea otters, and seafowl, as well as salmon, affording protection from commercial greed that was sweeping such marine species of the Northwest out of existence. There was no such classification as a wildlife refuge, so Grinnell and others settled for designation of the "Afognak Forest and Fish-Culture Reserve."[1]

Most remaining portions of the Alaska national forests were added during the Roosevelt-Pinchot conservation crusade of 1902–1909, a period when speculators were trying to control the natural resources and tie up land which might eventually prove of value. It was, in fact, Pinchot's effort to block the "Alaska Syndicate," composed of Morgan and Guggenheim Wall Street interests, that led to his celebrated feud with Interior Secretary Ballinger and to his dismissal.

The Move to Multiple Use

For roughly the first third of the century, management of national forests evolved slowly. Forestry, according to the saying, was 90 percent fire protection. Personnel were primarily custodians: fighting fires, making timber sales to settlers, issuing grazing permits to ranchers already grazing within newly established forests and to homesteaders struggling to make it on a 160-acre forest homestead. Their job was to win friends and supporters.

In 1908, Gifford Pinchot listed the objectives of administration as fire protection, timber harvesting, improvement in the growing crop of timber, protection of timber supply, utilization of forage crop (with betterment of range condition), and communication. During his period, there was ample room for all these functions to coexist.

Patterns began to change following World War I. Timber and range values of public lands became more evident. Agency leaders developed and endeavored to apply a principle of multiple use, predicated on the principle of sustained yield: the continuous output of resources without impairing productivity of the land. It implies an *even flow* and a limitation on the harvest of any renewable resource—whether timber, game, forage, or recreation—based on the ability of the resource to produce at least that much more. Especially in relation to timber, the application of sustained yield ensures the stability of a logging and milling community, in contrast to the old cut-and-get-out system.

The Forest Service had an established procedure to fulfill the multiple-use concept. Inventories were made—at national, regional, and local administrative levels—of present and potential use, condition of resources, and needs of the people. On this basis, multiple-use analysis of inventories was made for specific areas. Then, "action plans" for the national forests were carried out to meet approved coordinating requirements. Inspections, public reaction, and management experience were used to evaluate effectiveness.

In his annual report to Secretary Henry A. Wallace in October 1939, Chief Ferdinand A. Silcox wrote as follows: "National forests are administered on a multiple-use basis. Besides protecting from fire, insects and disease, Forest Service stewardship involves developing and administering these properties—including their land, water, timber, forage, wildlife, and recreational resources and the services they perform—in the public welfare."

With resource shortages at the end of World War II, the concept of multiple use came increasingly to the fore. "Grazing," Chief Richard McArdle declared in his annual report to Secretary Ezra Taft Benson in October of 1953, "must be integrated and coordinated with the multiple-use policy of management that recognizes water and timber production as paramount uses with equitable consideration for the interests of stockmen, recreationists, hunters and fishermen and the general public."

But this quickly was seen to require more attention and refinement. "The needs for water, timber and forage, for recreation, for wilderness areas, and for hunting and fishing, mount constantly," the Chief wrote in his 1955 report. "This places our multiple use principle of management under severe strain and tests our skill in both resource management and human relations."

This led to passage of the 1960 Multiple-Use–Sustained-Yield Act redefining the purposes of national forests to include the enhancement of outdoor recreation, soil, range, timber, watershed, wildlife, and fish, based on the "most judicious use of the land for some or all these resources." Critics would claim the Act gave the Forest Service legislative carte blanche to do almost anything it wished under the mantle of professional expertise. The Wilderness Society had pressed for protection of wilderness as one of the defined uses; in a last-minute compromise, this phrase was included: "The establishment and maintenance of areas of wilderness are consistent with the purposes and provisions of this Act."

Social changes and society's needs have made it difficult for the National Forest System to meet growing demands without stress on and sacrifice of one value to develop another, and difficult, also, to fulfill short-term goals and sustain long-term productivity. Consider the following example.

Worm grunting, or "fiddling," is an ancient activity in the southeastern bureaucracy, particularly in Florida, and there is even an International Worm Fiddling Contest. Basically it's a way to earn money, and at a pretty fair rate, by collecting worms in the woods and selling them to bait shops, which sell them to fishermen.

Approximately 600 permits per year are issued (at $30 per year in 1980) to worm grunters on the Apalachicola National Forest, near Panama City, in western Florida. The worm grunter, man or woman, hammers a stake in the ground, then uses a heavy file, saw, axe, or tire iron like a fiddle, rubbing it fiercely against the stake. Presently the earth begins to tremble. Vibrations grow to a tremor and evoke the desired result, with earthworms wiggling feverishly out of the ground.

The worms are easily taken. An adept grunter is said to fill ten buckets, with 500 earthworms per bucket, in a morning, which he will then market to the bait shop for a price somewhere between $7 and $11 per bucket. That is pretty fair business, especially when it helps to catch fish. However, Bobby J. Brady, ranger on the Apalachicola District in 1980, expressed a doubt. "Nobody really knows what we are doing to the soil," he said, "when we take worms out of it so fast."

The extraction of worms is one of the less significant uses of the national forests, yet in a way it represents them all. The national forests are the public estate, land wealth of people and nation, yet also a natural resource base of commercial enterprise to be either husbanded or exploited. And Ranger Brady's concern about knowing "what we are doing to the soil" might well be asked many times over.

The major categories of use of the national forests are covered in detail in separate chapters. The primary uses and the conflicts that they evoke are presented here in capsule form:

Timber. Since the end of World War II, greater pressure to increase cutting has been placed on national forests, notably in the Pacific

Northwest but in other regions as well. Mills built in a period of presumed lumber plenty now struggle to survive. Timber harvest and associated road construction conflict in various instances with wilderness, wildlife and fisheries, watershed protection, and recreation.

Range. Forest Service grazing lands, principally in the western states, originally were used by frontier settlers and their descendants. Ownership in recent years has shifted considerably to absentee and corporate interests. Grazing can be compatible with wildlife; private ranchlands, in fact, furnish extensive winter habitat. However, overgrazing has been a continuing problem, especially in the Southwest, where climate favors year-round use. Low grazing fees make national forests and BLM lands readily attractive for heavy use. Grazing is permitted in wilderness areas.

Oil, gas, and minerals. Most national forest land, outside of designated wilderness, is open to exploration and development. Funding for agency minerals work was boosted in the early 1980s in anticipation of major growth. Potentials even in wilderness have been tested, leading to conflict, as in the Bob Marshall Wilderness of Montana. Regulations have been applied to protect other values with varying degrees of success. Development of energy resources in national forests may impact nearby national parks, as in the case of geothermal supplies in the Targhee National Forest of Idaho, which scientists fear might critically affect geysers and hot springs in Yellowstone National Park.

Recreation. Americans have flocked to the national forests for many forms of outdoor leisure activities. In many cases commodity uses have been restricted, though recreation can be as damaging as any of them, with erosion and polluted streams the consequences of overuse and misuse. Recreation has built-in conflicts, as between riders of trail bikes and people who want to find relief from mechanical noises, and between snowmobilers and cross-country skiers.

Fish and wildlife. National forests provide important habitats for many game and nongame species. Deer in many cases are compatible with logging. Game departments in Wyoming, Idaho, and elsewhere have criticized roading and logging plans in elk habitat. Effects of logging on wild turkey was an issue at stake in the historic conflict in the Monongahela National Forest in West Virginia. The same is true in terms of the spotted owl in the Pacific Northwest.

Wilderness. The Forest Service has long been the principal federal land management agency involved in wilderness protection. Millions of acres have been set aside both before and since passage of the 1964 Wilderness Act. However, unclassified roadless areas have been a major field of conflict between wilderness advocates and commercial interests, principally timber and mining.

Watershed and soils. The 1897 Organic Act states that: "No national forest shall be established, except to improve and protect the forest within the boundaries, or for the purpose of securing favorable conditions of water flow, and to furnish a continuous supply of timber for the use and necessities of citizens of the United States." Likewise, protection of high elevation watersheds was a major factor in the Weeks Law of 1911 that brought national forests to the East. Many communities are dependent upon these watersheds for water supplies and protection from floods. Water and soil programs, based on sustaining the forest mantle, are compatible with safeguarding plant and wildlife species and some recreation, but are in conflict with any heavy disturbance.

What Went Wrong with Multiple Use?

As a process for achieving balance in land management and allocation among these conflicting pressures, multiple use proved wanting during the 1960s and 1970s. It is now conceded that logging was sanctioned on various sites ill suited to it—with tree species at the limits of their range, on infertile, shallow soils and rough terrain, where intensive operations proved costly and damaging to watershed, wildlife, and mountain scenery, and disruptive to recreation. Pressure for timber cutting proved unrelenting, both from without and within the agency. Foresters would cite logging as a multiple-use tool rather than an end in itself, because it provides the beginning of road systems and openings for sportsmen and wildlife; because timber roads were designed for heavy machinery, however, they could not avoid disturbing game areas.

Though public reaction was intended as a means of measuring multiple-use effectiveness, there was really no way of making it work in practice. In April 1970, Theodore Schlapfer, then regional forester of the Southern Region, in a letter to the author, wrote as follows: "We are hired as professionals and paid by the taxpayers to make decisions based on our education, training and experience. While the public must be involved if we make changes in management practices, we still have the managerial responsibility of making the final decision."

Senator Jennings Randolph, of West Virginia, complained in 1972 that: "Public comments are invited but the consultant atmosphere appears to be lacking. The prevailing feeling expressed by those after attending the hearings is that decisions have already been made and their expressed concerns have only been accepted as an empty polite gesture." Senator Randolph was referring to the long, bitter controversy over multiple-use management of the Monongahela National Forest in his state. In a letter published in the *New York Times* of February 29, 1972, he wrote:

> The Cranberry Back Country, located in the southern portion of the Monongahela National Forest, must be preserved in its unique and primitive state. The Forest Service, since 1964, awarded timber contracts despite pleas from the public that such cutting would destroy its chances for wilderness recognition.
> Earlier clearcutting in the Gauley Ranger District near Richwood nearly destroyed the local wildlife habitat, recreational potential, drainage systems

and the land. This tragic mistake prompted Chief Cliff in testimony before the Senate Committee on Public Lands to state:

"In 1964 and until recently we stated that even-aged management would be the basic system of management in the so-called general forest zone. This has been changed. Our policy now is to use a variety of methods, with no one method as primary."

Actually little has changed. Contracts awarded since Chief Cliff's testimony are significantly weighted toward even-aged management and un-even-aged management continues to be ignored on the general forest zone, which constitutes over 80 percent of the forest. This is diametrically opposed to the recommendations of the West Virginia Forest Management Practices Commission, established by the West Virginia legislature.

The practice of clearcutting, which Senator Randolph decried, involved large block cuts designed to achieve a forest of "even age," in contrast with the old system of selective logging in an "all-age" forest. Chief Cliff defended clearcutting as scientific forestry being properly applied. And he defended his organization, insisting that all uses and users were receiving their fair share.

In December 1971, the Chief took note of a critical editorial in the *New York Times*, as well as critical articles appearing in *Atlantic*, *Washington Monthly*, and *Reader's Digest* (written by Paul Brooks, James Risser, and James Nathan Miller, respectively, all considered journalists of merit), responding with letters to the editors. To the *New York Times*, he wrote: "Land-use planning is not a precise science, but the Forest Service system of multiple use planning is as advanced as any system, applicable to large land areas, in use today. As a result, the national forests are producing more goods and services for the use and enjoyment of the American people, and in greater variety, than ever before."[2]

He wrote much the same to *Washington Monthly*, adding: "If nothing else, this should prove that timber production is only one of the major responsibilities in the array of multiple uses." In the field, however, personnel wondered whether this was quite so. They questioned high, and perhaps unrealistic, quotas for logging on unproductive and marginal lands, the system-wide commitment to clearcutting without fully probing its consequences, continued reliance on hard pesticides, intrusion into roadless areas while reviews were pending, and the inadequate representation of biology and other disciplines to balance production-oriented foresters.

During the summer of 1970 a six-man Forest Service team spent the summer observing logging practices on four national forests in Wyoming (Bighorn, Shoshone, Teton, and Bridger) and conducting interviews with private citizens. Team members were research and management personnel not directly involved with these forests.[3] The issues already were well known. Concerned citizens had objected to various timber sales and had induced national conservation organizations to investigate. Representatives of the American Forestry Association, Izaak Walton League, National Wildlife Federation, and Sierra Club corroborated the substance of the complaints. They criticized the agency for laying out roads badly, conducting large cuttings that led to erosion, and endeavoring to log spruce in high elevations with poor prospects of regeneration. Their findings were substantiated by Senator Gale McGee, who toured portions of the Bridger National Forest in the fall of 1969. "Clearcutting in many areas of Wyoming's national forests has caused near devastation," he wrote to Russell Train, chairman of the Council on Environmental Quality. "I was appalled. Slash piles abounded, reforestation was meager, if not totally absent. It takes 150 years for grow-back to occur and in many places even forced planting of seedlings does not take."

The Forest Service team found much the same. In its report, issued in 1971, the team cited instances where road construction caused soil erosion and landslides—and, moreover, that knowledge to prevent mistakes was available but not used. "They [mistakes in road location and construction] cannot be dismissed on ground that 'we are no longer doing it this way.' Some of the cited roads were built during the past five years. There is still not enough quality control of road construction for watershed protection."[4]

The team also weighed the impact on relations with the public. "Our investigation confirms that a confrontation has developed between residents of the state and the federal agency charged with protecting the public interest in national forest lands," wrote the chairman, Dr. Carl M. Berntsen (who later became director of science programs of the Society of American Foresters). "In essence, the criticism we heard is that the Forest Service has given priority to timber production, whereas it should have given priority to watershed, wildlife, recreation, and scenic values; that is, to preserving the quality of the environment. Public confidence in the ability of the Forest Service to manage forests has suffered."[5]

Subsequently, Chief Cliff conceded that all was not right. Between 1963 and 1970 the proposed level of spending for timber sales administration and management was fulfilled to 95 percent of the total, while other phases of multiple use did not fare nearly so well. The planned levels of reforestation, stand improvement, recreation, public use, wildlife habitat management, and soil and water management were financed at levels of 40 percent, 45 percent, 62 percent, and 52 percent, respectively. "Our program is out of balance," the Chief declared in discussing the fiscal 1972 budget before the Senate Appropriations Committee. "Over the past two or three years we have increased the amount of money that we have for timber sales administrative activities, but not by sufficient amounts to do the top quality job needed. We are not adequately financed in other areas."

Achieving a balanced program has grown no easier with the years. "Why should conservationists accept unchallenged the concept that profits of lumber companies automatically take precedence over the interests of the nation as a whole?" demanded Nicholas Roosevelt in his book, *Conservation: Now or Never*, published in 1970. "The saving of scenery for the long-term enjoyment of present and future generations of Americans is more important to the nation as a whole than are profits which the lumber companies hope to make in cutting the last stands of virgin timber in the national forests which they can induce the Forest Service to let them cut."

That is one side of it. On the other, in 1969 at House hearings on the National Timber Supply Act, William D. Hagenstein, executive vice president of the Industrial Forestry Association, voiced the strong viewpoint of his industry:

> Eighty percent of America's homes are made of wood—a continuously renewable raw material through forestry. And the growing, protecting, harvesting and growing again of continuous timber crops is forest conservation at its best and its principal product is homes for the American people.
> That is why it is essential, if Congress really wants the American people to have the 26 million homes it has set for the next decade, that it must take the next step to provide the means. With the national forests having 54 percent of our total softwood timber supply, it's clear that timber production must be increased from this most important single source. And that production must be increased only by intensified forestry in every respect to safeguard the timber supply of generations yet unborn.[6]

This same view was emphasized anew in the "Agenda for Forest Productivity in the 1980s," prepared by the National Forest Products Association. Terming RARE II, the Resources Planning Act Program, and forest planning under NFMA as "major battlefields" in the conflict between wilderness and resource production, the Agenda declared that "righting the balance also demands strong direction to Forest Service field managers to take innovative and aggressive steps in national forest policy to blend timber management and environmental objectives, rather than give way automatically and blindly to non-timber objectives which tend to increase costs and decrease human and resource productivity."

Out of the West (and Elsewhere): The Sagebrush Rebels

Environmental laws adopted by Congress during the 1960s and 1970s, dealing with clean air, wilderness, preservation of antiquities, and threatened and endangered species, have affected national forests in more ways than one. They triggered an outburst of resentment among communities dependent on public lands. Perhaps the most significant law was the Federal Land Policy and Management Act (FLPMA) of 1976, giving permanence and authority to the Bureau of Land Management, an Interior Department agency, while relating to national forests as well.

Following enactment of FLPMA, livestock interests rose in wrath. In Arizona they forced the transfer of a field official when he sought to introduce a balanced multiple-use plan. In Nevada they induced the state legislature to pass a resolution demanding transfer of federal lands to state jurisdiction. Thus the so-called "Sagebrush Rebellion" was born. And in various quarters calls were raised for defederalizing national forests and transferring large portions to the states.

Pressures of this nature are as old as federal land-ownership. Stockmen have tended to resent and resist almost any form of governmental interference. Even at the first meeting of the American National Livestock Association, held in Denver in 1898, one serious proposal was to transfer federal lands to the states. During the 1920s, President Herbert Hoover supported this idea. The West had come of age, he insisted, and was fully capable of managing its own landed affairs.

Livestock interests are not alone in the current rebellion. The values of land in the West having risen appreciably, a variety of entrepreneurs have joined with political rebels against red tape to make it a thoroughly serious issue. During the election campaign of 1980 Ronald Reagan proudly and publicly labeled himself a Sagebrush Rebel.

"The power of the community cannot be restored simply by trading the power of big government for the power of big business," declared Governor Scott Matheson of Utah in denouncing the Sagebrush Rebellion early in 1981. Interior Secretary Watt endeavored to defuse the Rebellion, or at least to divert its energies, by asserting that federal land agencies henceforth would be more responsive, as "good neighbors" should be.

But yet another aspect emerged. President Reagan in February 1982 created the Asset Management Program, with the stated goal of selling $17 billion worth of federal buildings and property by 1987 to help reduce the national debt. Conservative economists and others argued in support of this idea, asserting that market forces responding to private ownership could induce high levels of productivity, promote the preservation of amenity and recreation values, encourage recovery of grazing lands, and reduce costs to the consumer as well as reduce the national debt—all this through the sale of "inefficiently managed" federal lands.

The "unneeded federal property and property of significantly higher value in private rather than public ownership" in the President's plan include: lands in and adjacent to urban areas, which limit community growth; scattered tracts which cannot be economically or efficiently managed, and "lands which have higher potential in private ownership, such as for raising crops or timber."

Thus in August the Department of Agriculture announced that the Forest Service would have land for sale, starting with 60,133 acres of forest and grassland restoration projects, unneeded administrative units and research centers, and certain parcels under 640 acres of possible public use to neighboring local governments. Looking beyond, all Forest Service land would be divided into two other blocks: (1) 51 million acres to be retained, notably wilderness, national recreation areas, national monuments and wild and scenic rivers; and (2) 140 million acres to be reviewed quickly to identify those with potential for sale.

With completion of this process, Secretary Block reported 6 million acres had been identified for further study to determine specific tracts to be made available for sale, dependent on Congressional authorization. The remaining 134 million acres were added to the 51 million already in the retention category.

Eventually, according to Secretary Block, 15 to 18 million acres would be identified for close study and sale, based on congressional authorization.

"Idaho is not for sale," retorted Governor John Evans amidst the heat of the 1982 election campaign, during which the fate of federal lands within the state became a key issue. The entire idea of divestiture stirred a furor of opposition. Environmentalists advanced figures showing that federal lands generate far more revenue than they cost to operate. Grass roots Sagebrush Rebels backed away from privatization: It wasn't really what they wanted, they said; their interest was only in making bureaucrats responsive. Over the long run the review of potentially disposable land might cost more than the government earns from any land actually placed on the auction block.

Nevertheless, the antifederal feeling reached beyond the West and beyond those interests utilizing public lands for private purposes. In 1979 the Indiana legislature established a study committee to consider revoking earlier legislation (of 1935 and 1937) which enabled the federal government to acquire land for fish hatcheries, wildlife preserves, and forest preserves. At public hearings most witnesses testified for repeal of the enabling legislation; these included conservation spokesmen critical of national forest management. Claude Ferguson, the retired Forest Service whistleblower, testified as immediate past president of the Indiana Conservation Council:

> I had hoped to see some sign that would signal change in the Forest Service management style. I have seen none. . . . The planning process of 1977–78 again led the public to hope, but saw intensive and expensive public participation for naught.
>
> Gone are the community-oriented local public servants—replaced by transients affixed to upward and onward mobility. State foresters are very capable and close to the land. And they are more responsive.[7]

For the Field: New Laws, Tasks, Opportunities

For those Forest Service employees on the ground, who want to discharge their responsibility to the fullest, life has become a sequence of dealing with conflicting demands and of trying to absorb new data.

Times have changed since personnel kept diaries of their daily activities (a practice which ended in the 1960s). Staff people and planners in the supervisor's office have assumed various functions the ranger formerly filled. The ranger is always conscious of the fact that 25 percent of earned receipts—from timber sales, grazing permits, recreation fees, mineral leases, and land use charges—is returned to the states in lieu of taxes. These monies are applied to support of public roads and schools in counties where national forests are located. And someone is always asking whether a local county could earn more revenue by administering the land itself.[8]

For years a large part of the district ranger's job was fire prevention and suppression. It was not uncommon for a district ranger to direct hundreds of men on a single fire. Chief Henry S. Graves in 1910 enunciated control as a primary focus when he declared: "The first measure necessary for the successful practice of forestry is protection from forest fires." This idea became entrenched as policy and aggressive initial attack standard in operation. The "10 a.m. Fire Suppression Policy" called for dispatching sufficient forces to control a fire by 10 a.m. of the following day.

During the 1930s and 1940s, Harry T. Gisborne, a respected researcher, questioned the concept. He reasoned that it didn't make sense to treat all areas alike—to attack a fire in the "goat rocks" with the same vigor as one in a white pine plantation. Aldo Leopold questioned it, too, pointing to fire as an historic constructive influence on ecosystems.

Since 1977, field personnel have been directed to follow a new approach in fire management: to allow some natural fires to burn, under prescribed conditions, when it is ecologically, socially, and economically acceptable, and to vary standards of suppression based on weighting resource loss versus cost of control. This is not to imply that personnel walk away from fire and let it go—there is still plenty of fire suppression in the old tradition.[9]

Among new laws, significant though seemingly obscure, which forest supervisors and rangers must keep in mind, are the National Historic Preservation Act of 1966 (with strengthening 1980 amendments) and the Archaeological Resources Protection Act of 1979. These require managers of federal lands to inventory their areas for traces of activities and life styles of past inhabitants for possible protection by listing on the National Register of Historic Places. This can mean redesigning a timber sale, rerouting a road, or posting to prevent vandalism until evaluation and knowledge can provide information about the best use and management of a given site.

This process is a burden to some, but an opportunity in management to others, since a site preserved becomes a means of expanding knowledge of past human cultures and adaptation to their environments. In an article in the *Journal of Forestry* (March 1981), Janet Friedman, formerly head archaeologist of the Forest Service, cited an example of what excavation discovered at Chavez Pass, in Coconino National Forest, Arizona:

There, 1,200 and more years ago, fields were terraced and cobbled to prevent erosion and to trap water in the arid land. Fields so treated are still greener and freer of erosion than those left in their natural state. Check dams, too, served the residents, controlling floods that could otherwise wash out their fields. Water storage facilities also have been uncovered by excavation.

Scientists are using archaeological information to record past climatic patterns, revealing weather cycles to provide insight into our own environment. Studies of fossil seeds and pollens, animal remains, and tree rings all contribute to knowledge about prehistoric and, potentially, future climate and environments.

A Basis for Decisions

The election returns of 1980, when Ronald Reagan and many Republicans were elected on a campaign against big government, showed that there was more to it than self-serving interests. Craig Rupp, regional forester of the Rocky Mountain Region (Region 2), in a speech in April 1980 recognized that many Americans were frustrated by "centralized government which prevents prompt and responsive decisions on local issues." As Rupp declared:

Both Congress and the American people want decisions of substance. Decisions that affect resources and people must be made by managers closest to the situation. They also want a uniform and open process that lays out how the decisions of substance are made.

The public does not want centralized or standardized decisions of substance, but they want centralized and uniform decisions of process.

The more decisions of substance that can be made at the local level with public participation, the easier it is to create and maintain productive harmony.

That is one way to look at it, though, as indicated, clear decisions of substance have grown increasingly difficult in the face of competing demands, overpowering political pressures, complex decision processes, and court challenges. Yet the Asheville (North Carolina) *Citizen-Times* expressed it a different way in an editorial of August 23, 1970, a simple statement that goes to the heart of the national forest idea: "To a large extent private timberland outside the national forests is still being thoughtlessly exploited with little regard for the future. The U.S. Forest Service is the agency that must set the example for others to follow."

THE PLANNING PROCESS
AND PROBLEMS

> Instead of having a comprehensive plan for the governing
> and protection of our resources, we have tended to focus on
> each problem individually, such as timber cutting, watershed
> management, livestock grazing, fish and wildlife resources, and
> wilderness. . . . It is quite obvious that all are related and
> therefore need to be considered as a totality rather than as
> individual items.
>
> —Senator Hubert Humphrey, of Minnesota,
> before the Senate Forestry Subcommittee,
> November 20, 1973, in advocating a National
> Forest Environmental Management Act.

Long-Range Planning: The Arguments

Timber harvesting on the national forests faced a considerable impasse
in the early 1970s as a result of court actions and court decisions. Two
of the most prominent cases of the period directly affected national
forests in Alaska and West Virginia, yet they also cast a shadow across
the entire National Forest System.

The Sierra Club brought suit against the Forest Service in 1970 to
block a proposed sale on the Tongass National Forest in southeastern
Alaska. That sale would have been the largest, and longest, in U.S.
history: spanning 50 years and one million acres, designed to allow
Champion International to harvest almost one billion board feet, enough
lumber to build a plank road from the Arctic Circle to the tip of South
America—though virtually all of it was intended for shipment to Japan
as lumber or pulp. The Sierra Club contended violation of the 1960
Multiple-Use–Sustained-Yield Act. The Club cited, among other things,
that only a small portion of Admiralty Island, focal point of the sale
and one of the foremost wild areas on North America, consists of
economically accessible commercial forests, while the better timber zones,

at the heads of bays and along valley bottoms, are critical for brown bears, bald eagles, salmon, and other wildlife species.

In 1973 attorneys for the Sierra Club and Natural Resources Defense Council (NRDC) went to court in behalf of the Izaak Walton League of West Virginia to challenge clearcutting on the Monongahela National Forest. Senator Jennings Randolph in 1970 had urged a moratorium pending a further open review. The Forest Service and Department of Agriculture had denied the request, asserting that such a moratorium would "inhibit the public understanding" of clearcutting, though effects of that system were already quite evident in West Virginia. "The Forest Service is employing a management practice of irreparable damage on the Monongahela," Senator Randolph declared on the Senate floor on June 8, 1973, "including a change of composition to that which the Forest Service believes to be 'a more desirable species.'" The litigants challenged clearcutting under the 1897 Organic Act, which stipulated that trees to be cut on national forests must only be "dead, mature or 'large growth' timber." A federal district judge ruled in favor of the Izaak Walton League, a verdict subsequently upheld on appeal in 1975.

Foresters were frustrated and timbermen were furious. In March 1973, rising lumber prices sparked an industry effort to obtain large increases in national forest output. The President's Cost of Living Council directed the Forest Service to increase sharply the volume of timber it planned to sell in order to curb rising consumer prices for new housing (although government statistics indicate that lumber accounts for only 17 percent of the construction cost and only 9 percent of the fully financed cost of the average single-family dwelling, and an even smaller percentage of multi-family dwellings, when weighed against costs of labor, land, and financing).

In order to undo the legal impasse, Senator Hubert Humphrey introduced legislation to provide for long-range comprehensive assessment and programming for national forests and rangelands. This was to become the RPA, or properly the Forest and Rangeland Renewable Resources Planning Act of 1974. The Act directed the Secretary of Agriculture to assess the entire nation's forest and rangeland scene every ten years and to submit recommendations for long-range Forest Service programs, these to be updated every five years. At Senator Humphrey's insistence, Congress ordered the Forest Service to prepare the first Assessment and Program in 1975, which the agency put together hurriedly, largely from existing information. The second assessment was due January 1, 1980, the third January 1, 1990, with a developed program to be delivered each five years.

This planning framework was not universally welcomed. The Office of Management and Budget felt that it would implicitly infringe on the power of the President in preparing annual budget requests. Environmentalists felt that nontimber programs would suffer since they would have to be justified in terms of specific programs and goals, and demonstrated costs—while timber sales in theory earn revenue, nontimber activities in theory cost money. For foresters in the field RPA meant an entire new set of paperwork. Nevertheless, in developing the first RPA program, management activities were grouped into six definable "resource systems": outdoor recreation and wilderness, human and community development, wildlife and fish, timber, range, and land and water. The report outlined accomplishments, with statistical summaries, in each system.

RPA was followed by further debate and legislation to allow timber harvest to proceed under a fair measure of environmental control and consideration of other forest uses. Senator Randolph introduced one bill (cosponsored in the House by Representative George Brown, of California), supported by environmental groups organized in the Coalition to Save Our National Forests, while Senator Humphrey introduced another, generally supported by timber and housing industries, and by organized labor. Foresters objected to tight restrictions on clearcutting in the Randolph-Brown Bill as "prescriptive forestry."

Both bills, however, called for revision of alleged outmoded limitations imposed by the 1897 Organic Act. Pinchot might have argued otherwise. As he wrote in *Breaking New Ground:* "The men who wrote and passed the rider of June 4, 1897, were taking no chances. They intended to see that all resources of the Forest Reserves were open to use, and what they tried to do they actually accomplished."[1] Proponents of revision in the mid-1970s complained specifically about the stipulation in the Organic Act that allowed only "dead, matured, or large growth of trees" to be cut. Pinchot might have cited the full context in which the phrase appears and argued that it encourages and allows plenty of latitude for the practice of sound management:

> For the purpose of conserving the living and growing timber and promoting the younger growth on national forests, the Secretary of Agriculture, under such rules and regulations as he shall prescribe, may cause to be designated and appraised so much of the dead, matured, or large growth trees found upon such national forests as may be compatible with the use of the forests thereon.

Ultimately, following lobbying and debate, a law was enacted in the waning days of the 94th Congress—the National Forest Management Act of 1976—combining features of the Humphrey and Randolph-Brown bills. Senator Randolph complained it was so watered down that it would do little to correct the problems one finds on the national forests.[2]

NFMA repealed strictures on timber harvesting as defined in the Organic Act and made national forest plans the basis for regulating harvest practices. Each national forest henceforth would operate under such a plan. More important, perhaps, the Act called for equal consideration for, and protection of, all the renewable resources in an optimal mix. NFMA directed that the public be involved in planning and the private landowner be informed of opportunities to improve his land; but the challenge mostly was directed to national forest administrators: to "recognize the interrelationships between and interdependence within the renewable resources" and "the fundamental need to protect and, where appropriate, improve air, water and soil quality." Through all this management would be based on analysis of cost and returns.[3]

RPA set a precedent for other agencies and other laws. The Federal Land Policy and Management Act (FLPMA) of 1976 became the charter for the Bureau of Land Management, though also calling for range planning on grazing allotments managed by both BLM and the Forest Service. The Soil and Water Resources Conservation Act of 1977 called for appraising and long-range planning for the nation's nonfederal soil and water resources, to be undertaken by the Soil Conservation Service.

The assessment of all the nation's forests and rangelands and the program of Forest Service activities required by RPA were delivered to Congress in June 1980, with an update scheduled for 1985. New requirements for national forest planning are at the core of NFMA. This has not been an easy system to develop or make operable. The NFMA process has shortcomings from everyone's viewpoint, raising more questions than it answers, seeming to defer rather than decide. Foresters have felt it takes initiative from land managers and transfers it to lawyers, economists and special interest groups. Yet it marks a step forward to address local problems in management through systematic planning. John F. Krutilla and John A. Haigh, of Resources for the Future, discussed this point in a 1978 article titled "An Integrated Approach to National Forest Management":

> The RPA appears to have been motivated in part by what to many seemed an increasingly chaotic process of managing the public forest lands in response to pressures from conflicting interest groups, which in innumerable instances resulted in embarrassing and disruptive litigation. The RPA as amended by the NFMA was seen as consistent with the Multiple Use—Sustained Yield Act; together with that and the Forest Service Organic Act of 1897, it was viewed as providing the maximum benefits by means of systematic planning and management.[4]

In implementing the planning process, the Forest Service delineated five major themes and ten basic steps in planning, which are worth noting to understand the process:

1. Consider all resources in determining optimal mix of resource use.
2. Improve data collection ability, and base Assessment and Program on resulting information.
3. Base decisions on analysis of costs and returns.

4. Coordinate programs with those of other federal, state, and local governments, and private landowners.
5. Involve the public throughout the decision-making process.

The ten basic steps in preparing the plans are:

1. Identification of public issues, management concerns, and resource opportunities, including those perceived by the public through meetings and questionnaires.
2. Development of planning criteria: "process criteria" to guide data gathering and analysis and "decision criteria" to guide selection of a proposed alternative.
3. Inventory the resources based on "best available data which may require that special inventories or studies be prepared."
4. Management analysis, heavily dependent on economic analysis, weighing cost against benefits of timber, wildlife, and other resources.
5. Formulation of alternatives, "to reflect a range of resource outputs and expenditure levels."
6. Estimation of effects of alternatives, of "physical, biological, economic and social effects" of each alternative and how each "responds to the range of goals and objectives assigned by the RPA program."
7. Evaluation of alternatives, identifying one or more to best meet criteria defined in step 2.
8. Selection of an alternative, with description of benefits relative to other options.
9. Implementation of the plan, contingent upon adequate budget appropriations.
10. Monitoring and evaluation.

Planning is divided further into three tiers: national RPA Assessment and Program, a planning guide for each of nine Forest Service regions, and a plan for each of more than 120 administrative units (national forests and grasslands). The regional guides link the national Assessment and Program with plans of individual forests, establishing the direction and distributing, or assigning, production targets and objectives.[5]

In principle, though direction comes from the top, there is also an information flow upward from the national forests to the region and the Washington office, but serious questions have been raised as to whether latitude actually exists. In theory, the regional forester performs the role of reconciling RPA assigned targets with the capabilities of the individual forest to produce them. In the case of timber, the nationwide plan charts production goals for each region, which then breaks them down into goals for each forest. When the allocated goal is considered too high, the forest is expected to bargain for a reduction through a process of negotiation called "iteration."

The RPA in Practice

The first draft plan in the country was completed in 1980 on the Lolo National Forest, which covers more than 2.1 million acres in northwestern Montana, a great stronghold in the northern Rockies of elk, mountain goat, moose, deer, falcon, grizzly, bald eagle, bighorn sheep, wolverine, and potential peregrine falcon sites. In recognition of wildlife and other noncommodity values, a draft Environmental Impact Statement (EIS) and the proposed Lolo Plan showed timber production increasing at a relatively slow rate, from 100 million to 129 million board feet over the next decade. But the draft EIS and proposed Lolo Plan had been released before the 1980 RPA assessment was completed and before Region 1 was given its timber goals from Washington. The timber industry objected to the content, economic analysis, and yield table use.

Under the pressure of time, the 1975 Assessment and Program documents were issued together. Since the Program is based on the Assessment, however, an Assessment was issued in 1979, one year in advance of the subsequent Program. Both were submitted to Congress in mid-1980, showing consideration of various alternatives for development of principal forest and rangeland resources: recreation, wilderness, wildlife and fish, range, and timber.

The idea of ranges—from sharply increasing to sharply decreasing resource outputs—was not used in the first go-round, nor in the drafts of the 1979-1980 documents. The Carter Administration, however, had ideas of its own, designed to make all parties happy, but ultimately satisfying no one; it directed preparation of a range, or "bounds," covering the spectrum of multiple uses. The "high bound" was based on the assumption of increased federal budget dollars and favorable economic conditions, the "low bound" on the assumption of constrained federal spending. "Bounds" were shown in the 1980 Program ("A Recommended Renewable Resource Program—1980 Update"), plainly rendering the process complicated and confusing.

How useful can such a program be, demanded its critics, with goals so broad, reflecting uncertainty and disagreement, with widely discrepant high and low bounds? Field personnel felt a forest supervisor would have to mount up and ride off in all directions at once. Congress in its wisdom in effect discarded the bounds idea, but possibly without simplifying things. In the Interior Department and Related Agencies appropriation bill for FY 1981 (Act of December 12, 1980—PL 56-514), the 96th Congress revised and modified the President's idea.

The planning process became laden with jargon and difficult for nonplanners to follow. The Idaho Environmental Council complained in its newsletter (of January 1981) that commodity production was getting the best of things:

> National Forest Region 4, which includes Idaho south of the Salmon River, is not a major timber producer, but is still being squeezed for timber by the RPA. National Forest Region 1, which includes Idaho north of the Salmon River, has a few more trees and it is being gouged.
>
> The Clearwater National Forest has cut between 130 and 180 million board feet for the past ten years. To get this, many miles of road have been built and much wildlife habitat has been damaged and lost. Even that level cannot be maintained without invading most of the remaining roadless areas outside the Selway-Bitterroot Wilderness. (The Bighorn-Weitas Roadless Area, designated nonwilderness in Rare II, furnishing a wealth of wildlife habitat, including known wolf habitat, is already slated for roads and logging.) Yet, the RPA dictates that the Clearwater National Forest must *double its annual timber output* by the year 2025.

Actually, the RPA program is not quite site-specific and doesn't give numbers for specific forests; regional foresters have the option of determining from which national forests goods and services will be drawn, though the Idaho Environmental Council may still have had valid basis for complaint.

The Committee of Scientists Report

On the other hand, after nearly three years of consultation and study, the Forest Service in September 1979 issued a set of regulations to encourage sensitive resource management and to prevent the misuse of clearcutting. These regulations were based on the report of a Committee of Scientists (a team of seven members mostly associated with forestry schools) appointed by the Secretary of Agriculture.

"From the initial inception of work to develop the regulations through to the present time, the Forest Service has maintained an open door policy with the public and interested groups to obtain information as well as to explain work and progress," the agency stated in publishing the regulations in the *Federal Register* of September 17, 1979. "Eighteen Committee of Scientists' meetings were open to the public, and a total of 737 individual responses containing 5,373 distinct references to various parts of the August 31, 1978, draft regulations were received, a number of which were elaborate, detailed, and explicit."

This committee (of which Dr. Arthur Cooper, of North Carolina State University, was chairman) gave the Forest Service a set of management guidelines based on biosystems rather than dollar systems or political pressures.

"Land management systems, harvest levels, and procedures must be determined," the regulations direct, "with due consideration for (1) their effects on all resources, (2) the definition of 'multiple use' and 'sustained yield' as provided in the Multiple Use–Sustained Yield Act of 1960, and (3) the availability of lands and their suitabilities for resource management."[6]

Equally significant was the call for an interdisciplinary approach to ensure coordinated planning of outdoor recreation, range, timber, watershed, wildlife and fish, and wilderness opportunities, specifying that "the interdisciplinary team shall be guided by the fact that the forests and rangelands are ecosystems and, hence, that management of goods and services requires an awareness of the interdependencies among plants, animals, soil and other factors that occur within such ecosystems."

The guidelines deal with clearcutting, declaring it may be used "only when it is determined to be the optimum method of silviculture for a specific forest type," and optimum is defined as the system "most favorable and conducive to the achievement of multiple-use goals specific in the forest plan." Resource managers are counseled to provide for diversity of plant and animal species, and to restrict timber harvesting to sites where it will not cause damage to watersheds and where streams can be adequately protected. The scientists also stressed the value of regional planning and of state personnel serving on planning teams. "Careful coordination among Forest Service and state planners is critical to the success of plans, particularly in areas of shared responsibilities, such as wildlife management."

For all the time and effort involved in preparation of these regulations, the work of the Committee of Scientists was virtually ignored by the Reagan administration early in 1982. Proposals for modification were announced in the *Federal Register* of February 22, presumably to streamline the land-management planning process.[7]

Deputy Assistant Secretary MacCleery on January 21, 1982 (in a speech before the Southeastern Section of the Society of American Foresters at Mobile, Alabama), explained the philosophy behind the modifications:

"In the last administration, there was a tendency to restrict or ban traditional forestry tools, particularly chemical control tools, based on cursory assessments of the risk and inadequate attention to the existence of economically viable and environmentally acceptable alternatives. Foresters will be freer to practice their profession, unfettered by interference from the federal government—and to use management tools and practices that their professional judgment tells them are legitimate."

Following publication of the proposed revisions, more than 2,000 letters were received. Most were critical. Chairman Cooper of the Committee of Scientists wrote Chief Peterson: "The fact that the regulations were changed by a new administration directly upon taking office creates the dangerous precedent that each new administration may do the same. If this happens the regulations will become a document for conveying the political wishes and aspirations of each new administration rather than a technically sound blueprint for governing national forest planning."

Representative James D. Weaver, of Oregon, chairman of the House Forestry Subcommittee, challenged the Administration's contention that regulations were only being "streamlined." He and others charged that the changes were designed to step up logging of old growth timber in the Northwest. In response to "substantial public interest," the period for comment was extended and a public meeting was convened at which members of the original Committee of Scientists were invited to testify.

Environmental critics of revision were especially apprehensive over the outlook for integrated pest management, interdisciplinary team planning, consideration of wilderness values, and public participation. However, when the *Federal Register* of September 30 carried the long awaited "Reagan version" of the regulations, the original standards had been weakened, but the Administration clearly had retreated and not gone as far as intended. Charles R. Hartgraves, director of land-management planning for the Forest Service, the career official most directly involved, told the media: "We cleaned up the language in the old rule and we eliminated philosophical passages which don't belong in regulations. But it doesn't mean we're not going to follow those principles."

Bringing the Public into the Process

Even at its best, planning has brought with it a combination of frustration mixed with hope. The 1970s and 1980s saw a manyfold increase in budget for planning, creating paperwork and problems for both managers and concerned citizens. There had been multiple-use plans, then unit plans, so general that a manager could do almost anything he wanted. Those processes were complex and constantly changing. Each national forest had its own approach and in many cases only a portion of each forest plan was completed. Though constraints appeared to make life harder for rangers in meeting timber-sale targets, they also found themselves looking at critical areas they had overlooked, with significant visual appeal, high priority wildlife, streamside values, and the like.

The National Environmental Policy Act had already changed many things. It mandated public participation; field personnel knew they had knowledgeable people watching. Long-range planners began to cope with problems addressed only in a cursory manner in the old multiple-use plans. Unfortunately, planning often was done by interdisciplinary specialists with little if any ground experience. Often forest supervisors and rangers were not deeply involved; they were still expected to meet timber sale objectives, regardless of the extent to which such objectives were weighed in the plans.

The National Environmental Policy Act directed consideration of environmental impacts of major federal activities, specifically through the process of preparing environmental impact statements. The Council on Environmental Quality, established under terms of the Act, in 1970 issued detailed guidelines on how the statements (or EISs) should be prepared; then, on May 24, 1977, President Carter issued an executive order directing the Council to replace the guidelines with regulations that would improve compliance and reduce paperwork. These new regulations, which became effective July 30, 1979, required that each agency adopt its own procedures, with provisions that actions be classified as one of three types: those normally requiring a full EIS, those normally requiring at least an environmental assessment, and those of least significance categorically excluded from the NEPA process.

For the three years of 1979-1980-1981, the Forest Service prepared an annual average of 12,600 environmental assessments. In November 1981, revised policies and procedures were announced, designed to exclude from analysis actions with little or no impact on the environment, enabling personnel to focus on potentially significant actions. But the idea was questioned forthwith. After all, as critics demanded, how can potential impact be reckoned, one way or the other, without an assessment?

Managing the RPA and NFMA procedures and paperwork represents a considerable challenge, which becomes complicated by inviting in the public or publics. Foresters tend to be reluctant about sharing ideas with nonprofessionals or listening to them, though some now see it differently.

"The graduate forester of fifty years ago was the best equipped person for making most of the national forest management decisions," Regional Forester Coston observed in an interview in 1981.

> Today we must realize that all of the resource capabilities and intelligence are not in the Forest Service.
> Some administrators in the past have felt public involvement was cumbersome and a stumbling block in the way of efficient management. That I can't accept. It has proven a great benefit to the national forests of the northern region. Participating individuals and groups help Forest Service personnel make better decisions and confirm the democratic approach to public land management. National forests don't belong to just those working in the Forest Service—they are part of our national heritage.

The degree to which field personnel involve the public, or seriously consider inputs received, has varied widely in the process prescribed by RPA-NFMA. In the case of the National Forests of Florida (administering the Apalachicola, Ocala, and Osceola), 1,800 packets were mailed in May 1980 to individuals and groups believed to be interested. This mailing listed more than twenty "preliminary issues for forest plan development," a serious effort to air concerns and conflicts and to elicit public response.

Besides issues that might be expected (relating to timber, recreation, wilderness, and land acquisition), the mailing posed questions to the public about deer hunting with and without dogs; deauthorization of the Cross-Florida Barge Canal (since the Oklawaha River borders Ocala National Forest); possible future of Choctawhatchee National Forest (which the military preempted in World War II for Eglin Air Force Base); existing and abandoned military ranges; impacts of mineral developments on other forest resources; proposed phosphate mining on the Osceola National Forest; and impacts of cattle grazing on other resources. Out of 1,800 questionnaires, the national forest received 282 replies identifying nine major issues, as follows:

1. Allocation of national forest lands: What lands are appropriate for special classification (such as wilderness, wild and scenic river), what lands appropriate for other uses?
2. Recreation: What types, amounts, and mix of recreation opportunity should be provided?
3. Silviculture: In what amount, by what methods, and in what relation to nonmarket products should timber be produced?
4. Minerals: To what extent and under what conditions should mineral resources be utilized?
5. Special use permits: What types and to what extent should permits for special uses of national forest land be granted?
6. Forest roads: What kind and how many miles of roads should be developed and managed?
7. Range: What is the appropriate level of range resource outputs, and what actions are needed to achieve that level?
8. Water: What is the national forests' role in producing quality water?
9. Land acquisition: What guidelines should be established to direct land acquisition through purchase and exchange?

In the case of the "Issues and Concerns" mailing of the George Washington National Forest in Virginia, of April 15, 1980, public comment was invited, but with an overtone of bureaucratic professional prejudgment. The timber issue was defined as follows: "Clearcutting as a method of harvest is supported by some, opposed by others. Some people are concerned about the impacts that harvesting timber has on other resources and/or values. Some people are concerned that too much timber is being cut from the national forest." Then the brochure contained a section of "Background Information," detailing benefits to be derived from accelerating clearcutting.

In October 1980, the Green Mountain National Forest in Vermont, following its initial questionnaire, identified twelve key issues by relative importance:

1. Availability of fuelwood   Medium
2. Effects of energy production   High
3. Effects of mineral exploration and extraction   High

4. Types of recreational opportunities to be provided — Medium
5. Decisions affecting wilderness values — High
6. Protection of special areas — High
7. Effects of even-aged management and clearcutting — High
8. Effects of chemical herbicides and pesticides — High
9. Effects of grazing program — Medium
10. Off-road vehicle and motorized access — High
11. Maintaining and developing the transportation system — Medium
12. Effects of land acquisition — Medium

Public participation, though it depends on individual skills and judgment, has been developed as a structured process. A *Public Participation Handbook* stresses that citizen participation is integral to decision making and that everyone in the Forest Service should participate on a continuing basis. The *Handbook* lists six goals in public participation:

1. To help the Forest Service reach better decisions.
2. To inform the public of Forest Service activities, plans and decisions.
3. To encourage public understanding about and participation in the planning and decision process by providing information.
4. To be aware of and responsive to the values of the publics the agency serves, and to evaluate how these publics will be affected by the decisions.
5. To ensure that the Forest Service understands needs and concerns of the public.
6. To broaden the information base upon which agency decisions are made.

Critics complain that citizen input isn't heeded, that agency officials go through the motions of public involvement and then reach decisions without regard for it. In one instance, however, involving planning for the Alpine Lakes wilderness in the Cascades near Seattle, 4,600 responses included a seventeen-page report submitted by an employee of the City of Seattle who had spent many years hiking and photographing the Cascade Range. One paragraph concerned a wetland ecotype in the East Fork of Foss River, most of which the Forest Service had excluded in the proposed boundary in the interest of identifiable topographic features. Yet his input resulted in a 600-acre addition to the wilderness.

The Land Use Planning Team then used this experience as centerpiece of a pamphlet issued in 1973 inviting average citizens to express their viewpoints. The text included this appeal:

We have found your participation to be much more than an exercise in democracy or satisfaction of some policy requirement to include the public in governmental decision-making. Many individual expressions we have received are best described as just plain common sense. While in a democracy numbers certainly count, you may be the only one to express a certain good idea. . . . We stress that this is not an elective process, but rather an opportunity to share your opinions and thoughts with the administrators of your forests before a proposal is developed.[8]

Citizen groups have been discouraged to the point of despair in dealing with some national forest officials. The Mount Hood Forest Study Group, however, was formed in 1971 when Mount Hood National Forest, located near Portland, Oregon, was starting its comprehensive land-use planning. Members of the Study Group (about a dozen teachers, architects, lawyers, students, social and physical scientists) worked with the multidisciplinary planning team, providing data and reviewing land plans for each portion of the national forest, and extending to larger issues of management.

The group urged broadening the list of "public issues and management concerns" and pressed for strict use of economic criteria in identifying lands unsuitable for timber harvest. Members developed a "multiple use alternative" for inclusion in the draft environmental impact statement, declaring that water, plant communities, wildlife, and fisheries are primary forest resources, and that "uses" such as recreation and timber harvest must be planned to maintain and enhance those primary resources.

Water, in fact, was defined as the single most important resource, since nearly one million people depend on the national forest for their domestic supplies. Maintaining old-growth corridors along streams and rivers, reasoned the Study Group in its proposed alternative, would prove the most economic and efficient means of maintaining water quality. To maintain and enhance anadromous fisheries, wilderness designation was advanced as the means of assuring stable soils on steep slopes and preventing sedimentation of spawning beds.

In many cases the Forest Service hears different citizen voices. On the Mount Hood National Forest, a series of public workshops was conducted to develop offroad-vehicle regulations. At the final public meeting, the public and Forest Service agreed on a general prescription, but there were still areas of conflict among hikers, horsemen, four-wheel drivers, motorcyclists, snowmobilers, and cross-country skiers. Someone from the audience suggested the Forest Service convene a session of one representative of each of the six major user groups. These people among themselves devised a workable plan of seasonal and partial closures. One road, for example, would be managed for use by snowmobilers during certain periods and by cross-country skiers during others. One trail would be closed to motorcycles in early spring to allow the tread to dry out. Thus discussion and openness led to common ground and a degree of harmony, at least for a time.

Planning into the Eighties

"Alternative Goals," issued in December 1981, was part of the process of the 1985 RPA Program update. It was intended to achieve the following:

1. More directly tie the Forest Service program to assessment findings.
2. Orient the program toward specific goals and objectives.
3. Improve data from national forest, state, and research planning.
4. Make sure government policymakers were involved early and frequently.

This early document presented needs, opportunities, and alternative national goals for the following "opportunity areas": Timber Supply, Range Productivity, Recreation Use, Wilderness Use, Wildlife and Fish Habitat, Minerals and Energy Development, Rural Communities and Human Resources, International Forestry, and Protection and Support. The purpose of the document was to obtain public views on preferred goals in the opportunity areas to be addressed in the EIS for the 1985 update. The recommended program in that statement, following review by Congress and selection of a single national goal for each of the ten opportunity areas, will then be used to set policy, budgets, and output targets for the years 1985 to 2035.[9]

There weren't many surprises evident in "Alternative Goals." As in years past, it was suggested that timber supplies can be sharply increased by accent on fertilization, use of genetically improved trees, vegetation control, and intensive management on private lands. The alternatives were well publicized and 8,000 "Goals" documents were mailed to individuals, interest organizations, and government agencies and officials for review and comment. The 430 written responses ranged widely. In general, business interests favored greater timber production and less emphasis on wilderness, while conservation groups urged the opposite.

The American Mining Congress advocated actions "to open or maintain access to mineral resources." Additions to wilderness, it contended, would "adversely affect all other opportunity areas." The National Association of Manufacturers took the same approach, asserting that "Recent statistics support the concept that the present wilderness base is sufficient."

The Wilderness Society and Sierra Club in a joint letter to Thomas E. Hamilton, Director of RPA, dated March 31, 1982, asserted that the document was "seriously flawed, both in substance and in process."

They said the wilderness section was basically antiwilderness; that all alternatives under the minerals and energy section called for rapid and massive development, and that sharp increases in timber cutting on the national forests were projected with little accounting for justified needs and costs to the taxpayer.

Region Plans were criticized as well. In a letter to R. E. Worthington, Regional Forester in the Northwest, V. Alaric Sample, of the Wilderness Society, wrote that calls for steep overall increases in timber outputs would put tremendous pressure on forest supervisors. "The stage has been set," wrote Sample, "for inevitable and explosive conflicts."

For all the complication and criticism, the planning process mandated by law, despite its loopholes, may still be the best way to ensure wise use of the national forests. This can only happen, however, if the inevitable conflicts are utilized to evoke factual data; if administrators are openminded, rather than locked in by old ideas and by inflexible production quotas; and if the basic options in land allocation and management remain open through the planning process.[10]

7

TIMBER CUTTING: BENEFITS AND COSTS

> Multiple use seems a bit more of a reality when you consider that our timber program annually makes significant returns to the U.S. Treasury, as well as to the states and counties. In addition, increasingly important wildlife and other resource money is set aside through the Knutson-Vandenberg (K-V) sale area betterment program. Then, there's the bottom line: the timber program operates to the net benefit of the taxpayers—not as a net cost.
>
> —John E. Burns, Supervisor, Targhee
> National Forest, Idaho, in a letter to the
> Idaho Environmental Council, March 1981.

The Timber Program: An Overview

Government Timber Sale No. 1, the first regulated cutting on any land owned by the United States, was begun in the Black Hills Forest Reserve early in 1898, soon after the Organic Act of 1897. The sale was made to the Homestake Mining Company, which needed a continuing supply of timber for mine props. In the eight-year contract period, 15 million board feet were removed; the ponderosa pine forest was heavily cut, but the sales contract called for leaving two large trees on each acre for seed purposes. This was the beginning of Pinchot's conscious forest management. The Black Hills, where once every gulch had its portable sawmill freighted in by bull team, now promised a future of sustained yield—with the year's cut limited to a reasonable, sustainable level and, it was hoped, with better timber as the stand improved.

From that sale onward the goal of timber management activities, in principle, has been to produce continuous flows of wood products in perpetuity. Timber management is practiced on nearly 90 million acres of the National Forest System. Over the years, timber resources of these federal lands have become important to communities supported in part or whole by logging and milling, as well as to other industries, including paper, lumber, furniture, and agriculture, and to consumers. The national forests now provide approximately one-fifth of the nation's total consumption of lumber and other wood products.

Timber values are demonstrable in dollars and cents. Since 1905 the Forest Service has earned $9.4 billion in gross income, with 92 percent derived from sales to timber operators. In the four years of 1977 through 1980 receipts for this "stumpage," or standing timber, amounted to $3.285 billion; the highest year was 1979, when timber receipts reached $968 million. These funds are deposited in the federal treasury, with 25 percent transferred to the states for support of public roads and schools in counties where harvested areas are located. Through administrative practice and NFMA provisions, purchaser road costs are deducted from gross stumpage prices. Another 10 percent of timber sale receipts is spent on construction of national forest roads and trails in states producing the revenues.

This system is implicit with mixed blessings. It links the volume of timber cut, and earnings from it, to justification for appropriations and places the Forest Service under continuing pressure to bring in more money. Higher employment and larger federal rebates are always welcome to a Congressman and beget his support.

The authorization for sale of national forest timber "at not less than appraised prices" was written into the 1897 Organic Act. The 1976 National Forest Management Act renewed this directive, adding specific requirements regarding cutting methods and bidding practices. Further authority for alternative methods of timber sales was also included in the 1978 Timber Sales Bidding Act.

Tree planting and other cultural activities aimed at improving growth rates and tree quality are associated aspects of the timber program. Funds for these purposes, to which critics have been addressing increasing attention and interest, include regular appropriations and an offset of timber sale receipts or stumpage payments as authorized by the Knutson-Vandenberg Act (K-V funds). In 1978, approximately 411,000 acres were reforested; 420,000 acres were treated to timber stand improvement (thinning, pruning, fertilizing, and release cutting), while twelve Forest Service nurseries produced 120 million trees.

Timber Sales: Processing and Problems

The general operating procedure in planning and processing a timber sale depends to a large degree on the district ranger, but he receives guidelines and a target, or quota, from the forest supervisor, based on the approved management plan for the forest. He draws upon the expertise of the supervisor's staff and of his own, not only of timber specialists but of other disciplines as well.

The regulations of September 17, 1979, in pursuant with NFMA, even as altered in 1982, are apt to make it difficult, or challenging, for the ranger to meet an assigned target. Management prescriptions that involve vegetative manipulation of tree cover, according to the regulations, must not be chosen primarily because they will give the greatest dollar return or the greatest output of timber, although these factors should be considered.

Individual cut blocks, patches, or strips are required to conform to maximum size limits established by the regional guide, according to geographic areas and forest types. This limit may be less than but not exceed: 60 acres for Douglas fir of California, Oregon, and Washington; 80 acres for Southern yellow pine of Alabama, Arkansas, Georgia, Florida, Louisiana, Mississippi, North Carolina, South Carolina, Oklahoma, and Texas; 100 acres for hemlock–Sitka spruce of coastal Alaska; and 40 acres for all others, "except as provided." In addition:

> Special attention will be given to land and vegetation for approximately 100 feet from the edges of all perennial streams, lakes, and other bodies of water. . . .
> When trees are cut to achieve timber production objectives, the cuttings will be made in such a way as to assure that lands can be adequately restocked within five years after final harvest. . . . Five years after final overstory removal in shelterwood cutting, five years after the seed tree removal cut in seed tree cutting, or five years after selection cutting.

Once the sale is decided, a timber specialist surveys the area. He may do this by helicopter, studying topography and distance to roads. He consults the district engineer, whose job is to design the main haul road, recommend road standards, and write construction specifications for inclusion in the sales contract. Next, the timber specialist leads a crew in closely surveying (or "cruising") a percentage of the timber on the ground, in order to gather data for an appraisal of value. Since each species brings a different market price, the crew estimates, by species, volume, and quality of timber to be sold.

The appraised value is arrived at by this formula:

> Estimated selling price of timber products,
> minus harvesting expenses (costs of removal, road building, fire protection, disposal of logging debris),
> minus manufacturing expenses,
> minus reasonable profit and risk estimate,
> equals appraised value.

This approach has been questioned at various times over the years, but investigations by Congress, the General Accounting Office and various commissions have upheld it, recommending only minor changes. The profit and risk estimate is based on what a firm of average efficiency might expect by harvesting the timber at the time of sale. The Forest Service collects product-value and manufacturing-cost data from timber purchasers, verifies the data, then averages them for local areas.

Bids are opened at appraised prices through advertisements in local newspapers and post offices and by notices to interested firms. Official figures show that the top twenty purchasers of national forest timber buy about one-third of the volume offered for sale in any one year. About 54 percent of timber volume in 1979 was purchased by small businesses. Only 9 percent of major sales (707 of 7,655) during that year were for more than five million board feet, while 3,991 were over $2,000 in value. A total of 65,418 minor sales went for under $300 apiece.

Bidding is usually conducted at the office of the forest supervisor or district ranger, with appraised value as the starting point. The minimum acceptable bid is required to include adequate funds for reforestation. A procedure resembling an auction begins at the appraised price, with successively higher bids recorded on a blackboard until the winning high bid is reached. Oral bidding has been questioned as a device by which firms hold prices down, but sealed bids can present the same difficulty. A larger problem arises when a company purchases a timber sale, then defers logging for a long period until the market price of lumber rises appreciably.

Speculation and Slippage

Speculative bidding became widespread during the 1970s.[1] Assistant Secretary Crowell explained it simply in a talk before the Natural Resources Workshop of the American Bar Association at San Francisco on January 18, 1982:

> Many purchasers have been betting that inflation would continue, and have bid excessively high rates for longer-term sales of four, five or six years. Some sales were bid so high, particularly in California, western Oregon and, to a lesser degree, western Washington, that they could not be profitably harvested even under the best markets we have ever experienced. The only way such sales could be cut would be if an easy money policy continued to fuel raging inflation.

This hangup became so serious that in May 1980 the Forest Service waived normal criteria for sales extensions on more than 1,400 individual sales, and in October 1981 granted additional contract extensions. Crowell directed that national forests in western Washington and Oregon offer as many sales as possible in the first three months of 1982 in the hope that they would be purchased at depressed market prices and promptly operated. In January 1982, the Forest Service proposed a set of new procedures to reduce speculative bidding and provide for prompt and orderly timber harvest. These were placed into effect in April, reflecting various degrees of public and industry comment. Partial payments at mid-point of sales would be required, but purchaser road credits could be used as cash. Several provisions would apply only to sales of more than three years and more than $10,000. Providing discounts for prompt logging of a sale would be tested on selected western forests for at least two years before being offered throughout the National Forest System.

Once the sale is awarded, Forest Service personnel work with the logger and his crew, making sure that road building meets specifications concerning grade of the road, materials for the base, drainage, angle of slopes on cuts and fills, and seeding of side slopes to protect them from erosion. Where logging is done by the selection method, government timber cruisers mark trees for cutting. Where staff biologists have been involved, they are apt to mark specific trees or areas within the sale to be left for wildlife.

Procedures that appear foolproof on paper are not without imperfections in practice. There is room for laxity and slippage between policy and performance. Some rangers insist on strict compliance in every detail, with continuing inspections during logging and afterwards, but this is not always the case. Various charges have been made over the years of arbitrary changes in timber appraisals by field personnel and of designing timber sales for particular buyers at low figures.

The timber purchaser's aim in removing wood from sales areas, of course, is to obtain the maximum profit. It is an understandable goal, particularly for a small operator paying off expensive equipment. His concern is with his own immediate sale, rather than with efficiency of forest management.

The General Accounting Office made this point in a 1973 study on how the increased use of felled wood would help meet the demand for lumber, plywood, and fiber wood products, while at the same time reducing environmental damage on federal forests (administered by the Forest Service, Bureau of Land Management, and Bureau of Indian Affairs). GAO found enormous waste: the purchasers were simply extracting the most profitable timbers and leaving abundant quantities that would meet minimum standards, but which conveniently were not included in the selling price.

Timber sales controls have been revised during recent years to spur the removal of additional materials. In the Pacific Northwest, the Forest Service makes the point that required payment on a per-acre basis of material *not meeting* utilization standards on timber sales (actually dating from the late 1960s), has spurred greater utilization and, when chip markets are good, the utilization of "cull" material. Besides such controls, the Wood Residue Utilization Act of 1980 authorized a five-year pilot program to determine the commercial potential of residue use.

In Congressional discussion preceding passage of the Wood Residue Utilization Act, it was estimated that 190 million tons of wood and wood residues are left across timber harvest areas of the nation, with far more of it, however, on public than private forests. The cost-conscious landowner recognizes the benefits of removal of waste from his own property. Public forests tend to contain residual old-growth stands with high levels of defect, but that's only part of it. Private lands may be closer to processing plants, making it more profitable to limb, buck, load, and haul the timber. Further, nonindustrial private forest owners often sell a block of timber for a lump sum. The purchaser thus owns the entire tree and stand of trees and is inclined to take all material on which he can do even slightly better than break even. By contrast, Forest Service timber is sold on a pay-as-cut basis, an inevitable temptation to high-grade.

Laxity in enforcing sales contracts results in increased profits for the timber purchaser, inefficient forest management, and damage to other resources in the multiple-use spectrum. Felled wood remaining in sales areas plainly causes delays in planting new trees, creates fire hazards, causes air pollution from unintentional fires (or intentional ones needed to facilitate replanting), and generally leads to increased costs—instead of the revenues that might have come. As to effects on other resources, Walter Kirkness, then Commissioner of Fish and Game in Alaska, wrote, in a memorandum dated May 29, 1964, titled "Logging Damage to Streams": "Probably the greatest damage which takes place to the watershed is from loggers who do not follow stipulations in the contract. They attempt to effect savings in their operation either in time or money, to the great detriment of the salmon. It is here that we are having our most trouble. . . . The Forest Service does not condone these practices but, depending upon the individual forester in charge, in many cases does not maintain enough control over the loggers' activities."

Much later, on November 5, 1979, Ernst W. Mueller, Commissioner of Alaska's Department of Environmental Conservation, wrote Chief R. Max Peterson: "I am particularly dismayed by the apparent view of a few of your key agents that they feel no real need to cooperate effectively with the State of Alaska in preventing damage to anadromous fish streams from logging and related activities."

Elsewhere in the country, the South Fork of the Salmon River, once the most important salmon and steelhead stream in Idaho, lies in the heart of a geologic zone known as the Idaho Batholith, a composition of granitic bedrock and erodible soils. During the 1960s, clearcutting along the South Fork, in the Boise and Payette National Forests, led to extensive damage and to a ten-year moratorium within the most heavily impacted section of the drainage. When the Forest Service proposed to resume logging, the Idaho Wildlife Federation and other groups raised a vigorous protest, citing threats to the fishery, and also to habitat of waterfowl, peregrine falcons, bald eagles, and sandhill cranes.

An Associated Press article from Boise, appearing in the *Lewiston* (Idaho) *Tribune* on July 18, 1983, brought the issue up-to-date. It noted that Forest Service officials had lately tried to persuade environmentalists and other segments of the public that logging would now be safe because of advancements in erosion control and lessons learned from mistakes. Despite such assurances, however, the Idaho Wildlife Federation has fought and caused deferral of almost every plan and timber sale in the area. To quote the AP report on Forest Service consideration of increasing timber harvest by about 50 percent a year:

> Although officials were confident it would not harm the river or fish, opposition by environmentalists was among reasons the proposal was scuttled, Payette National Forest Supervisor Ken Weyers said.
> "One of the major reasons is that it is quite expensive and quite controversial to develop timber-sale environmental analyses," Weyers said.

The Advent of Clearcutting

For many years foresters had preached the virtues of selection-cutting, or all-aged forest management, in which individual trees are logged when they reach maturity and younger trees continue their growth. Under this system, timber is grown on long rotations of one hundred to two hundred years, with cautions taken to protect the soil. Vernon Bostick recalls in a letter to the author how things went in the West before World War II, when he served in the Forest Service:

> In those days sustained yield and selective cutting were our guiding principles; clearcutting was synonomous with "destructive logging." The forestry profession lobbied most legislatures in forested states into passing laws prohibiting clearcutting on *private* land.
> I remember that one of my first acts as a forest ranger was to bring charges against a lumberman for clearcutting a piece of private land in violation of New Mexico's forestry law. He beat the rap, but the point is that in those days the Forest Service tried to get the state to prosecute a lumberman for destructive logging on private land. And I had full backing from my supervisor.

The selection system of that period was designed to follow and fit into nature's pattern of growth, maturity, and decline by selecting individual trees, or very small groups of trees, in order to favor species tolerant of shade, or possibly larger groups up to quarter-acre clearings to favor species intolerant of shade. It wasn't the only system. When Pinchot surveyed the Forest Reserves of the West for the Department of the Interior before the turn of the century, he noted a number of specific instances where clearcutting would be desirable. He may not have conceived it in the same scale as it was practiced later, but he used the exact term.[2]

Clearcutting had been accepted as a silvicultural practice mostly in certain short-lived forest types which reproduce easily, such as aspen, jack pine, lodgepole pine, and some southern pines. It had been applied in small patches, exposing mineral soil necessary for reproduction of these "shade intolerant" species. Patches were of a size that forest edges were able to supply the seed for regeneration and some intermittent shade cover for young seedlings.

Following World War II, lumber company foresters in the Pacific Northwest applied research reports which claimed that partially cut stands of Douglas fir and hemlock would revert in second growth to the less valuable hemlock. They felt that Douglas fir, as a species relatively intolerant of shade and requiring sunlight to reproduce, should be clearcut in patches, followed either by natural seeding or by artificial planting.[3]

On this basis, the era of clearcutting began. Thanks to new and heavy machinery, it became possible to upend as many as 1,500 to 1,800 trees in the course of a day, thus leveling a timber stand of hundreds of acres in short order. And road-building equipment could be used on steep hillsides where it had never been before.

Chief Edward Cliff, in addressing the National Council of State Garden Clubs at Portland, Oregon, on May 25, 1965, endeavored to explain the values of this system:

> Patch-cutting is something like an urban renewal project, a necessary violent prelude to a new housing project. When we harvest overmature, defective timber that would otherwise be wasted, there is bound to be a temporary loss of natural beauty. But there is also promise of what is to come: a thrifty new forest replacing the old. The point is that there often must be a drastic, even violent upheaval to create new forests. It can come naturally—and wastefully—without rhyme or reason as it has in the past, through fires, hurricanes, insects, and other destructive agents. Or it can take place on a planned, purposeful and productive basis.

Chief Cliff compared clearcutting to growing crops; a forest "field," as he said, with its crop recently harvested is not much different from a farm field at the same stage. "For the young, 'citified,' articulate part of our citizenry," he declared before the Pacific Logging Congress of 1966 in dismissing the critics of clearcutting, "it is especially easy and natural to get stirred up about outdoor beauty, recreation, wilderness, vanishing wildlife species and environmental pollution. It is not likely that very many know or even particularly care much about how timber is grown, harvested and used to meet their needs."

From the Northwest the clearcutting method spread throughout the national forests. It was said to be the best method of regenerating commercially valuable eastern hardwoods. In the South, clearcutting stands up to 200 acres in size would make logging easier and cheaper, while southern trees would reproduce and grow best in full sunlight. Once a major program is launched by a federal agency, or any large institution, and directives are issued from the top, it becomes difficult to reverse course, or to question it from within the ranks. When an agency recommends a program to Congress and receives approval, the direction becomes fixed.

No one has written a more incisive or lucid explanation of the growth of clearcutting than Sidney Weitzman in "Lessons from the Monongahela Experience," an in-service analysis completed in December 1977. Weitzman was a veteran Forest Service official who had retired as Director of Southeastern Area State and Private Forestry. He undertook his study at the request of Chief John McGuire "to learn what the Forest Service as an agency might do differently in the future to prevent similar controversies from developing."

Weitzman interviewed many people inside and outside the agency. He analyzed the decision-making process, reward system, personnel policy limitations, traditional attitudes to land administration, organization loyalty, and the directive system. He was not out to find fault or fix blame. "Foresters on the Monongahela National Forest made their even-aged clearcuts no less scientifically than foresters across the nation," he wrote. "But those cuts were generally applied to foster maximum timber production—and there lies the controversy."

Noting the internal strength and influence of timber management staffs, Weitzman recorded these happenings:

> The picture that emerges is of forceful and articulate specialists in the Washington office and field staffs being convinced that even-aged management is a sound silvicultural practice. Then, through force of conviction, they become the motivating force for establishing even-aged management across the eastern United States. This conversion from all-aged management was accomplished primarily through direct field contacts with timber management specialists, discussions and training sessions for field-based timber management staffs and later by Manual direction. Since all management plans for the national forests were approved in the Washington Office at that time, general compliance was assured. Thus, a sound technical decision automatically became an administrative decision. As later events showed, what may have been a sound timber management decision wasn't automatically sound politically, aesthetically, socially or good multiple-use practice. . . . The decision to proceed with the change in policy was based on the needs of timber management production goals and objectives rather than a coordinated decision of all disciplines.[4]

When that decision was reached, the Forest Service leadership stood solidly behind it. Despite later misgivings and self-examination over the direction pursued in West Virginia, the agency stood firm in its position during the heat of the controversy and implementation of clearcut logging. The upsurge of clearcutting generally paralleled rising industry demand for an increase in timber supply from the national forests. Until recent years such pressures had been light, for industry had available to it commercial forest lands held in private ownership and generally more accessible. However, since the post–World War II building boom, steady depletion of resources on private acres led the industry to press for greater productivity on the national forests.

"The tight log supply situation in the West might be eased somewhat by reducing exports of unprocessed logs," Chief Cliff declared in testimony before the Senate Committee on Banking and Currency on March 21, 1969, referring to another significant factor. During 1968, actions by Congress and the Administration had placed a ceiling on log exports from federal lands, but there were no similar restrictions on most state lands. This enabled industry to fill foreign orders (principally from Japan) for unprocessed "round logs" from its own sources, while insisting on greater access to national forests in order to keep the domestic mills going.

In 1950 the annual allowable cut of sawtimber was 5.6 billion board feet. Ten years later it was up to 9.2 billion, and in 1969 to 12.8 billion. Demands for increased cutting were greatest in the West, but the effects were felt throughout the National Forest System. Politics and conflicts grew steadily sharper. In testimony before the Senate Committee on Banking and Currency in March 1969, Chief Cliff cited the need to meet responsibilities other than timber:

> In addition to preserving an adequate wilderness resource, we need to manage our National Forest System to protect watersheds, scenic beauty and fish and wildlife habitat, and to provide forage and outdoor recreation opportunities as well as to maintain future timber harvests in perpetuity. This means there will be areas of the National Forest System other than in designated wilderness where timber harvesting will be foreclosed or restricted. Roadside and streamside zones and developed recreation sites are examples of areas where timber production is limited.

The Chief stressed that timber harvest was close to the maximum allowable cut supportable under current levels of management. But two months later, on May 23, Edward C. Crafts, a former deputy chief, told a House Agriculture Committee hearing that the Forest Service had made a grave error by permitting the term "allowable cut" to be substituted for sustained yield. "Allowable cut used to be the ceiling above which the cut would not be allowed to go. Then it became the floor below which the cut would not be allowed to fall."

Another procedure used to increase timber harvests was to combine sustained-yield management units, called "working circles," into fewer larger areas, making it possible to justify heavy cutting on the strength of growth estimates. In some places, the increases were based on reclassifying forestlands from noncommercial to commercial. Consequently, in order to make their allowable-cut quotas, or targets, supervisors and rangers were obliged to conduct sales of marginal species growing in scattered stands, on steep slopes, or in thin rocky soils.

Such high elevation forested land, as in the Rocky Mountains or Appalachians, often is not suited for timber harvesting. Soils are shallow and terrain makes operations costly. But the qualities that make an area a liability for commodity production make it a natural for human enjoyment: a variety of species of plants and trees; ridges affording distant vistas and cool climates; fishing streams and wildlife.

One such location, the Bitterroot Mountains of western Montana, became symbolic of steep-slope clearcutting, carried to extreme with costly mechanical terracing. Citizens protested with vehemence and a local newspaper, the *Daily Missoulian*, carried an extensive series of illustrated articles by Dale Burk. Although he did not intercede, Neal Rahm, the regional forester, later substantiated the public criticism, listing varied professional mistakes, including: wrong silviculture system choice, poor choice of site preparation measures, allowing logging and road equipment in streams, and allowing logging debris in stream courses.[5]

In neighboring Wyoming, concerned citizens objected to timber sales on the western flank of the Wind River Mountains adjacent to the Bridger Wilderness. The immediate justification advanced by the regional office for clearcutting these mountain slopes was an epidemic-size infestation of bark beetles. But this only stirred alarm and consternation among sportsmen, dude ranchers, stockmen and others. They challenged the value of timbering for bark beetle control on grounds that it would disturb thin mountain soils, uproot huge rocks, and offer only putative benefits to recreation. In view of marginal qualities of timbering at high elevation with a short growing season, they argued that the Forest Service was serving the expanded mill capacity of a large, out-of-state firm, U.S. Plywood-Champion, on a short-term basis. Natural regeneration in such a circumstance would be a slow, painful process, and replanting would be necessary. Since no private industry would bear the cost, responsibility would therefore be shifted to the public.

Who Pays the Bill—and for What?

Does it, in fact, take a public subsidy to construct expensive, extensive road systems to log remote unroaded and marginal timberland? Does it take place at the expense of other legally mandated uses? These questions were in sharp focus early in the 1980s, even more than in the late 1960s, as a result of increased concern over federal spending.

Under the Knutson-Vandenberg Act of 1930, the Forest Service may calculate reforestation costs and deduct them from the receipts of an individual sale. Congress has authorized the Forest Service to require timber purchasers to build and maintain roads needed to remove the timber. The Forest Service explains that the purchaser pays the same stumpage price whether or not K-V funds are taken for reforestation. K-V is considered an "offset" from stumpage before the receipts go to the Treasury and after the purchaser pays for "stumpage." Road costs, however, do reduce receipts or stumpage value.

Critics charge this is part of the subsidy system. However, Forest Service officials make the point that while some timber sales show a "loss on paper," this is due to lack of accounting for both near-term and long-term benefits provided by roads, timber stand improvement, fire protection, wildlife habitat enhancement, watershed improvement, and other positive, but monetarily unmeasurable, effects on all forest resources. Furthermore, they reason, if each sale were required to show revenue greater than the federal costs of selling and administering the sale, as critics contend should be done, timber in some areas would be priced above acceptable market levels and opportunities to develop access to forests and the enhancement of other resources would be foregone. Thus the public would lose the use of other resources and availability of reasonably priced timber into the future.

On the other hand, Thomas W. Power, chairman of the Economics Department at the University of Montana, reported following an economic analysis in 1978 that the Forest Service was ignoring fundamental economic principles by not considering costs or benefits associated with different types of forest sites and by pricing the commodity in a way that ignores the costs of management:

> If in allocating its forest management budgets and forested lands the Forest Service took into account the costs of management as well as the value of the commodity flows, it would never enter most of the remaining roadless areas in Montana but would instead concentrate its efforts on the more productive already roaded areas. If the timber industry had to pay stumpage prices which reflected the full costs to the Forest Service of managing the sites for timber instead of a price which guarantees a "normal profit," they would not be interested in further sales in most of the remaining roadless areas.[6]

The Natural Resources Defense Council claimed in a 1979 publication, "An Analysis of Forest Service Sales Below Cost," that direct Forest Service costs of planting, thinning, administration, planning, and most road construction are completely ignored in its system of appraisal and pricing. Sales of old-growth timber in the Northwest and California show an actual profit above appraised stumpage, according to NRDC, but they are the exception rather than the rule, and sales in more than half the national forests between 1974 and 1978 were money losers. "No roadless area," argued NRDC, "should by opened for development until the economics have been examined."

Chief Peterson responded to the NRDC charges in a letter of October 28, 1981, to Representative Jim Weaver, Chairman of the House Subcommittee on Forest, Family Farms, and Energy:

> A considerable portion of the current dollars spent for national forest management are related to long-term investment. The return on investment may be in just a few years or it may be 20 to 100 years away, depending

on the needs of the timber stands and the forest property as an integral management unit. It is important to understand that these investment costs will not necessarily be recovered within the same year of investment, or even within a 5-year period. . . .

The NRDC also overlooks the proportion of the timber funding required for purposes other than timber sales and timber management. Notable examples are the costs of managing the free fuelwood program and the funding for nontimber specific items, such as land management planning or the RARE II effort.

While denying the scheduling of "below-cost" sales, the Chief acknowledged that:

> Current receipts from thinnings in the Eastern Region do not always cover all costs of road access and other current management costs. In the long range, the high value hardwood (and softwood) timber stands that will be produced through management will more than adequately justify the current investment in management and access construction costs.
>
> In the Alaska, Intermountain, and Rocky Mountain Regions the initial costs of providing access are higher than in other regions on a per thousand board feet basis. The timber is generally overmature and highly defective. Much of the timber harvested is already dead, a condition which lowers the value and the net volume. To improve timber growth in these regions it is important to replace the old, decadent, dying or dead timber stands with young, vigorous, healthy timber stands. Costs are high for all facets of timber harvest and large unroaded areas must be provided with access. As a result, current timber harvests may not always cover current costs. Again we can expect the higher value of the future harvests, resulting from today's management, to yield receipts that justify our current management efforts.[7]

A major element of the NRDC criticism has been that sales of timber from national forests compete unfairly with private, nonindustrial ownerships, discouraging management of those timberlands. Chief Peterson, in his letter to Rep. Weaver, noted that national forests are concentrated in the West, while more than 90 percent of private, nonindustrial timber holdings are in the Northeast and Southeast. In 1976, he wrote, nearly 68 percent of all timber harvested in North and South came from private nonindustrial ownerships, while national forests supplied less than 5 percent of the timber harvested in the same areas.

Nevertheless, NRDC in 1982 released a report, "Reforming the Unfair Federal Timber Magnet: Proper Pricing to Prevent the Penalizing of Private Industry." It renewed the charge of below-cost timber sales, asserting that the Forest Service was "(1) wasting considerable taxpayer dollars and (2) denying price levels to private and corporate tree farmers needed to induce proper management of their timberlands."

Another area of contention revolves around the definition of commercial forestland. Writing in the *Journal of Forestry* for November 1981 ("An Economic Classification of U.S. 'Commercial' Forests"), Marion

Clawson, of Resources for the Future, notes that "to an economist, the term *commercial* is particularly unfortunate, implying as it does that these forests have economically profitable possibilities for wood growing. . . . In fact, much of the 'commercial' forest area cannot be managed economically to grow wood under present or foreseeable conditions."

The Forest Service defines commercial timberland as forestland currently producing or capable of producing crops of industrial wood and not withdrawn from timber utilization by statute or administrative regulation (as, for instance, for wilderness, recreation, or streamside protection). Areas qualifying as commercial timberland must have the capability of producing in excess of 20 cubic feet per acre per year of industrial wood in natural stands. Inaccessible and inoperable areas currently are included. On the other hand, noncommercial forestland is defined as forests incapable of yielding crops of industrial wood because of adverse site conditions and productive forestland withdrawn from commercial timber use through statute or administrative action.

Critics have felt the figure of 20 cubic feet per acre per year to be unrealistic and unscientific. Clawson in his article proposed a classification system as follows:

Class A: biologically and economically productive forestland, with no serious conservation or environmental hazards to timber harvest, with capacity to grow 85 or more cubic feet of potential industrial wood per acre annually under extensive forestry, and with ability to return 10 percent or more on any incremental investment in timber management and production.

Class B: forestland capable of growing 50 to 85 cubic feet of potential industrial wood per acre annually under forestry, and that does not develop serious environmental problems when timber is harvested.

Class C: forestland with biological and economic productivity too low to warrant investment in timber management, or with environmental or conservation hazard too great to permit timber harvest; specifically, land incapable of growing 50 cubic feet per acre annually, especially if poorly located, with high road costs, steep terrain, poor soils, or other conservation hazards.

In general, Clawson urged, the Class C forests should not be cut, even if some return could be secured from existing old growth. "Under the National Forest Management Act of 1976" he wrote, "the Forest Service is probably excluded from harvesting those stands, both by the 'unsuitability' clause and because the costs of reforestation required by law would exceed the value of the timber that might be removed."[8]

Another point was made by Roland Cheek, of Back Country Horsemen, a Montana outdoors group. In reviewing the 1977 budget request of the Flathead National Forest, he noted that 60 percent, or $2.8 million, of the total request of $4.4 million, was for timber harvest activity. Recreation received $159,000 and wildlife and fisheries management, $64,000. Yet 10 percent of recreation and 80 percent of fish and wildlife funds were for personnel working on timber sales.

Whether or not these allocations make sense depends on one's own perspective, philosophy, and objectives. In 1981 the Forest Service again found itself at loggerheads with a substantial body of citizens in western Wyoming, this time over the proposed Jack Creek timber sale on the Bridger-Teton National Forest near Pinedale. The decision to proceed with the sale after extensive review and public hearings evoked protest from the Sublette County Commissioners and an investigation by the Wyoming congressional delegation.

State Senator John Turner, whose family has been in the guest ranch business for many years, charged that the sale was intended primarily to support two logging mills. In a letter of August 16, 1981, to Forest Supervisor Reid Jackson, he wrote that the agency's environmental assessment had carefully considered employment numbers of the region's timber mills, while disregarding recreation and resort employment and economic impacts. Senator Turner proceeded with this further criticism:

> I generally support the multiple-use philosophy and yet we have witnessed timber programs render many areas in this region of Wyoming into almost single-use land units. We have seen scenic values decline, recreational opportunities whittled away, resident wildlife populations pushed out, big game migration routes destroyed, watersheds impacted, and opportunities for small timber operators decline. Considering that this region possesses some of the finest outdoor resources left in North America, such management programs and priorities must change or these unsurpassed values will be lost forever.

Turner raised the question of cost versus benefits, calling for an economic review of the Jack Creek sale, to include all outlays of effort and public funds for silvicultural work to analyze and prepare sales, road reconnaissance, preparation of environmental assessment, harvest design and timber cruising, bidding and sale contracts, road construction credits, harvest management, slash treatment, replanting, thinning, rodent control, and long-term road closure program.

Timber interests view things differently, of course, and so do some Forest Service officials, who reason that deficit sales not only serve other forest uses, but benefit the consumer and the labor market: that a reduction of just one dollar per thousand board feet in the price of softwood lumber, induced by an increase of national forest timber output, can save consumers $40 million per year; while on the other hand, a one-million board foot decrease in national forest timber output can be translated into the loss of three American forestry jobs, even after accounting for substitution from nonfederal timber producers. In 1981 a sale proposed near Estes Park, Colorado, on the Arapaho and Roosevelt National Forests, led to controversy and criticism on grounds of cheating the taxpayer while subsidizing the industry. The *Denver Post*, however, quoted a Forest Service spokesman, John Windsor, defending the deficit sale as a legitimate aid for an ailing industry. Windsor said he hadn't heard environmentalists complain about the $1.5 million spent for recreation programs on that national forest alone.

Whether poor timber lands are being harvested at a net loss to the Treasury is one question that may be answerable in economic terms. Is it cost-efficient forestry to build road systems in order to log remote, marginal timber? If such a timber program benefits other forest uses, the answer may be affirmative. Or does it take place at the expense of legally mandated multiple uses, degrading recreation, fish and wildlife, soil and water protection, and wilderness? If deficit sales give timber an unfair advantage in weighing it against nontimber values, against long-term public interest, then something must be wrong in the formula.

Challenges to Clearcutting

The Forest Service has made headway since the 1960s. Foresters themselves have challenged clearcutting as a system that always works, even the idea that Douglas fir must be cut in large blocks.

"Clearcutting tends to minimize diversity and makes it almost impossible to avoid damage to the site, to streams, and to aesthetic qualities," wrote Dr. Leon Minckler (who worked in Forest Service research for twenty years, then began another career in forestry education).

Most of all, it eliminates the forested character of a particular area for a long time. Ecologically it is a major disturbance. When harvesting mature stands, clearcutting is a cheap and effective way of extracting timber, but the sacrifice of other values may be a poor trade-off for cheap timber harvesting. In immature or partially mature stands, clearcutting may not even be the cheapest way of harvesting timber.

For integrated uses (such as timber, wildlife, watershed, recreation and aesthetics), management should aim toward maximum diversity and minimum damage to the environment. This can be accomplished by single tree selection, group selection, small patch cutting, of a few acres, or a combination of these.

There is no doubt that the Forest Service has undergone critical self-examination. Following the controversies of the 1960s, Chief Cliff appointed a multidisciplinary team to conduct a nationwide review. The resulting report, "National Forest Management in a Quality Environment—Timber Productivity," was released in 1971. It found that knowledge, skills, and abilities of silviculturists needed to be improved. An action plan has been in effect since 1972.

The plan provides for each region to send silviculture specialists for refresher courses at selected forestry schools. They study a variety of subjects and hopefully become more adept in dealing with ecological values. The Northern Region led the way by sending personnel for a month's training at the Universities of Idaho, Washington, and Montana.

About 170 trainees per year have upgraded their skills and been designated as certified silviculturists, in accordance with national standards. In an article on this program in the *Journal of Forestry* for April 1981, Carl R. Puuri (silviculturist, Washington office) and Raymond G. Weinmann (director of timber management, Pacific Southwest Region) wrote: "The silviculturist has more to consider than just ensuring perpetually productive stands and maintaining a timber management program based on sustained yield. Once the silviculturist could be primarily oriented to timber. Now, though, he or she must be concerned with management objectives that also include regulation of water yields, enhancement of wildlife habitat, and improvement of recreation values."[9]

Soil and Water Resources—Unmeasured Benefits

The need for broadened awareness in silvicultural planning and practice has been raised by Forest Service hydrologists and soils specialists. Erosion under natural conditions is a slow process, steadily wearing away the earth's surface. It never stops. The action of water is continuous, whether freezing, thawing, dissolving, or transporting sediment. Over long periods of time, humid regions have been changed into desert, and desert regions into humid. The process of change, however, has often been accelerated by man and his activities, sometimes positively, at other times with drastic destruction as the consequence.

When conditions in the forests are altered by wildfire, grazing, logging, mining, and road construction, erosion rates are likely to increase sharply. Flooding is a visible effect. Damage to fish and fish habitat is another, since increased sedimentation (of clays, silts, and sands) destroys spawning areas, smothers eggs deposited in gravels, and reduces insect foods upon which fish depend. But there is also the invisible effect, significant over the long term and directly related to the silvicultural goals of "ensuring perpetually productive stands and maintaining a timber management program based on sustained yield" (to quote again from the article by Puuri and Weinmann). Compaction of the soil due to man's activity reduces air and water normally available for plant growth. When the topsoil is removed, nutrients are lost, no longer contributing to forest nourishment and health.

Plants in natural systems generally anchor the soil in place. Forest ecosystems absorb rainfall, releasing stored water gradually. They filter impurities and improve water quality. On the other hand, water can exert tremendous force on exposed soils. The lush heartland of America is now losing its fertility because of the excessive loss of topsoil from planted croplands. It is estimated that Iowa alone each year is losing 260 million tons of soil from its cropland.

The protection of soils and water supplies and the prevention of erosion is implicit in the mission of the Forest Service. "'Watershed management' is not the exclusive domain of the hydrologist," wrote

Clifford Benoit, hydrologist of the Northern Region, in a paper on "Forest Hydrology" in 1973. "The quest for quality land management requires that soil and water stewardship through the application of sound principles and techniques be practiced by all to the highest degree possible."

By this, Benoit added, he meant minimizing by all means possible the impact of commodity production on soil and water resources. "Many foresters know little about solar radiation and find it difficult to comprehend. The subject generally lies outside their training and experience. However, the manner in which the forest absorbs, reflects, radiates, and transmits radiant energy is basic to all land management concepts of ecosystems, habitat types, silvicultural practices, etc."

Hydrology is virtually as old as civilization. Developments to utilize and move water through engineering devices go back to ancient Greece, Egypt, and Mesopotamia, as well as to the Anasazi of the American Southwest, with their wells, canals, collecting tunnels, cisterns, and dams. The Romans engineered and built aqueducts that are standing to this day. The unified concept of soil, water, and forest, however, came into its own in the United States through the work of George Perkins Marsh, Bernhard Fernow, Franklin Hough, Gifford Pinchot, and Theodore Roosevelt. Thus the watershed mission of the Forest Service was first enunciated in the Organic Act, with renewed emphasis in more recent legislation, such as the Multiple-Use–Sustained-Yield Act, National Environmental Policy Act, Federal Water Pollution Control Act, Forest and Rangelands Renewable Resources Planning Act, and National Forest Management Act, as well as various executive orders and federal regulations.

Watershed management focuses on drainage basins for protection and also on production of water supplies and water-based resources; this includes the control of erosion and floods, protection of fisheries, and protection of aesthetic values associated with water. Over the years various efforts have been made to increase water yield through either complete or partial removal of vegetative cover—the idea being that what plants don't consume can be transferred for human use. Research experiments and studies at Wagon Wheel Gap, Colorado; Coweeta, North Carolina; San Dimas, California, and elsewhere have strongly indicated the feasibility of this approach. The outcome has not always fulfilled the hope. "Results of individual treatments vary widely and for the most part are unpredictable," reported A. R. Hibbert, a Forest Service specialist in watershed research in 1965, after reviewing all the experiments to that time. The chemical spraying of chaparral at Globe, Arizona, treated elsewhere in this book, was designed in part to increase the water yield. Ranchers in the foothills of the Bitterroots of western Montana complained during the height of the clearcutting controversy there that all the extra water they were told to expect drenched them in a massive outpouring and when they didn't want it.

120   TIMBER CUTTING: BENEFITS AND COSTS

"We need to elevate the erosion control job to a higher level of recognition," urged Don Boyer, soils scientist on the Watershed Management staff in the Pacific Northwest Region, in 1979 (in a paper on "Special Road Considerations—Planning Ahead"). "Everyone should be concerned but someone should be responsible for the total job at all levels. At present, the responsibility is often fragmented into two or three disciplinary areas and coordination is difficult. I realize that the job does not lend itself easily to a 'production measuring stick' concept but we can call attention to the idea that a successful stabilization program is the reflection of our organizational concern for the soil and water resources and it's out there for everyone to see. In a sense, our land ethic is showing."

Boyer was then facing, in his region, covering national forests of Oregon and Washington, the harvest of 700,000 acres of timberland yearly, with disturbance, by his estimate, of 100,000 acres of soils. "It is much wiser to prevent soil-water damage than to attempt to restore to the original conditions," he declared in urging coordinated interdisciplinary presale planning. "Complete restoration to the natural state is often very difficult, if not impossible, to achieve."

Benoit, Boyer, and others have pinpointed serious erosion and stream sedimentation resulting in national forests from poor road construction, inadequate drainage facilities, skidding up and down steep slopes, vegetation removal, accumulation of debris in stream channels, working mechanical equipment in stream beds, and activities which change natural stream channel alignment or gradient.

Road construction has been identified as especially damaging to fish and fish habitat, based on the following: location of roads too close to stream channels; siltation from road drainage and unstabilized slopes; landslides triggered by poorly located roads; removal of vegetative cover on stream banks (resulting in higher water temperatures, accelerated bank erosion, and diminishing food supply); disturbance of stream bed and banks at stream crossings during road construction; and improperly installed culverts that inhibit or prevent fish passage.

How should values be assessed and priorities determined? Which rank highest: economic returns from commodity production or the fishery and the quality of water and soil that it represents?

F. Dale Robertson, while Supervisor of the Siuslaw National Forest in the big timber country of western Oregon, endeavored to respond to these questions. In a letter to all personnel of the national forest, dated August 2, 1974, he wrote:

As many of you know, the Siuslaw experienced a lot of mass soil movement this past winter. Because of the detrimental impact this has had on water quality and fishery habitat, the Oregon Wildlife Commission has been particularly critical of some of our land management and road building practices. Although many good things are happening on the Siuslaw, several of our own people have also expressed concerns to me about the effect mass soil movement is having on our quality of land management. In view of this feedback, and based on some of my own observations, I felt there was a need to re-examine the basic direction of the Siuslaw National Forest.

A ranger/staff group, wrote Robertson, discussed the erosion issue and drafted a policy statement. It was reviewed by many people, internally and externally, and the final draft was now being placed into effect. The supervisor asked for full commitment and support from all personnel to keep it from becoming another meaningless document, "forgotten and filed away like so many others in the past."

The Siuslaw policy statement is well worth citing here,[10] in part if not in full. It reaches to the core of the conflict and controversy of how national forests are to be used and managed:

> In the long run, we will strive toward a more balanced budget to reflect the total resource management job. In the short run, we will strive for quality forest management in meeting production goals with a budget favoring the timber resource over recreation, fish, and wildlife. However, an unbalanced budget is no excuse for doing the timber job at less than "acceptable quality standards" or at the expense of mass soil movement or other resource values. If a choice has to be made between production and quality, after putting forth our very best effort, the decision will be to accept a fall down in production.
>
> Responsible land stewardship includes the protection of soil, water quality, and fish and wildlife habitat. Timber sales and related road construction, reconstruction and maintenance will be planned and administered to leave the "land whole again." We cannot afford to cut corners in our timber sale program and permit additional degradation, risk to health or safety, or other undesirable and unintended consequences such as log jams, debris slides, or mass soil movement. If a proposed timber sale cannot be harvested within acceptable environmental constraints and leave the "land whole again," it will not be made. Timber sales will be planned with the intent to satisfy all environmental objectives so that the areas will be free from future problems or liabilities which would degrade watershed values or require correction with appropriated funds. Short-run economics will not override long-term needs of high quality land management.

The policy statement stipulates ten specific safeguards before reaching a summary of classic stewardship and public forestry:

> a. Leaving timber that is more valuable for watershed.
>
> b. Leaving "long corners" that cannot be reached from an acceptable road or landing or impractical to log with a helicopter or other system not requiring roads.
>
> c. More use of helicopter or other logging systems to harvest timber in areas which cannot be safely roaded.
>
> d. Better and more timely road maintenance to prevent mass failures.

e. Incorporation of the science of water management in road location and design to neutralize the often adverse effects of changes caused by road construction to surface and subsurface water movement.

f. Using available timber sale contractual flexibility to prevent and clean up debris in streams.

g. Keeping fuel accumulations down to a safe level through better utilization and fuel treatment.

h. Reclaiming material over the side of landings and along roads when it is likely to trigger mass soil movement.

i. Constructing stable roads which will withstand the worst winter storms.

j. Disposing of road and landing right-of-way debris when it is likely to affect slope stability, hinder fire protection, or not meet visual management standards.

Finally, when there are no alternatives for logging an area, except at the expense of mass soil movement or creating a future liability, the timber will be left until an acceptable logging method is developed. Such high risk areas will be identified and classified as marginal in our Timber Management Plan and the Forest's allowable cut calculation will be adjusted accordingly.

8

RANGELAND USE AND ABUSE

Diversity of plants is the underlying factor controlling the
diversity of other organisms and thus the stability of the world
ecosystem.
—Peter H. Raven, Director, Missouri
Botanical Garden.

A History of Overuse

The rangelands of today are the grasslands of history, a part of inner
America richly endowed by nature but undone by Western man through
stormy generations until now they are caught in the process of desertification. The course of history, as written in lands bordering the
Mediterranean and in Africa, has arrived to make the New World old.

Until the 1870s the grasslands were ruled by the Indians, the "Plains
Indians" as we now call them, living in concert with buffalo, antelope,
deer, and a host of other animals that thrived on grass. Then entrepreneurs
sent their cowboys north from Texas, disgorging vast herds of domestic
livestock. Within a few years the buffalo, which had numbered in the
millions, were virtually extinct and their species had to be saved in an
eastern zoo. The Indians without the buffalo were left like the sky
without the sun. Then came the homesteaders, plowing submarginal
grassland ill suited for farming. This pattern inevitably led to the Dust
Bowl era of the 1930s, when meadows turned to desert, cattle starved,
and furious storms of dust lashed the plains and prairies.[1]

"Synopsis of Range Conditions on the Tonto National Forest," a
1979 internal review, portrays with openness and candor a scene in a
major Arizona grassland. The Tonto covers almost three million acres,
ranging from the Sonoran desert at lower elevations to conifers below
the Mogollon Rim (or "Under the Tonto Rim," to cite one of Zane
Grey's titles). The "Synopsis" reports:

> Presently the majority of the forest has serious grazing problems that
> have been compounded through many years of misuse. The Sonoran Desert

123

and associated grasslands are producing at just a fraction of their former level of productivity. Areas in the Tonto Basin are described by early settlers as producing native grass to cut for hay. These areas are now dominated by thorny shrubs and annuals. Perennial grasses have been almost completely eliminated over large areas.

Historically the Forest Service has tried to bring livestock numbers in line with grazing capacity. The Forest Service was successful in significantly reducing numbers from the maximum that occurred near the turn of the century. Generally, these early reductions were only stopgap measures that did not have lasting results. Many times, rather than fight for the total reduction needed, there was a compromise for something less. When this happened the range continued its downward trend at a slower rate.

Bringing stocking in line with grazing capacity has met with only limited success during the past 30 years. A primary reason is that through the years low grazing fees and the unrecognized buying and selling of forest grazing permits have built up a current false value of about $1,000 per animal unit. Any effort to adjust livestock numbers to grazing capacity is regarded as a financial loss to the permittee. Rather than accept a reduction in livestock numbers, many livestock operators seek political assistance in their behalf. Under political pressure, the Forest Service either capitulated or increased the intensity and number of studies determined necessary to support anticipated grazing appeals. . . .

By the nature of the permit system, speculation in buying grazing permits has been high. Since 1959, about 25 percent of the ranches with Forest Service permits have transferred ownership each year. . . .

Where water was developed on overstocked allotments in lighter utilized areas, grazing pressure shifted to these areas with resulting range deterioration. Current studies indicate that many allotments are overstocked now to a greater extent, depending on initial degree of overstocking, than they were before money was expended on them. Hundreds of thousands of dollars invested by the Forest Service and the permittees have been wasted because a cardinal rule of grazing was not adhered to—that no grazing system will work, or range sustain itself, if the system is overloaded with too many livestock, and plants are not allowed to meet their physiological growth requirements.[2]

The chronic effects of intensive grazing are widespread on public lands, likely even more so on areas administered by the Bureau of Land Management than those administered by the Forest Service. The Portland *Oregonian* of April 3, 1982, reported on a study made by the Bureau of Reclamation of stream problems along 360 miles of the Grande Ronde River system in northeast Oregon. The study singled out cattle grazing as a major factor leading to streambank erosion, siltation of gravel beds, and high summertime water temperatures currently blocking efforts to revive the once plentiful steelhead and salmon runs.

"While no individual resource practice is solely responsible for the loss of streamside vegetation," according to the *Oregonian* account, "livestock grazing, farming practices and timber removal have done the most damage. The bureau also cited road building, stream channelization, mining and recreational developments." In the words of the Bureau of

Reclamation: "Of all the factors affecting the quality of riparian habitat, livestock grazing has perhaps had the greatest impact on streamside vegetation."

Many Forest Service officials have sought to bring grazing under control, struggling with political pressures from without and a willingness to compromise from within. Earl D. Sandvig recalls how Regional Forester "Major" Evan W. Kelley vigorously sought to stop overgrazing throughout the Northern Region during the 1940s. Sandvig had been supervisor of the Beaverhead National Forest in Montana, then was promoted to Chief of Range Management for the Region. He completed range surveys and ordered large reductions, with Kelley's complete support. "There was no appeasement to cover poor range management or poor resource management of any kind under Kelley," wrote Sandvig in a personal letter in 1982, continuing his concern to control the range long after his retirement. "It was a good example of positive leadership from a regional forester."

C. N. Woods, Regional Forester of the Intermountain Region, conducted a running battle with the Washington office over the grazing issue. In a letter to Chief Lyle Watts on December 4, 1942, he wrote:

It is unnecessary to go into detail as to range conditions on R-4 forests. The WO and RO know and admit they are, over a big acreage, not what they should be. Why do these conditions exist after 35 years of grazing administration? The answer can probably be found in one or more of the numbered paragraphs below.

1. Not enough rangers and administrative guards.
2. Forest officers not spending enough time on RM [range management].
3. Forest officers not practically and technically qualified for RM.
4. Forest officers did not do what they knew should be done relative to reductions in numbers of stock and length of seasons, handling of livestock, and personnel action.
5. Grazing regulations and policy wrong.
6. Lack of encouragement and support of the lower levels by the higher levels.

There is no more difficult job on a national forest than that of RM on districts with much depleted range, which inevitably presents big grazing problems. It requires, in a high degree, patience, diplomacy, familiarity with RM, courage, aggressiveness and persistence to handle big grazing problems successfully. . . .

We have administered these forests for 35 years. I feel strongly we could, without any blow-up, have gotten down to carrying capacity in not more than 10 to 15 years at any period during the 35 years of administration with little if any exception.

I know we have been told that good PR made it imperative that we go slow. I do not concede for a minute that we could not practically have gone much faster than we have gone, and I want to emphasize what I said in the last sentence of the preceding paragraph.

Grazing permittees constitute but a small percentage of the people of the Region. Their numbers and interests in the national forests are small, as compared with water users, recreationists, and others. Surely, it's poor PR to countenance depletion of the natural resources. Besides, I believe most of our permittees are reasonable people. It is felt that the best PR is proper management of the national forests, the prevention of depletion. This is surely true in the long run.

Washington officials refuted and rebuffed Woods, with personal criticism implicit in their correspondence. Officials in high places, after all, are expected to have passed the stage of boat-rocking. But on May 6, 1943, Woods wrote:

I am much disturbed, for the good of the Service, and for the cause of conservation, at the attitude of the WO. It is my judgment, after much consideration, that the WO is in certain very important things showing a decided timidity, fear, attitude of appeasement. Besides, there is some very good evidence of bureaucracy, of lack of sympathy, or of failure to give more than very superficial consideration to some things I consider of great importance, things presented by this office in good faith and advisedly.

The level of grazing fees bears strongly on the issue. When regulations were written and permits awarded early in the century, the idea was to protect the homesteader, setting roots in the land, from unfair competition of the large corporate owner and cattle baron. The initial fee for grazing was 5 cents for an animal unit month (or AUM), based on a calculation of one cow, one horse, or 5 sheep in one month. This is part of a legacy that still lingers. As Earl Sandvig wrote to Representative Denny Smith of Oregon, on March 15, 1982:

Beginning with the inception of selling forage to ranchers 75 years ago, the owners of the land—the public—have never received fair market price for the grazing that was sold. Numerous attempts to correct the situation have failed because of the opposition brought on the Forest Service and BLM by that segment of the livestock industry receiving subsidy of very cheap grazing from public land. Actually, the general public is not informed of the loss of revenue to pay administrative costs and the loss to public land counties from their share of fees from grazing and other products sold from the public lands. In eastern Oregon subsidized grazing has never contributed a fair amount to school and road funds.

Who Controls the Range?

Range management covers about 102 million acres of national forest and national grassland, divided into 11,000 range allotments in 31 states. A total of 16,000 paid permits enable permittees to graze 1.6 million cattle and 2.1 million sheep. (In addition, 87,000 free-use permits have been issued for horses used in forest recreation and for small numbers of cattle and sheep.)

Most grazing is in the West, where use of high meadows and fields of the national forests during summer months is considered essential to many ranches. But grazing also extends across the plains and in a lesser way to tidal marshes and piney woods of southern national forests. Receipts from grazing in 1978 totalled $11.4 million, of which 25 percent was returned to the states.

Four basic goals theoretically apply in Forest Service range management: (1) utilization of the kind or class of livestock best suited to the range, (2) grazing in proper numbers, (3) grazing in the proper season, and (4) getting good distribution. Range planning is undertaken jointly on the ground by the district ranger, a range technician, and individual stockmen holding permits. It is based on technical studies of such factors as soil, condition of forage, and topography. Under choice circumstances, the objective is to ensure sustained production and use of quality forage consistent with other uses. For example, if elk are known to forage in higher reaches, some of the allotment will be reserved for them. Ideally, sagebrush will be left to provide food and cover for grouse and other birds and mammals instead of being converted to grass. Where recreation visitors show preference for a shady grove as a picnic site, it will be withheld from the allotment, at least through the popular visitor season, and campgrounds are fenced to keep cows out.

That is how it works—at times. "Let me be first to admit that we still have many problem allotments on the Gila," wrote Ray Swigart, of the Range and Wildlife Staff of the Gila National Forest, in a letter responding to an informal report prepared by Earl Sandvig for the National Wildlife Federation in 1971.

By the same token we have some very good allotments, as good as any I've observed through the years. On the majority of our ranges we are making considerable progress. Everything I can find in our files, including range surveys of the 1930's, old photographs, and other old studies, indicates this is generally true. Perhaps much of this country was harder hit in the early days when miners grazed herds of goats and burros, as well as the usual overstocking by sheep, cattle and horses.

In many respects I believe it can be compared to the alpine country of Colorado, Wyoming and other portions of the West. Unfortunately, though, in the Southwest the climate allowed livestock to use the range year-long—and this arid country just couldn't, and still won't, stand up under extreme abuse. This complicated administration when the national forest was created. As late as the 1940's major trespass was a common thing and we have had problems with a few individuals within the past six years. In these instances we developed what we thought were good sound cases, but they were all appealed. In most cases the Secretary's Board of Forest Appeals has sided with the permittee.

Around the turn of the century overgrazing was common, and there was no regulation of use. "The general lack of control," reported the Public Lands Commission of 1905, "has resulted, naturally and inevitably, in overgrazing and the ruin of millions of acres of otherwise valuable grazing territory. Lands useful for grazing are losing their only capacity for productiveness, as of course they must when no legal control is exercised."

A measure of legal control was initiated over key Arizona rangeland in 1905 with establishment of the Tonto National Forest. Sheepmen and cattlemen had been embroiled in bitter feuding, particularly over winter ranges along the Verde and lower Salt Rivers. The steadily deteriorating range only increased competition for what good forage remained. Because the national forest was intended to administer watershed and grazing areas in the public interest, efforts were directed to bringing the number of domestic stock in line with capacity.

This had its pitfalls, not without a humorous side. "We have done a good job on the Tonto," officials would declare, tongue in cheek. "In 1915 we had 50,000 head and started to reduce. Now in 1919 we have reduced to 82,000."

Gifford Pinchot was extremely critical of damage done by sheep, which he thought ten times worse than cattle damage (John Muir had called sheep "hoofed locusts" and wanted them banned altogether). But Pinchot felt that if grazing were completely prohibited, the western ranchers would not quit fighting until they killed off the forest reserves. He also believed, perhaps wishfully, that under strict control grazing could proceed without harming soil and vegetation.

The result was the revolutionary permit system of 1905. Albert F. Potter, a pioneer Arizona stockman whom Pinchot recruited into the Forest Service, is credited with early policies and regulations by which permits for grazing were issued on a scale of priorities. First, permits were issued to small owners who lived in, or close to, the national forest and whose stock had regularly grazed on the range, then to others in the surrounding area who had also used the range, and finally to stockmen who had never used it. Fees were instituted, and the Service required permittees to own a certain amount of cultivated land and water, eliminating at once nomads who had used the public domain for many years without supervision. This was strong medicine for stockmen and, although overgrazing was not eliminated, the institution of these regulations stands among the monumental achievements of the Forest Service.

Today, permits are issued based on guidelines from the supervisor's office and in accordance with the Federal Land Policy and Management Act of 1976. FLPMA directed that an Allotment Management Plan be developed on virtually all allotments in consultation, cooperation, and coordination with grazing permittees. The plan must specify the number of head to be grazed, indicating specifically where, when, and under what special considerations, as well as what range improvements are to be installed and maintained, and by whom.

This range improvement program marks yet another attempt to restore overgrazed lands and achieve a greater balance of uses with the participation of stockmen who have enjoyed low-cost grazing for many years. Such progress has come slowly.

In 1966, the Forest Service and Bureau of Land Management contracted with the Statistical Reporting Service of the Department of Agriculture to obtain economic data on grazing use within ninety-eight national forests, nineteen national grasslands, and fifty-five Bureau of Land Management districts in the western states. These data were to serve as a basis for evaluating current fee structures and determining grazing values.

Interviews were conducted with federal permittees and with ranchers leasing private grazing lands. The analysis and evaluation showed that fees were much higher for grazing nonfederal lands. Because of the considerable difference, permits had assumed a value, and buyers of ranch property or livestock associated with a Forest Service permit had to pay a premium because of it.

Despite problems in this aspect of grazing, improvements had been made on the national forests. The public domain lands under jurisdiction of the Department of the Interior, by contrast, had remained uncontrolled for years. In 1934, the Taylor Grazing Act was adopted to "stop injury to the public grazing lands by preventing overgrazing and soil deterioration, to provide for their early use, improvement and development, to stabilize the livestock industry dependent on the public range." But only in recent times has BLM succeeded in establishing so-called priority grazing periods and in getting funds for range improvement. Politically powerful livestock associations blocked management at every turn, while the condition of millions of acres deteriorated, and all values—soil, water, wildlife, and forage, too—suffered. Stockmen received further subsidy through predator control programs, by which federal trappers and "gopher chokers" of the Fish and Wildlife Service were paid to exterminate wildlife that might (or might never) interfere with domestic stock. In 1969 permittees objected to the raising of fees and demanded proprietary rights not granted to other users of the public lands. The 1966 joint study of fees had resulted in a plan to adjust them upward gradually to fair market value by 1978. Then FLPMA provided that fees not be increased until completion of another joint study and report by the Interior and Agriculture Departments. A report to Congress followed in October 1977, recommending phased increases over a period of years until fees reached a par with fair market value—but action was postponed by congressional and Administration action.

Permittees insist that payment for certain improvements on their part refutes the allegation of subsidy. On the other hand, it also tends to reinforce their claim on land they don't really own, adding to the tribulations of the resource manager and of other resource users. Grazing fees today are determined by a formula which Congress established in 1978. It includes an index for "ability to pay," which is affected by beef cattle prices and cost of production. Application of the formula in 1981 resulted in a steep reduction in federal grazing fees, while the average lease rate for private grazing land rose. In 1982, the public land grazing fee, set according to the formula, was $1.86 per AUM, compared with $8.83 on private land.

FLPMA gave permanence to the BLM and a directive to manage the lands in its care in behalf of all the public. This sparked the Sagebrush Rebellion, led by livestock interests fighting federal regulations. In like manner, livestock associations had sought for years to dominate the national forests, or even to destroy them. Following World War II, a political push reached its peak in 1953 with introduction of the Uniform Land Tenancy Act, known as the "Stockmen's Bill." It was pressed in Congress by a handful of Westerners (including Senator Pat McCarran, of Nevada, who denounced Forest Service officials as "swivel-chair bureaucrats," and Representative, later Senator, Frank Barrett, of Wyoming), who designed it to grant their stockmen constitutents monopoly control of public lands. The bill was finally defeated through unified efforts of national conservation organizations.

Facing the State of the Range

In a speech before the American Society of Range Management in 1967, Chief Edward Cliff discussed the need to rehabilitate millions of acres of deteriorated range. During both World Wars heavy pressures "to meet the war effort" had led to increasing the numbers of livestock, followed by reductions when peace had come, but the cutbacks proved inadequate. "Watershed values are becoming more and more important in area after area," the Chief declared. "Neither the livestock industry nor the Forest Service can live with grazing practices that result in damage to watersheds. Maintaining an adequate plant cover must be one of the measures of our performance."

Two years later the Chief again referred publicly to the problem of overgrazing. In a letter to the National Wildlife Federation, on June 20, 1969, he provided these figures: of a total of 106 million acres included in allotments to grazing permittees, 50 million acres (47 percent) constituted suitable range, while the balance of 56 million acres (53 percent) was unsuitable. "The suitable land," he wrote, "essentially is what can be grazed by livestock under reasonable management requirements without damage to itself or to adjacent land areas." As to the state of the suitable range, the Chief reported that 18 million acres, or 36 percent, was in poor condition; although 64 percent was considered fair or better, only 18 percent could be classed in good condition.

In November 1973, Regional Forester William D. Hurst advised of range conditions in the Southwest. Of 19 million acres within grazing allotments in Region 3, only 13.2 million were considered suitable (the remainder were classed as unsuitable because of unstable soils, steep topography, dense timber, barrenness, or inherent low forage-producing potential). He provided the following data.

Condition of Suitable Range			
	Poor	Fair	Good
Acres	5.3 million	6.6 million	1.3 million
Percent	40	50	10

Individual allotments with stocking problems, Hurst conceded openly, had not been identified, but adjustments were being approached in three ways: completion of improved management plans for all allotments; construction of improvements and range revegetation to meet requirements of these plans; and reduction of permitted numbers where management improvements fail to work. Various management systems have been employed in the Southwest and elsewhere. These include rotation grazing (dividing large units into small units and rotating use in order to avoid grazing the same area at the same time each year); deferred grazing (delaying grazing during the vegetative growing season, as in the high meadows); deferred rotation (rotating deferment of two or more units); and rest-and-rotation (completely resting parts of the range during certain years).

These intensive management systems have been tried with varying degrees of success. Earl D. Sandvig, however, has remained unconvinced of the efficiency of these techniques. "A majority of critical condition areas can never be improved under the pressure of use," he declared in a personal letter in 1980. "Only removal or heavy reductions of livestock will restore them. The longer corrective action is delayed, the longer and more difficult becomes the job of restoring sick land to health. Range unsuitable for domestic livestock may be valuable for wildlife. Yet choice areas for non-game and game species have been sacrificed."

During recent years, small economic ranches have been consolidated into large holdings. Absentee corporations and urban-based hobby ranchers in search of tax shelters have become permittees. Local communities that once supplied and serviced homesteaders and small ranchers have disappeared in many locations. A far cry from Pinchot's vision of democratic land use.

A provision in FLPMA directs that 50 percent of monies received as fees for grazing on the public lands (including national forests in the eleven contiguous western states) be made available for range rehabilitation, protection, and improvements. The Public Rangelands Improvement Act of 1976 extended coverage to the sixteen contiguous western states; it also provided for incentives or rewards to ranchers improving the rangelands.

While stockmen generally would welcome this approach, those concerned with other multiple uses might be hesitant to cheer certain aspects, such as chemical brush control and weed control, seeding the range with species intended for cows rather than wildlife, and predator control, all of which have been tried, tested, and found wanting. Nevertheless the Oregon Range Evaluation Program (ORVP), begun in 1976, is intended to increase grazing capacity 25 percent by the year 2000. The project has involved the Forest Service, other state and federal agencies, and twenty-five private landowners, proving that the old order perseveres.

"Thoughtful Vegetative Management" in Arizona

Land managers increasingly recognize the place of plants both in their immediate areas of concern and in spheres extending far beyond them. Plants are at the base of food chains. They generally anchor the soil in place. Each organism, no matter how humble, performs an essential role in the ecosystem, and natural ecosystems influence global patterns of air circulation and climate. That is why Peter H. Raven, Director of the Missouri Botanical Garden and a leading plant scientist, considers diversity of plants the underlying factor in the stability of the world ecosystem and feels constrained to warn that every plant species that goes extinct takes an average of ten to thirty species of other organisms with it.

"The Need for Thoughtful Vegetative Management on the Tonto National Forest," the report of a study completed in 1978 by Bruce Hronek, the supervisor of that national forest, brought the plant factor into an equation with overgrazing in the arid Southwest. The report was prepared as heavy reductions were being made in grazing allotments on the national forest, largely through the efforts of Hronek and William E. Pint, Jr., in charge of range and wildlife programs.

The report revealed that until the mid-fifties, most permits on the Tonto were held by descendants of the original settlers, or by those whose principal business was producing and selling livestock. They lived on their ranches and directly managed their own operations, and income from the sale of livestock produced on the national forest contributed to the economy of communities. Following World War II, a new kind of rancher, the absentee owner, arrived. His numbers multiplied in time. For each year between 1959 and 1975 approximately 25 percent of the ranches on the Tonto National Forest changed hands. The resale value of the ranch, because of the grazing permit that went with it, overshadowed the value of the ranch for producing income from the livestock operation.

Hronek and Pint decided it was time to take decisive action in defense of soil and vegetative cover. As Hronek discussed the problem in his paper:

> Overgrazing tends to increase surface runoff by reducing vegetation ground cover, causing a change in plant composition and compacting surface soils. A reduction in vegetative ground cover exposes more bare soil to raindrop impact which can dislodge numerous soil particles (soil movement quantities greater than 100 tons per acre have been measured). Rainfall on bare soil partially seals the soil surface. In addition, poor vegetative ground cover does not slow surface runoff. This results in less infiltration on site and a greater potential for erosion.
>
> Grazing by livestock can cause significant changes in vegetative composition. Often, overgrazed ranges that were once dominated by perennial grasses are characterized by annuals which do not have a root system capable of adequately stabilizing the soil. The protective mulch of the more sandy soil is lost by erosion or by compacting into the clay subsoils. The loss of soil mulch allows cracks in the clay to open. Loose surface material falls into the cracks and a churning action begins when soil moisture changes. Nature does not have a way of correcting this problem once it begins.
>
> Increased runoff on overgrazed ranges usually results in accelerated sheet and gully erosion. Severe erosion, in turn, can cause a total loss of site productivity. In desert ecosystems, where soil development is extremely slow, the loss of surface soils is essentially irretrievable and must be avoided at all costs.[3]

Grazing on the Tonto National Forest had long been conducted throughout the year, thanks to the favorable climate. Ranchers had no need to produce or purchase feed to support their stock. Field personnel in charge of the national forest, however, felt convinced that such continuous grazing without rotation caused the biggest problem, and reduced grazing to give the earth a chance.

National Grasslands—Another Side of the Story

The Bankhead-Jones Farm Tenant Act of 1937 may now appear obscure, but it ranks among the classic land-use laws. Superseding the Homestead Act, it provided for relocation and rehabilitation of those who wanted to move to better farming areas and for restoration of the lands they left behind, which should never have been farmed. Or, as the law stated its purposes: "To promote more secure occupancy of farms and farm homes, to correct the economic instability resulting from some present forms of farm tenancy, and for other purposes." One section, Title III, gave the Secretary of Agriculture authority and direction to develop a program of land utilization and conservation. The following year lands purchased under this authority were administratively organized as Land Use, or LU, projects. It was the beginning of the national grasslands, a distinctive system within the National Forest System.[4]

Land Use projects emerged as an attempt to meet a desperate problem. People who had farmed the Midwest, with its rich soils and dependable climate, or who had never farmed at all, had moved westward onto the plains. They had pressed homesteading beyond its limit. The soils, climate, and moisture would not support agriculture. They had come in too great numbers for the land to absorb. Any serious reading of John Wesley Powell would lead one to understand that nature would impose its own necessity on people, and with them on the economy and government.

Farming drought-prone grasslands didn't work. During the 1930s "black rollers"—devastating hot, dry winds—darkened the skies at midday with dirt blowing from the plowed lands. When it did rain, "gully washers" would cut deep scars in the earth, carrying silt down creek beds. People went broke, packed up, and abandoned their farms in the Dust Bowl. In 1935, after the fact, the federal government responded. President Franklin D. Roosevelt established the Resettlement Administration with authority to purchase up to 10 million acres of such submarginal land. Subsequently, the program was shifted to the Farm Security Administration and then to a new agency, organized in 1935, the Soil Conservation Service.

SCS agents worked with farmers and ranchers to implement conservation practices, hoping to remove soil erosion from the list of environmental threats. On the plains they seeded to grass. The region is too dry for trees. To survive, plants must have thin, shallow root systems adapted to wind, heat, and drought.

In this environment, growing forage has appeared to be the most efficient way to get production out of the land. SCS formed cooperative grazing associations to provide economic impetus and to demonstrate proper management practices for landowners who remained. In the late 1950s and, finally, in 1960 the Land Use projects were transferred to administration of the Forest Service, as a land-management agency, and were renamed the national grasslands.

Today nineteen national grasslands cover almost 4 million acres. Most are on the Great Plains—the Dakotas, Kansas, Oklahoma, Texas, Wyoming, Colorado, and New Mexico—with one each in Oregon and Idaho. They are outstanding volumes in the library of public lands, rich in history and the ecosystem of grass. The Cimarron National Grassland, in southwest Kansas, embraces a portion of the Santa Fe Trail and also the unique habitat where the lesser prairie chicken performs its elaborate spring courting ritual. The Pawnee National Grassland, in northern Colorado, provides nesting habitat for 85 percent of the country's mountain plovers. The national grasslands hold an abundance of antelope and deer, prairie dogs and the rare black-footed ferret that preys on them, as well as grouse, pheasants, game birds, and song birds. During June, low-rolling hills, coated with grass, are decorated with varied species of colorful wildflowers.

"The Rolling Prairie has a uniqueness of its own," the 1975 Management Plan for the Rolling Prairie Planning Unit of the Little Missouri National Grasslands, in North Dakota, declared with a touch of poetry. "It commands the awe of its vastness beyond the interpretation of being boring. The beauty of the prairie, the open, undefined, seemingly unrestricted landscape is its inherent characteristic. The openness of the prairie has for contrasts scattered buttes that stand like sentinels throughout the grasslands. Some buttes are well known, such as Blue Butte, Chimney Butte, Sheep Butte, Black Butte, and Tracy Mountain. They are the landmarks of the area and are the mountains of the prairie."[5]

The government does not own solid blocks of grassland; its holdings are intermingled with private land. Grazing associations and the Forest Service play a significant cooperative role in administration of both. The associations are responsible for range improvements and for limiting the number of cattle. As living standards rose in grassland country and the economy stabilized, grazing associations came to support the federal role, opposing disposition—in contrast to Sagebrush Rebels elsewhere.

While national grasslands are designed to demonstrate sound practical conservation principles of land use, the decade of the 1980s now finds them called upon to yield energy resources. Most are being explored and exploited for oil and gas, and some for uranium. Vast deposits of coal are located under the Little Missouri National Grassland in North Dakota and the Thunder Basin National Grassland in northeast Wyoming.

Thunder Basin lies between the Bighorn Mountains and the Black Hills, covering 1.8 million acres of federal, state, and private land in the Powder River Basin. It provides habitat for antelope, deer, elk, mountain lion, red squirrel, Merriam's turkey, sage grouse, lesser sandhill crane, bald eagle, owls and hawks, and forage for livestock. But it also holds coal, oil, gas, and uranium. Every acre of the grassland is under lease, except where mining companies have purchased ranch lands outright. Virtually each new development will require its own railroad spur, road system, powerline, gravel source, water supply, and telephone hookup. Large population increases are likely. For example, a coal gasification project north of Douglas, if constructed as once planned, would have a peak work force of 2,600. Conflict with ranching would be inevitable. According to a Forest Service assessment:

> Exploitation of these resources is having a tremendous impact on traditional uses, specifically livestock and wildlife. Sizeable acreage is taken out of availability. For each mine there will be about 200 additional acres disturbed each year and the total time required before the land will again be available for livestock grazing will be 13 or more years (1 year for mining, 1 year for rehabilitation, and 10-20 years of protection). Based on

a conservative 13-year cycle, there will be 2,600 acres per mine out of availability at any one time in the future. The mines will cut through a large number of allotments in an ever-changing pattern. In addition, transportation corridors of all types will change many pasture and allotment lines.[6]

Balancing Resource Values and Needs

In determining the best uses of land, public and private, are all factors fully considered? Commodity production seems automatically to come first, and production of energy commodities to come foremost. Is that because of actual public need and want, or because of the power and political influence of corporate energy producers as compared with livestock growers, or any other group? Are such uses compatible or incompatible—and if the latter, which should be required to yield in short-term and long-term public interest? How significant are the lessons of history in shaping policy and practice?

Or perhaps one should question whether any lessons are being learned at all. In 1977 the General Accounting Office found the nation experiencing 25 percent more erosion than during the Dust Bowl days. In 1982 the Soil Conservation Service reported (as part of its survey under the Soil and Water Conservation Act of 1977) "high erosion" on 140 million of 370 million acres of cropland. Inherent productivity is falling because of excessive topsoil loss. Food production has increased through intensive cultivation, but topsoil productivity has decreased through abandonment of conservation practices and traditional rotations.

Western rangelands generally are not suited for raising crops and it takes a lot of open space to provide for a single cow. More and more livestock are being raised in feedlots, concentrated areas where they are provided with grains, soy, and other nutrient sources harvested specifically for this purpose on productive lands. In the late 1960s Frances Moore Lappé asked the hard question: "How close are we to the limit of the earth's capacity to provide food for all humanity?" In her classic little book, *Diet for a Small Planet*, she reported her findings. The amount of humanly edible protein fed to American livestock, she wrote, and not returned for human consumption approached the protein deficit of the whole world. "By relying more on non-meat protein sources we can eat in a way that both maximizes the earth's potential to meet our nutritional needs and, at the same time, minimizes the disruption of the earth necessary to sustain us."

Lappé cited harsh reminders of the limit of the earth's agricultural capacity, which should be self-evident to students of history, ecology, environment, and resource management, and then went on: "An acre of cereals can produce five times more protein than an acre devoted to meat production; legumes (beans, peas, lentils) can produce ten times more; and leafy vegetables fifteen times more."

She noted that "feed" consists not only of highly nutritious grain and soybeans, but milk products, fish meal, and wheat germ. Evidently we feed about 90 percent of corn, oats, barley, and unexported soybean crop to animals. As for the return: the average conversion ratio for U.S. livestock (excluding dairy cattle) is 7 pounds of grain and soy feed to produce one pound of edible meat. According to this estimate, of 140 million tons of grain and soy fed to beef cattle, poultry, and hogs in 1971, one-seventh, or only 20 million tons, was returned in meat. The rest, almost 118 million tons of grain and soy, became inaccessible for human consumption.[7]

American eating habits have undergone many changes. At the beginning of the 1980s consumption of red meat was definitely on the decline. *The National Food Review*, a quarterly issued by the Economic Research Service, an Agricultural Department agency, on facts, figures, and trends related to food choice, showed that during 1980 the average American ate 1,402 pounds of food—about the same as 20 years before—but the "mix" of food was considerably altered.

According to this report, consumer concern with nutrition and food safety were major considerations, along with price. A growing number of shoppers were aiming to restrict intake of sugar, salt, and fat and were watching their weight. From 1960 to 1980 red meat consumption increased only 11 percent, whereas consumption of poultry and fish rose 84 and 28 percent, respectively. Most of the red meat gain was in the 1960s, very little in the 1970s. Beef consumption in 1980 was 13 pounds below the 20-year trend and nearly 18 pounds below the 1976 peak of 95 pounds per person.

There is yet another part of the picture. In an area like Jackson Hole, Wyoming, ranchers wonder why they continue to run cattle in the face of soaring real estate values. As their properties are converted to subdivisions and condominiums, does the public gain or lose? Or does the fate of private lands matter when considerable land is federally administered?

These questions are discussed in a report titled "Jackson Hole: Protecting Public Values on Private Lands," completed in 1981 by two planners, Jean Hocker and Story Clark, as part of a project of the Izaak Walton League. They focused attention on 50,000 acres of private land in the southern part of the greater Yellowstone region, or Yellowstone Ecosystem, which embraces two national parks (Yellowstone and Grand Teton), five national forests (Bridger-Teton, Custer, Gallatin, Shoshone, and Targhee), national wildlife refuges, state lands, and private lands.

Hocker and Clark observed that ranch lands, while relatively small in acreage, are strategically situated, a critical buffer for the federal areas, an important element in the scenery, rural western flavor, and wildlife habitat of the region. The same could be said of ranch lands bordering other national parks and national forests. They combine open space and development generally complementary to public values. Being located on valley floors, with warmer temperatures, more fertile soils, and lower snow accumulations than are found in the federal areas, ranches furnish essential winter habitat for many wildlife species, including moose, elk, deer, and bald eagles, which summer in the high country. Thus, urged the Izaak Walton planners, "the overall goal must be to keep the land in open space, wherever open space is necessary to protect scenic and wildlife values. This can best be done on private lands by encouraging the continuation of ranching, a land use that is not only very compatible with the valley's other resources, but which also perpetuates the special western heritage for which Jackson Hole is known."[8]

This study included a number of recommendations that might be applied elsewhere as well. They include encouragement of ranching by public or private funding for purchase of conservation easements; changes in federal tax laws to encourage donation of easements, including estate tax and/or income tax credits; improved communication between federal agencies and ranchers about permit procedures and grazing allotments; establishment of policies that recognize ranching values to local tourism and enhancing the visitor experience to federal lands, and establishment of a land trust (the Nature Conservancy has, in fact, begun a program to accept donations of conservation easements in Jackson Hole). Other recommendations called for emphasis on wildlife in designing and approving, or disapproving, developments, with special emphasis on protection of the Snake River corridor and its riparian ecosystem.

New concepts may be hard to digest or accept, particularly on the rangelands, a stronghold of individualism, where tough and determined men have pursued their own ideas, for better or for worse. Now, however, they are no longer alone in a region that once seemed open and free. Change is coming faster than anyone can visualize, let alone cope with in a democratic society. Yet this change brings with it entire new horizons of opportunity in resource ideology and administration.

9

RE-CREATION IN THE FORESTS

> The rustic, or primitive, tends to lead us closer to nature and natural values. The challenge is to offer primitive experiences with only minor sacrifices in efficiency, while maintaining health and safety standards, and to move away from rather than towards urbanization of our recreation sites.
> —Roy Feuchter, Director of Recreation
> Management, Forest Service, 1980.

The Scope and Role of Recreation

In 1980 Congress designated the River of No Return Wilderness in central Idaho, covering two million acres and thus becoming the largest wilderness area south of Alaska. The River of No Return is larger and perhaps wilder than Yellowstone, with a greater variety of plants, fish, and wildlife. It embraces rugged peaks, flowery alpine meadows, highland lakes and forests, portions of the beautiful Salmon River and some of its tributaries flowing through deep granite gorges. The wilderness is not a park like Yellowstone, Glacier, or Grand Teton, but a composite covering major portions of six national forests: Bitterroot, Boise, Challis, Nez Perce, Payette, and Salmon.[1]

The River of No Return epitomizes "national forest country," where recreation at its best comes in low density and low-key dimensions. This means hiking, backpacking, trail riding, ski touring, snowshoeing, and river running. Hunting and fishing have been important traditional activities, too, but times and tastes are changing; many of today's sportsmen would just as soon *see* a mountain lion as kill one.

The Forest Service administers an abundance of areas comparable to the River of No Return, no longer quite unspoiled, yet still vestiges of the original America. These cover a wide spectrum across the highest mountains of New England, the Southern Appalachians, the Ozarks, Rocky Mountains, Southwest, Sierras, Cascades, and southeast Alaska, as well as the Northwoods of the Lake states and Piedmont and Coastal Plains of the Southeast. They range from wilderness to legendary wonders like Lake Tahoe, astride the California-Nevada border, and the Boundary Waters Canoe Area, on the border between Minnesota and Ontario; the natural spectacular of Mt. St. Helens; historic and prehistoric points of interest; most of the slopes of the country's major ski areas, including Sugarbush, Vail, Aspen, Sun Valley, Squaw Valley, and Mount Hood; Blanchard Springs Caverns in Arkansas; plus many of the wildest rivers, toughest trails, and quietest corners.

Virtually the entire National Forest System is open to the public. National forests are said to furnish more outdoor recreation use than any other system, public or private. During fiscal 1980, a volume of 233 million recreation visitor days was recorded, compared with 87 million visitor days in the national parks. National forest facilities include 111,000 family picnic and camping sites and 100,000 miles of trail. Major portions of the Appalachian Trail and Pacific Crest Trail are in the national forests, and so are 300 national recreation trails. There are boating and swimming sites, summer homes, interpretive sites, organization camps, lodges and resorts, and 800 outfitting and guiding enterprises—all of which provide a capacity of more than a million visitors at one time.

The National Forest System provides hunting virtually everywhere; fishing on 83,000 miles of stream and 2.7 million acres of lakes and reservoirs; hiking, cross-country skiing, horseback riding, bicycle riding, and motor travel on 205,000 miles of forest roads; offroad-vehicle use where it is deemed feasible and appropriate; and visitation to historical, cultural, and scenic areas.

Nevertheless, the first RPA program submitted in 1975 recommended increasing the supply of outdoor recreation opportunities and services, especially those emphasizing "dispersed recreation," a particular national forest specialty. In the 1980 assessment of renewable resources, including supply and demand projections for the next fifty years, the Forest Service reported these statistics:

The greatest increases have come in snow and ice activities. A decade ago the United States imported 12,000 cross-country skis a year; now we import 50,000, and one U.S. company alone produces another 250,000.

Americans bought nearly 60,000 new boats in 1977, and membership in the American Canoe Association climbed fivefold in eleven years. Demand for water activities is expected to increase 106 percent, with sailing and canoeing leading the way.

The Appalachian National Scenic Trail saw 75 percent more use in 1976 than in 1974; land-based activities in such areas are expected to grow by 60 percent.

Downhill skiing and snowmobile use have both slowed in growth, but participation will still increase 140 percent by 2030.[2]

A Look Back—and Around

Recreation use of national forests began long before the areas were established as such. During the nineteenth century, visitors came from the East and Europe to hike mountain trails, seek adventure and big game, and to study nature in an unspoiled state. Gifford Pinchot touched on recreation during his tenure as Chief Forester, but only lightly. In 1915 the Service published *A Handbook for Campers in the National Forests of California*. A few years later it promoted leasing of summer home sites in order to stimulate interest in remote regions (a practice that outlasted its purpose).

Camping in the national forests during the early years was primitive, without designated areas; facilities were simple or nonexistent. National parks, on the other hand, were like museums meant for organized tourists, definitely with more development, roads, and rules. In national forests, even today, with minimal restrictions semiprecious stones may be sought and collected, pine cones and other common plants taken home, game animals, birds, and fish harvested according to state laws, and dead timber, forage, and such materials used for camping or taken home as fuelwood—all of which would be forbidden in the national parks.

By and large, recreation management remained quite casual for many years; there was no reason for it to be otherwise with millions of acres of elbow room. By 1958, however, pressures caused by increasing population and leisure time were clearly evident. That year Congress created the Outdoor Recreation Resources Review Commission to conduct a nationwide appraisal and to offer appropriate recommendations. In the same period the Forest Service launched the National Forest Recreation Survey, a resource inventory aimed at more effective decision making. Recreation was aided further when it was specifically authorized in the 1960 Multiple-Use–Sustained-Yield Act. And the 1976 National Forest Management Act provided guidelines and further authorization to include recreation in multiple-use management.

The Forest Service is involved in studies and management of outstanding free-flowing streams, as directed by the Wild and Scenic Rivers Act of 1968. Under terms of the Act, certain rivers flowing at least partly through national forests (Clearwater and Salmon Rivers, Idaho; Feather River, California; Rogue River, Oregon; Rio Grande, New Mexico; Wolf, Wisconsin; Eleven Point, Missouri) were classified as wild or scenic or recreation—the wild to be kept nearly primitive, restricted to nonmotorized craft; the scenic to be developed modestly at most; the recreation to provide somewhat more intensive use. These and other rivers, subsequently classified after study, are to be kept natural and free-flowing, without dams or other obstructions, with water quality maintained.

The National Trails System Act, also adopted in 1968, recognized the values of long-distance hiking by designating the Pacific Crest and Appalachian Trails as the initial national scenic trails and providing for the study of others. This law and others that followed reflect the exciting grass-roots ferment among hikers craving more trails and willing to work for them.

The 2,000-mile Appalachian Trail, though administered by the National Park Service, winds through eight national forests between Maine and Georgia. The Pacific Crest Trail, for which the Forest Service has management responsibility, crosses twenty-five national forests on its 2,500-mile journey from the Canadian border to Mexico. Other trails are taking shape, including the Continental Divide Trail, administered by the Forest Service, which will cover 3,100 miles between the Canadian border in Montana and the Mexican border in New Mexico and embrace existing trails in Glacier, Yellowstone, and Rocky Mountain national parks and in twenty-five national forests through some of the most spectacular mountain scenery in the West. Then there are the Lewis and Clark Trail, tracing the steps of the explorers of 1804—1806; the Bartram Trail, commemorating the exploration of colonial naturalist William Bartram in the mountains of the Southeast; the Florida Trail, of which the first 66-mile section crosses Ocala National Forest; the Daniel Boone Trail, following the route of the pioneer from North Carolina to Kentucky, and the North Country Trail, designed to cross half the country, ultimately to link the Appalachian Trail with the Lewis and Clark Trail in North Dakota.[3]

A large part of the Forest Service activity involves working with trail clubs, especially in the East. Those in the Appalachian Trail Conference maintained the trail for years without much federal help; in fact, they insisted on grass-roots control until commercial development and road construction made them recognize they couldn't do it all alone. The Appalachian Mountain Club (or AMC), the country's oldest outdoor club, operates a system of huts, or hostels, across the Presidential Range in the White Mountains. AMC offers guided summer hikes utilizing the huts, as well as a year-round program of climbing, canoeing, orienteering, skiing, and hiking.

The National Trails System Act also provided for establishment of national recreation trails within reach of urban areas, designed variously for use by bicycle, foot, horse, snowmobile, wheelchair, canoe, and cross-country ski. The Forest Service made the most of it. By September 1981 there were 329 national recreation trails, totalling 4,200 miles, in the national forests. The Forest Service furnishes considerable helpful material to visitors, including maps, guides, and leaflets (dealing with such activities as birdwatching, fall colors, safety in the woods, trip planning), yet there is question as to whether recreation values are fully understood or appreciated by agency personnel, a point which will be discussed later in this chapter.

Total receipts from admission and recreation user fees, and from charges for special use, amounted to a little over $18 million during 1981, or only 16 percent of the $111 million Forest Service recreation budget. Even on sites where charges were made, only an estimated 55 percent of average operating costs per visitor was recovered by fees.[4]

Does this make recreation a net loss? Stated differently, to what extent should government underwrite the cost of outdoor recreation? Or, what is the government's appropriate role?

Many other nations have already answered these questions. In Europe, government-sponsored mountain hostels, vacation villages, and holiday centers are designed variously for vigorous outdoors people, families, and working people. In Canada, the Province of Quebec reorganized its parks and reserves to become part of a new Ministry of Recreation, Fish, and Game. Its purpose, simply, is "to support conservation, physical fitness and relaxation in healthy outdoor leisure."

And in the United States? "Despite its importance," an industry group, the American Recreation Coalition, warned in 1982, "recreation is receiving precious little attention in federal policy-making today."[5] The Reagan Administration, in fact, had abolished the Heritage Conservation and Recreation Service (formerly the Bureau of Outdoor Recreation) and buried its functions elsewhere. The word "recreation" did not appear in the title of any federal department or agency. The Administration blocked federal grants from the Land and Water Conservation Fund so that states, counties, and cities all were pressed to reduce their recreation programs.

Recreation contributes significantly to the economy, as travel and recreation industry groups are quick to point out. It also sustains the quality of life and public health. Outdoor recreation spans a variety of interests, tastes, and goals. Theme parks like Disneyland, Six Flags, and Busch Gardens fill particular niches, as demonstrated by their popularity. So do commercial resorts and campgrounds. They enable those who prefer, and can afford to make the necessary investment, to bring conveniences of urban living into outdoor life away from home, or to find these conveniences waiting for them. There is scant emphasis on self-reliance or on the need to respect the natural environment.

Public parks and forests, on the other hand, may be said to provide an antidote to the physical and psychological pressures of urban living. Expansive areas like the national forests afford a chance to exercise mind and body in harmony with the great outdoors. Keeping things simple and direct, as national forests generally do, enables individuals to understand the mechanism of the land as a basis for desiring and supporting protection of the nation's resources.

What's It All Worth?

National forests are endowed with some of the greatest and most varied recreational opportunities on earth. The Nantahala National Forest, in mile-high country in western North Carolina, and surrounding national forests (Pisgah, Cherokee, and Chattahoochee) embrace remnants of the ancient Cherokee hunting ground. Joyce Kilmer–Slick Rock Wilderness includes within its 15,000 acres a magnificent virgin hardwood forest (with patriarchs five and six centuries old) abundant in wildflowers and flowering shrubs. The national forest is laced with trails (including 81 miles of the Appalachian Trail from the edge of the Great Smoky Mountains across the high Nantahalas). And the most challenging wild river in the East, the Chattooga, begins under the shadow of Whiteside Mountain, then flows through other national forests forming the state boundary between South Carolina and Georgia.

In Utah, the Ashley and Wasatch National Forests hold the high Uintas, this country's major mountain range running east to west (most ranges run north to south). Most of the Uintas are above 10,000 feet, with trails opening the way to riding, hiking, fishing, and hunting, and to spectacular Western scenery—a world of broad basins, rolling plateaus, and richly colored rocky canyon walls reaching 13,528 feet at the summit of Kings Peak, the highest point in Utah.

The Tongass National Forest is a composition of scores of islands and a mainland portion of Alaska cleft by bays, inlets, and rivers, all highly scenic, misty and moody. On Admiralty Island the visitor can travel by canoe for days over freshwater lakes and streams, crossing paths with few others, looking at snowy mountains, and at eagles, too, since Admiralty has the highest density of bald eagles in the world. Misty Fjords, in Alaska's southernmost corner, covers two million acres of rain forests, steep granite fjords, hidden glaciers, alpine meadows, and snowfields, accessible via Ketchikan, where outfitters and guides operate with floatplanes and boats.

National forests and other public lands in a sense maintain the opportunity for successive generations to learn firsthand the conditions that shaped the culture of the nation. Contacts of this type instill the vital spirit of pride in being an American. But shutting off young people from healthful outdoor recreation contributes to the tide of violence and crime. The decline in recreation facilities contributes inversely to rising public health costs. That is why other governments consider recreation a rational, economic national investment.

"Public lands serve as the basic reservoir for meeting outdoor recreation demands in America by persons of all economic levels," declared the American Recreation Coalition. "Federal lands, in particular, must continue to be important sources of recreational opportunities. Adequate management must be provided on these lands to assure protection of the resource base.

"No government which manages public land can legitimately abrogate its resource stewardship responsibilities by denying public recreation opportunities in the name of fiscal constraints."

The Coalition statement was made in response to Administration proposals for the 1983 budget. Unlike the favored treatment given to timber, oil and gas, and mineral developments on public lands, recreation evoked official concern about the need to "control the cost of government," the need to close facilities where "costs per visitor-day are high" (which could easily mean for the handicapped and wilderness user), and "reducing unfair competition with the private sector."

The Administration charted a multisided course to increase recreation fees (which Congress firmly and continually rejected); to "privatize" national resource lands, turning them over for management to commercial entrepreneurs; to dispose by sale or gift of "excess" federal lands; and to reduce recreation budgets.

Cutting funds could hardly be expected to keep visitors away from the national forests. But it could mean higher expenditures over the long run to deal with a backlog of effects of overuse. The agency, for example, estimated that the 1983 budget would prevent maintenance on 50,000 miles of trail.

The attractiveness and scenic beauty of the national forests have caught up with them. They are no longer remote or difficult to reach. The White Mountain National Forest in New Hampshire is the major playground of New England, popular even in winter for hiking, camping, skiing, and climbing. The same is true in one season or another, or through all the seasons, across the continent. Consequently, everything is more organized than it used to be, while the pressures of use often exceed the capacity of sites.

Crowds, Closeness, and Conflict

In timber management, as discussed earlier, the higher the cut the greater the appropriation to cut more is likely to be. The same principle can be applied to recreation: the higher the volume of visitors, the greater the appropriation to handle more of them.

Because national forests (and other federal and state lands) are public, the tendency is to open them to visitation with scant restraint. Currently, people come in droves to accessible areas. Coping with numbers gives the once wild forest an organized, urban texture. Litter and vandalism emerge as problems (the Forest Service spends 15 percent of its operating budget to repair and restore damage). So does crime, which leads to replacement of the friendly ranger by an armed patrol. Marijuana growers have made remote areas unsafe through cleverly concealed booby traps, like a pit full of sharpened stakes or a wire snare, or a shotgun aimed at the trail with a wire tripper tied to its trigger.[6]

In other ways, too, overuse or misuse by recreation visitors can prove as damaging as the uncontrolled bulldozer. The trampling effect eliminates vegetative growth, creating problems of erosion and water runoff. A concentration of people can be just as bad as a concentration of cows.

Conflict Over ORVs

In the early 1960s, the beginning of the snowmobile era, there were few machines and they were used primarily for transportation by game wardens, timber cruisers, and search-and-rescue teams. Their use and popularity have grown. Twenty years later, 17,000 miles of national forest trails are groomed, primarily by "cooperators" in snowmobile clubs, for snowmobile use. Thousands of miles of ungroomed trails are also available to snowmobiles. Critics complain that the focus of attention is on riding the machine as an end in itself, rather than on nature or scenery, and that it represents a mechanical intrusion in a natural setting, disrupting other forms of recreation (principally cross-country skiing) and damaging the environment.

Snowmobiles and other offroad vehicles—four-wheel drives, motor-cycles and dune buggies—have matched logging, mining, or grazing as a source of controversy. The vigorous proponents of ORVs, however, are not corporate interests (though they, too, are deeply involved), but citizen users, often articulate and organized in clubs and associations.

To trace the controversy: a 1972 Presidential Executive Order directed federal land-managing agencies to "develop and issue regulations and administrative intructions . . . to provide administrative designation of the specific areas and trails on public lands on which the use of ORVs may be permitted, and areas in which the use of ORVs may not be permitted, and set a date by which such designation of all public lands shall be completed."

The National Wildlife Federation charged forthwith that the order would open 600 million acres of public land to destructive use, while ORV groups decried the "lockout." Wilderness advocates criticized plans to permit continued ORV use in wilderness study areas, claiming this would be used as a criterion for excluding them in the inventory of roadless areas.

The issue persisted to the point that President Carter, in his environmental message of 1977, asserted that offroad-vehicle use "has ruined fragile soils, harassed wildlife and damaged unique archaeological sites." He directed the Bureau of Land Management, Forest Service, and National Park Service to act immediately in closing areas where ORVs cause "considerable adverse effects on the environment." The Council on Environmental Quality, responsible for coordination of such programs, took up where the President left off and went further: where adverse impacts are anticipated, CEQ warned, closings are mandatory, whether or not areas are open to offroad users, and field officials must not wait until damage is demonstrable before taking action.

A 1977 report by the Geological Society of America gave supporting evidence to charges of lasting damage to soil, plants, and wildlife. This report followed field study mostly in California. It strongly indicated that ORV-induced compaction and erosion lead to a self-propelling cycle of deterioration, with fewer and fewer plants able to take root.

"Damage to the land frequently becomes so severe that even the vehicle user no longer derives pleasure from the activity," according to Dr. Robert J. Stull, geology professor at California State University. "Consequently, new areas are sought and the damage becomes more widespread."

"Offroad vehicles are clearly out of control in many parts of our country," added Dr. Robert C. Stebbins, University of California zoology professor. "They must soon be more properly managed or we will suffer extensive losses of unique biological materials, irreplaceable soil and the visual appeal of many of our wild lands."

Throughout this running controversy ORV users and manufacturers have spoken with strong voices, asserting the right to their share of the public lands. In some circumstances they have demonstrated increased responsibility and respect for natural resources. As the International Snowmobile Industry Association points out, newer snow machines are much quieter and wildlife harassment has sharply declined.

Nevertheless, a 1979 report prepared for the Council on Environmental Quality, "Off-Road Vehicle Use on the Public Lands," termed the ORV "one of the most serious public land use problems." David Sheridan, the author of the report, charged the two major land management agencies—BLM and the Forest Service—with being slow to grapple with the challenge. Most decisions seemed to him to be based on subjective management experience which tended to legitimize the status quo, without sufficient analytic work to determine the actual impact of ORVs on natural ecosystems. "Too few federal land managers are effectively representing the interests of the land and the plants and creatures who live upon it."[7]

The Forest Service took exception. In a memorandum to Assistant Secretary Cutler, on August 10, 1979, three months after publication of the CEQ report, Chief Peterson criticized Sheridan for an "unbalanced approach" and noted steps taken to improve enforcement, public involvement aimed at environmental awareness, studies to determine public demand for ORV use, monitoring, and rehabilitation. The Chief wrote: "The perspective expressed by the report tends to exclude recent positive Forest Service actions and the growing cooperation and assistance being offered by ORV organizations to facilitate management efforts. This omission seriously reduces the validity of the report and unfairly biases the perceptions of an uninformed reader."

Through Cutler's influence, the Department of Agriculture in cooperation with the University of Michigan sponsored a three-day conference on the "management challenge" of ORV use. Subsequently he asked the Forest Service to review controversial areas and prescribe restoration areas where and if needed. The agency designated a committee of specialists—in wildlife, soils, hydrology, ORV recreation, surface rehabilitation, and cultural resources—to conduct an on-the-ground study.

The same scene gives different impressions through different eyes. In its report on one of the most controversial of areas, the task force commented as follows:

> Observations of Ballinger Canyon [a part of the Los Padres National Forest, California, referred to in Chapter 3 in connection with the "whistle-blower" case of Monty Montagne, a biologist] indicate that while impacts on wildlife by motorcycle operations could not be assessed by the inspection team, it is apparent that a wide variety of wildlife is present. It was observed that soil movement due to ORVs is slight versus soil movement caused by geologic process. . . . Hill climbing was thought by some members of the team to be a compatible use in Ballinger Canyon, but it was agreed that hill-climbing needs more intensive management than normal trail operation.

On the other hand, Howard G. Wilshire of the U.S. Geological Survey, in a paper presented to the Department of Agriculture–University of Michigan management conference, commented on Ballinger Canyon as follows:

> It has been shown that even the most resilient soils in the area—those determined by the 1972 forest soils report to be most suitable for motorcycle use—are highly vulnerable to accelerated erosion, and that the more sensitive soils have undergone erosion on a scale greatly exceeding even the high natural erosion in the area. In addition, wildlife habitat that the forest sought to protect by fencing has undergone degradation by erosion resulting from excessive runoff from ORV-denuded slopes and by burial by debris eroded from those slopes.
>
> The potential for rehabilitation, a useful criterion for any ORV area, has not been demonstrated for any soil-vegetation association present in Ballinger Canyon. No data have been adduced by the Forest Service to support the belief that rehabilitation is possible or feasible. No standards have been established for what rehabilitation is expected to achieve because no long-range plans for use of the land after rehabilitation have been made.[8]

The 1980 task force offered what might be called a moderate approach, placing the burden of judgment on national forest field personnel. "ORV use which minimizes adverse effects and promotes vehicle operation in harmony with the natural terrain and social setting and enhances appreciation of an outdoor forest environment, is generally appropriate," declared the task force report. Outside of recommending that emphasis for motorcycle use be placed on trail riding rather than hill climbing, it generalized by suggesting that: "Good ORV management requires a sound inventory of conditions, management plans, monitoring, educational programs for users, and adequate budgeting."

Budgeting could be a deterrent to an effective management plan, considering the nine steps proposed by the task force to implement it:

1. Determine soil-slope-moisture-landform relationships for specific sites.
2. Develop resource reclamation techniques to ORV use.
3. Study the sociology of ORV users.
4. Determine the impacts of sound level and other ORV activities on game and nongame animals.
5. Set standards and criteria for acceptable levels of resource damage for site-specific areas based on surrounding control areas.
6. Determine simple and useful inventory and monitoring techniques that would provide statistical validity and quantification of actual soil losses.
7. Find new trail hardening techniques.
8. Determine vehicle characteristics as they affect other resources, such as tires, weight, horsepower, sound levels, and muffler type.
9. Develop a handbook of recommended ORV management techniques, e.g., trail location, design, and construction; vegetative cover needs, trail hardening techniques, soil and water conservation systems.

In many areas forest administrators have worked hard to solve ORV problems, often with cooperation of ORV organizations to eliminate conflicts. When the Boundary Waters Canoe Area was being considered for expansion and clarification of purpose, the Eastern Region proposed the elimination of snowmobile recreation. Considering that Minnesota is the heartland of snow machine country, this stirred considerable opposition. Snowmobile groups and resort owners lobbied against this provision, but when Congress voted to classify the BWCA as wilderness, snowmobile use was sharply curtailed.[9]

Along the Mirror Lake Highway, in the Wasatch National Forest of Utah, certain winter trails have been designated for use by snowmobiles and certain others by cross-country skiers. While the question of resource impact may not be resolved, in many such circumstances both groups of users have learned to tolerate the other.

Downhill Skiing Delights and Dilemmas

Alpine, or downhill, skiing is by far the most advertised and commercialized winter sport. Of 700-some ski areas in the United States, 170 depend upon slopes on national forest land. Sometimes these recreational facilities also cause serious conflict.

Because the demand for downhill skiing in California doubled between 1970 and 1980 (and likely elsewhere as well), the State Resources Agency and the Forest Service Pacific Southwest Region agreed to review ways to accelerate expansion of existing areas and development of new ones. Ski areas, after all, provide recreation, pay fees to the U.S. Treasury based on the number of skier visits, and stimulate the local economy.

Considering these pluses, the Forest Service generally has encouraged development of downhill ski areas. In the case of Crystal Mountain, on the Snoqualmie National Forest near Seattle, the Forest Service went so far during the 1960s as to make public land available for private condominiums so the project would be attractive to the group of developers; or, as officially stated, "to provide needed accommodations that seemed otherwise unavailable."

In California during the same period, the Forest Service sought competitive bids for a major ski center in Mineral King Valley, a 15,000-acre portion of Sequoia National Forest surrounded on three sides by Sequoia–Kings Canyon National Park. The successful bidder, the Walt Disney entertainment conglomerate, planned for 2.5 million visitors throughout the year, with peak use bringing 16,000 persons at a time, with 5,400 cars requiring 82 acres of parking. The National Park Service expressed its concern over the effects of automobile exhaust in a bowl circled by peaks 3,000 feet above it and by the proposal for the access road to cross a corner of the national park.

Members of the Sierra Club in California (its home base and stronghold) were divided over the Disney proposal. Some argued the need for expanded downhill skiing opportunities, while others expressed concern over the potential of serious environmental impact, particularly on the national park. The club initially approved development of Mineral King, but reconsidered and ultimately expressed its opposition. Then it campaigned successfully for legislation, passed by Congress in 1979, to transfer the area to the national park.

Despite the delights of downhill skiing, foresters have increasingly noted dilemmas as well. In 1976 the Green Mountain National Forest issued a draft environmental impact statement for management of the Deerfield River Area, including the possible expansion of commercial winter sports facilities on national forest land in the Dover-Wilmington region. The impact statement aired potential effects on multiple-use management of the national forest and on communities around it.

Further development, for instance, would result in stream siltation, nutrient enrichment of streams, soil disturbance, removal of timberland from production, limitations on mineral extraction, increase in fire risk, and adverse visual impact. "All or parts of the area are utilized by wildlife at some time," the national forest noted in its draft EIS. "This modification will alter the wildlife populations. For every acre of urban development, an acre plus a larger area of influence is lost as game habitat and modified concerning nongame species."

Besides these impacts, the project would "generate increased use of both public and private lands, with the accompanying need to provide a complete range of complementary services and facilities," leading to more highways, traffic, taxes and rising consumer costs. On the other hand: "With further development of private land within the area almost inevitable [even without commercial ski expansion] the national forest will become more of an 'oasis' for those seeking an experience of remoteness and solitude."

From another perspective, conceiving and operating a major ski center is no small feat. In the case of Copper Mountain, Colorado, the developers were heartened by the Forest Service Recreation Evaluation Report they received in 1970 concerning their site: "If there was a mountain that had terrain created for skiing, it would be Copper Mountain. It is probably the most outstanding ski area in Colorado. The area will lie adjacent to Interstate 70 at Wheeler Junction and has been proposed as a possible Olympic site. The base area is extensive with the possibility of handling 10,000 skiers per day."

That report was encouraging, but only the beginning. The developers had acquired a substantial property at the base of Copper Mountain, 90 miles west of Denver totally surrounded by Arapaho National Forest. It had the physical attributes to support at least 14 or 15 lifts serving 40 to 50 trails and slopes, with grades varied enough for all classes of skiers from beginner to expert. On this basis, a Forest Service study permit was issued for preliminary work, followed by a 30-year special use permit allowing for development of the mountain facilities. Three years then were spent in land planning and utilization; studies of snow, wind, and sun; installation of utilities (electricity, telephone, gas, water, sewage); fire and police protection; marketing studies; financial feasibility and analysis; hydrologic and geologic studies; and planning summer activities.

It was determined to plan for approximately 10,000 visitors per day, the average projected for the thirty highest volume days during the season, with half as overnight or destination guests. This led to a year-round recreational design ultimately to include 1,200 hotel rooms, 1,000 condominiums, 200,000 square feet of commercial space, housing for 500 employees, and parking for 6,000 cars. Also included in the planning were such recreation facilities as golf course, tennis courts, play fields, swimming pools, and stream fishing facilities.

While the planners recognized that easy access with the new Interstate, built in the mid-1970s, might result in increasing population pressures, they also felt the isolated nature of the real estate would enable a comprehensive development with due respect for the resources. Developed land, including parking, would amount to less than 50 percent of the available base land area, the rest prevailing in open space. At the same time, they obtained zoning under a PUD (Planned Unit Development) ordinance in Summit County, allowing flexibility "to evolve and respond sensitively, within a set of overall constraints, to competing market factors, deepening environmental concerns, and shifting consumer preferences inherent in a recreational/resort development to be implemented over a period of years."

In 1981, with Copper Mountain operating prominently on the skiing scene, Chuck Lewis, then chief operating officer of Copper Mountain Corporation, in a personal interview complained about the lack of flexibility allowed by the Forest Service, especially in dealing with public health and safety:

> Master development plans are a dynamic device which must remain flexible, and any design criteria numbers utilized are solely for planning purposes and should not be used to implement use-limiting systems.
> The Forest Service Manual should clearly establish that the responsibility for public health and safety within the ski area boundary is the permittee's. The Forest Service's role is to see that the permittee is fulfilling these responsibilities. The Forest Service should exercise this role by ensuring that the permittee develops an operation plan. The Forest Service should monitor as necessary to see that the established procedures and schedules are being followed.
> The Forest Service does not have to, and should not attempt to, discharge any responsibility with respect to the review and approval of design or construction engineering for any ski area permittee's improvements provided such is prepared by a recognized licensed professional.

That is an aspect of the business of recreation on the national forests. The ski facilities at the most popular centers are operated by corporate entities with needs and political influence. Their primary desire in winter is to keep those lift lines moving and to add more runs on the slopes. But there is a difference in doing business on public land and on their own land, involving adequate parking, sanitation and regulations, and concern for noncommercial public values.

The Need for a "New Profession"

The Forest Service may not be quite prepared to meet its future in recreation. "Foresters are generally timber oriented, not people oriented," Roy Feuchter, Director of Recreation Management, declared before the annual meeting of the Society of American Foresters in Boston on October 16, 1979. "Timber pays the freight in most wildland management activities. Lack of recognition of recreation values exists everywhere, including within the Forest Service. I think our biggest training job is to convince the wildland managers, not the public, of the value of outdoor recreation."

There is no standard under classification of the Office of Personnel Management (formerly Civil Service Commission) for the recreation professional in the Forest Service, which means that timber-trained foresters are in charge. A program of three-week-long short courses, initiated in 1979 at Clemson and Utah State Universities, enables them to develop recreation skills. But there is still a long way to go. Recreation professionals scarcely are found in decision making, although colleges of forestry in Washington, Oregon, Idaho, and elsewhere include departments specializing in recreation.

"What we need to do is define the true importance of recreation," according to Roy Feuchter. "It is insufficient to recognize the importance of recreation in our own minds. We must also establish its importance in the minds of our civil servants, wildland managers, corporate presidents, professional societies and politicians—not for the sake of the profession or the professionals, but for the sake of the public and our country's natural resources."

Feuchter may be trained in forestry with emphasis on timber, but he expressed serious and progressive concepts in recreation:

> As the population becomes increasingly urbanized, our goal should be to help the public to know, experience and enjoy the *natural* values in outdoor recreation—a whole experience in contrast to urbanization. Also, too often we address the participation phase of their experience, overlooking anticipation and recollection values.
>
> Local public transportation networks should be encouraged and supported. So should upgrading rail facilities from cities to urban areas; expanding metropolitan bus schedules to nearby national forests that can reach out to inner city residents, including the disadvantaged and handicapped, and expanding packages with transportation companies for national forest and other recreation destinations.
>
> Energy-efficient awareness will make a significant contribution to public well-being, support goals of energy efficiency and bolster rural economies. At the same time we will advance the importance of outdoor recreation and professionalism in it.

Chief Peterson expressed the same idea in 1980: "We've trained a cadre of experts in silviculture for decades. Not so with natural resource recreation managers. We need to produce a trained cadre of recreation specialists to develop a technical base for planning and management and put it to use. Programs in educational institutions across the nation are now building the needed font of knowledge, and with it the status of a new profession."[10] Nevertheless, recreation has a long way to go.

For example, managers must still master issues of capacity and appropriate use. The thing to do, according to one school, is to educate the public to an acute consciousness of nature by keeping them off the grass. Paved nature trails, displays, interpretive centers, and a staff of interpreters are seen as the route to follow. Whether the lessons will last, however, is questionable when interpretation before large numbers must be impersonal and superficial. Spoon feeding at its best does little to encourage or stimulate understanding of biological process. Outdoor recreation, as another school of thought argues, must provide direct contact with nature. The best lessons come in quiet moments, while experiencing weather, at dusk, darkness, and early dawn. The consequence of such moments is the development of a personalized holistic outdoor ethic—respect for the environment and for one's self.

In some places the Forest Service has taken steps to limit use in order to protect a choice area so it can be properly enjoyed. The Boundary Waters Canoe Area requires visitors to use specific entry points on given dates, which they reserve in advance. Similar systems are in effect on the Middle Fork of the Salmon River in Idaho and in the California wilderness areas.

Feuchter reported promise in another approach: a prototype program teaching the simple basics of hiking. "The reception was tremendous. Most people don't want to admit they don't know. The average person may even be fearful about venturing into the woods. So we presented the information where they could participate easily without having to admit they really didn't know. Such basic knowledge is the real beginning point in giving people knowledge of how to enjoy and appreciate the natural experience."

How much is the "natural experience" worth? How does one cost-account the value of recreation and the price of management to sustain it? The cost of roads constructed by logging companies is listed by the Forest Service under the heading of "Income," but its own costs in behalf of recreation under "Expenditures." As capital investment, which yields the best long-range return? When timber cutting overruns and depresses recreation, by degrading the resource and polluting the environment (and, consequently, affecting the economic return to communities dependent on scenic and recreational values), should the investment in logging roads be credited as "income" without deducting depreciation of other values?

Forest Service timber and recreation values involve different systems of data gathering and cost accounting. Timber is measured in board feet, recreation in visitor days and satisfying experiences. The sale of timber is for commodity use, definable in dollar values, which are acceptable and comprehensible. Recreation, however, is designed to make life livable, to promote health and fitness.

These individual and social values are no less important than timber. Unfortunately, they still await definition and acceptance. Forecasting patterns in recreation has proven difficult if not impossible ever since the end of World War II. Nevertheless, it seems clear that allocation of trained personnel for recreation planning and management must at last become a Forest Service priority.

10

FOREST WILDLIFE: CONSERVATION AND QUANTIFICATION

> The preservation of forests and game go hand in hand. He who works for one works for the other.
>
> —Theodore Roosevelt.

Recognizing Wildlife Values

The grizzly bear, once common in the Colorado Rockies, was believed to have been extirpated. What was thought until recently to be the last grizzly in the area was shot by an outfitter or government trapper in the San Juans in 1951. Scattered reports of tracks and claims of sightings in the years since then were discounted by state and federal officials—that is, until a thirteen-year-old grizzly female was killed in September 1979, on the headwaters of the Navajo River in rugged southwest Colorado. The great bear, it now appears, survives in Colorado in very low numbers. If it makes a comeback, or holds its own, wildland of the national forests—notably the San Juan, South San Juan, La Garita and Weminche Wildernesses—will provide the sanctuary required for doing so.

The Rocky Mountain Region of the Forest Service, covering portions of Colorado, Kansas, Nebraska, South Dakota, and eastern Wyoming, is home to an abundance and variety of species. These include the black-footed ferret, the rarest mammal in North America, confirmed in a 1981 sighting in Wyoming,[1] and the endangered whooping crane which rests in Nebraska on its long migrations between northwestern Canada and the Texas Gulf Coast. Big game species like mule deer and whitetail deer, antelope, black bear, and elk are common. There are beaver, bobcat, coyote, song birds, game fish in mountain streams, game birds, and small numbers of moose, bighorn sheep, and Rocky Mountain goat, as well as lynx, cougar, and raptors like the peregrine falcon and bald eagle.

It is surprising in a way that these still survive in the face of continuous pressures of roading, logging, grazing, dam building, predator control, and the sheer spread of human population. This is true not only of the Rocky Mountain Region but the entire National Forest System.

In one part of the country or another, national forests furnish the habitat of the California condor, the largest North American landbird; Bachman's warbler, the rarest songbird; the ivory-billed woodpecker, if they are still found anywhere; Kirtland's warbler; red-cockaded woodpecker;[2] prairie chicken; Eastern panther; gray wolf; and red wolf. One of the world's truly great resources, whether measured in economics, aesthetics, or sheer natural wonder, the family of anadromous salmonids—sea-run salmon and trout—depends on many streams and rivers flowing in national forests from California to Alaska; while the Atlantic salmon is supported by waters of the White Mountain National Forest in New England.

Of 236 domestic species on the federal endangered-species list in 1981, 70 were found to occur on national forests, some *only* on national forests. But if such species are to survive much longer, their habitat must be perpetuated through conscious effort and commitment.

In 1980, Dale Jones, Director of Wildlife and Fisheries of the Forest Service, spoke of the need of "ecological perceptions" in administering the national forests, and of responding to "a demand for the peace of mind that comes from protecting and preserving wildlife because its users see it as a treasured gift." Chief Peterson supported this idea, calling for greater emphasis on fish and wildlife, with concern for whole ecosystems and not just game species. This represented a change in direction from the old approach that wildlife must be used like a commodity to be considered useful.

The wildlife role of the national forests differs from that of the national parks. The latter came into being as reservations by which threatened species were withdrawn for safekeeping. National forests complement national parks where they are adjacent to them. The Forests are open to hunting so long as the kill is limited to the natural increase of the game. The responsibilities are shared between state fish and game departments, which set license fees and bag limits, and the Forest Service, which is charged with protection and enhancement of habitat (basically food, cover, and water).

About 90 percent of lands in the National Forest System are sufficiently covered with trees or brush to attract game animals. National forests have been said to account for approximately 30 percent of the nation's annual big game harvest, including more than 80 percent of the elk, bighorn sheep, and mountain goat, nearly 60 percent of the mule deer, and 35 percent of bear and blacktail deer. In addition, the

System contains 128,000 miles of fishing streams, more than 2.2 million acres of lakes, and millions of small game animals, upland game birds, and waterfowl.

In years past, immense areas that now constitute national forests were abused by domestic livestock to the detriment of game animals. In Utah, for example, millions of head of cattle and sheep caused deterioration of the range to such a degree that hunting of deer, elk, antelope, and bighorn sheep was prohibited by the state from 1908 to 1913. As recently as 1969 an estimated 50 million acres of national forest land in the eleven Western states were considered to be in unsatisfactory condition for wintering game.

The 1960 Multiple-Use–Sustained-Yield Act gave thrust to wildlife management as a Forest Service program in its own right. It did not yet provide the means of dissolving the symbiosis between timber-oriented forestry and game-oriented wildlife management. Aldo Leopold defined game management on the opening page of his textbook as "the art of making land produce sustained annual crops of wild game for recreational use." Leopold personally advanced to a broader ecological concept, but many in his field focused on manipulating food and cover to produce favored game species. State wildlife agencies were tied to this approach, since they derived their operating revenues largely from the sale of fishing and hunting licenses and from excise taxes on the sale of guns, ammunition, and fishing gear. Thus logging generally was considered the means of creating more habitat for game species, and sportsmen were called upon to harvest the surplus. Government wildlife programs were (and essentially still are today) weighted toward game species, such as deer, pheasant, and trout, and hunting and fishing as industries and sources of revenue, rather than toward holistic ecosystem management for game and nongame species.[3]

Yet the Forest Service was assigned a major ecological responsibility by the Threatened and Endangered Species Act of 1973, as amended in 1978 and 1979, which established a national policy to protect species of native fish and wildlife threatened by extinction and to protect their habitat as well. The law was passed in response to a sense of public urgency and concern that no portion of the nation's heritage should be lost heedlessly to future generations.

Various positive actions have since been taken throughout the National Forest System to enhance the habitat of endangered species (where they are currently, where they might be, and where they formerly have been) and to give more recognition to wildlife in the framework of multiple use. Though such progress has not been uniform, it reflects the initiative of administrators and individuals, both inside and outside the agency. The public has been made more aware of wildlife values and opportunities through such publications as "Discover Birding in the National Forests," "Cavity-Nesting Birds of North American Forests," and "Rare and Uncommon Wildflowers"—a contrast with former materials like "How Even-Aged Management Benefits the Sportsman."

Examples of successful, or encouraging, wildlife projects include the following:

The Condor project on the Los Padres National Forest, California, is probably the best known, though not the most hopeful wildlife project. The condor, a member of the vulture family, has a wingspread of nine feet and weighs up to twenty-three pounds. Once it covered a wide range across the continent, but habitat loss, food scarcity, shooting, poisoning, egg collecting, and harassment have reduced the condor population to twenty-five or thirty. Two designated sanctuaries plus additional "critical habitat" on the Los Padres contain all known recently active condor sites. Condors face competition for their range, which contains oil and gas potential, deposits of gypsum and gold, and geothermal resources.[4]

Forest Service condor activities are coordinated with the U.S. Fish and Wildlife Service, Bureau of Land Management (which administers adjacent public land), National Audubon Society, California Department of Fish and Game, and the Santa Barbara Museum. The work includes sanctuary patrol, population monitoring, artificial feeding, and efforts to coordinate minerals and engineering activities. But there is a serious question if all of this is adequate or even right.

Since 1965 the condors have produced less than one chick per year. In 1979 the late Carl Koford, who studied the condors for years, proposed a natural-based recovery plan. Rejecting the concept of captive breeding, he urged improving environmental conditions: assuring food supply in all seasons, decreasing the use of pesticides, and evaluating possible reduction of competition from golden eagles and turkey vultures. But the recommendations of a panel of ornithologists for a "hands-on" program prevailed. This included trapping the remaining birds, holding them long enough to affix small radio transmitters and conduct laboratory studies, and captive breeding (assuming they would lay pesticide-free eggs and increase the hatch) for subsequent release. In June 1980, unfortunately, while two biologists of the Fish and Wildlife Service and one from National Audubon were endeavoring to take a chick from the nest, the chick died. It was a tragedy for the slow-breeding condor population.

The Three Sisters Project, on the Klamath National Forest, Oregon, maintains and protects a roost in the Lower Klamath Basin for 70 to 100 wintering bald eagles, one of the largest such roosts west of the Mississippi. In 1980 the Forest Service acquired an additional 960 acres for the area and instituted measures to restrict and control access.

The Golden Trout Wilderness, on the Sequoia National Forest, California, established by Congress, with the management program conducted in cooperation with Sequoia National Park and the Fish and Wildlife Service, is designed to protect and restore the Little Kern golden trout.

The eastern timber wolf is protected on the Superior National Forest, Minnesota. Though once it ranged as extensively as the white-tailed deer, the wolf has been reduced to a very low population. The Superior and nearby Isle Royale National Park comprise major remnants of its former habitat. However, where the national park is isolated and wolf protection goes unchallenged, the national forest is obliged to contend with continuing local hostility in order to defend the highly intelligent predator.

The Puerto Rican parrot is the focus of a joint project being conducted by the Forest Service, Fish and Wildlife Service, and World Wildlife Fund in the Luquillo Mountains, the last stronghold of this spectacular bird, located about twenty-five miles southeast of San Juan. Once these birds flew in huge flocks, but they were shot for food and taken for pets. The Caribbean National Forest protects the last remnant of their habitat. The numbers are down to approximately twenty, but efforts are being made to breed the parrots in captivity and return them to the forest. Dr. Frank Wadsworth, former director of the Institute of Tropical Forestry, the champion of the parrot, in 1967 successfully prevented a military maneuver involving thousands of troops from intruding into the area.

Every region has undertaken such projects, but they have largely been considered as isolated elements rather than integral in the process of management, and some of the best ideas for wildlife protection have been overlooked or inadequately applied. The Southern Region, for instance, commissioned George Lowman, a consulting biologist, to review "Endangered, Threatened and Unique Mammals of Southern National Forests," the title of his report published in 1975. His impressive study summarized distribution, status, habitat requirements, and life histories, and recommended management procedures covering forty-six mammals. Lowman specifically urged complete protection as "the most necessary step" for management of the endangered panther in the eastern United States. "As much of each national forest as possible," he advised, "should be maintained in unbroken, undisturbed tracts." National forests in the White Mountains, Green Mountains, Southern Appalachians, Florida, and the Ozarks could help save the panther; yet there are no programs evident in accord with Lowman's recommendations.[5]

A companion study, "Rare and Endangered Birds of the Southern National Forests," published in 1974 in order to "help guide habitat management of these species on the national forests," dealt specifically with Bachman's Warbler, a secretive little bird. It was last sighted in 1937 and may already be extinct, though it is known to go undetected for years in large swamp areas. "Because of the present lack of knowledge," advised the report, "about the only management program which can be undertaken is protection of any known birds and preservation of habitat."[6] On this basis, Jay Shuler, a professional naturalist of Mc-Clellanville, South Carolina, in 1976 sought to block a proposed logging sale in the area of Ion Swamp, a unit of Francis Marion National Forest, where 32 of 40 known nests of the warbler have been identified. The forest supervisor and biologist responded to Shuler's petition by asserting that forest openings and "dispersed cuttings" would prove more helpful to the warbler than dense swamp.

Examples of the Problem: Grizzly Bears and Mountain Goats

In 1981 a pair of yearling grizzly cubs from Glacier National Park was transplanted into the wilds of Kootenai National Forest, a similar area in northern Montana. It was part of an interagency effort to save the grizzly in fulfillment of the Endangered Species Act. Dr. Charles Jonkel of the Border Grizzly Project, a key figure in this effort, describes essential grizzly habitat as a combination of spring-fall feeding areas, travel corridors, denning sites, and other space-related and food-related requirements. When any of these needs is threatened by activity such as logging or mining, according to the interagency program, these activities should be tightly controlled.[7]

Nevertheless, a proposed logging sale on the Flathead National Forest in 1980 stirred question and controversy. The sale was intended to eliminate stands of trees infested by the mountain pine beetle. Critics claimed it would threaten habitat of both grizzly and northern gray wolf (and also that financial costs would outweigh the benefits). Jonkel himself advised against a sale in the Ketchikan Creek area, citing rising pressures against the bear in Montana and neighboring British Columbia, such as logging and roading, oil and gas exploration, and imminent coal development. "I fear," he declared, "that many individual bears are depending on the Ketchikan area as undisturbed habitat within their ranges, without many other options." (This sale, incidentally, was withdrawn in 1982 when there were no takers.)[8]

In 1980 the Forest Service, National Park Service, and fish and game departments of Montana, Idaho, and Wyoming issued detailed "Guidelines for Management Involving Grizzly Bear in the Greater Yellowstone Area." It was a means of recognizing that the grizzly has been losing ground, as a consequence of slow reproduction rate and rising mortality caused by human pressures. "Land uses which can affect grizzlies and their habitat will be made compatible with grizzly needs or such uses will be disallowed or eliminated," according to the guidelines. "Grizzly-human conflicts will be resolved in favor of grizzlies unless the bear is determined to be a nuisance." And even the "nuisance" bears are given a break, since they may be removed or relocated only when all measures have been taken to protect the bear and its habitat.[9] Critics claimed that even after the guidelines were published, the Forest Service allowed grazing of livestock in grizzly habitat and that it approved development of a winter resort, Ski Yellowstone, where it would impact grizzly activities adjacent to Yellowstone National Park.

In an article titled "Living with Mountain Goats," published in the August 1974 issue of *Defenders of Wildlife Magazine,* Douglas and Elizabeth Chadwick recount an experience in northern Montana with significance reaching far beyond. The question it poses is not whether management in this particular case chose the right option, but whether wildlife is factored properly into the multiple-use equation.

When the mountain goat is hunted, write the Chadwicks, it doesn't disappear into a thicket, but looks for safety by climbing to a steeper but equally exposed cliff face. The "challenge" of a goat hunt, in fact, consists almost entirely of climbing to reach the animal's home. In the Chadwicks' area most valleys already were marked with major roads and logging clearcuts within rifle shot of the goat cliffs, so that climbing wasn't really necessary. From such valleys they would watch groups of two, three, or four goats where local residents reported 30 or more prior to logging. And they continue:

> Now that logging roads are being pushed into the last remote timbered areas, the few remaining refuges for goats and grizzly bears and other wilderness species are becoming accessible. The last strongholds, the mountain fortresses, are under attack. In 1971, the two of us found between six and eight different grizzly bears regularly using Bunker Creek and the adjacent valley to which the goats fled during road building. By 1973, intensive surveys of these same areas by five people revealed only one resident bear, and that bear, a small female, was shot during the first week of hunting season. . . .
>
> The Forest Service continues its policies of road building and logging in those remote high mountain areas and other *de facto* wilderness habitats which harbor those species. Few and insufficient prelogging wildlife inventories are conducted. Of the nearly 200 people employed by the Flathead National Forest on a permanent basis, there is only one fisheries biologist and a single wildlife biologist employed only this year. The remainder are foresters, engineers, and administrative personnel. To their credit, the Forest Service has recently begun to close roads in a few critical areas, but not on a comprehensive basis, and not rapidly enough to keep pace with their search for cheap virgin timber in wild mountain valleys.[10]

Protecting Ecosystems

Many things have changed, and are continuing to change. The budget for Forest Service fish and wildlife programs rose from $9.2 million in 1975 to $41.6 million in 1981. In the 1970s some rangers thought biologists were getting in the way; in the 1980s rangers find their input contributes to better management and helps to avoid appeals by critical citizen groups. Various regions issue publications discussing questions formerly avoided, with such statements as: "Roads necessary for timber management alter habitat and allow increased human activity. Some wildlife, including elk, move out of areas while such activity is going on. Roadbuilding and timber harvest can adversely affect spawning areas for salmon and steelhead."

Under authorization provided by the Conservation and Rehabilitation Program on Military and Public Lands Act of 1974, known as the Sikes Act, agreements have been developed with state fish and game agencies to underwrite habitat improvements on the national forests. Though funds have not been forthcoming, the inventories mark a step in the right direction. "K-V" funds (provided by the 1930 Knutson-Vandenberg Act), however, are being applied to wildlife as well as to timber goals, through such means as planting browse, seeding patchcuts, installing watering ponds, and planting cover along streambanks to benefit fisheries.

The most significant change has come in recognition of wildlife ecosystems and how to protect them. Virtually all funding has gone into programs to enhance game species, though most wildlife is classified as "nongame," including 3,000 vertebrate species (birds of prey, songbirds, reptiles, and small mammals), many of which have long been subject to open season and poison programs as "pests" and "varmints." In 1980 Congress passed the Fish and Wildlife Conservation Act authorizing a modest allocation not to exceed $5,000,000 for reimbursement to the states for development and implementation of comprehensive nongame management plans. Some states already were funding nongame programs through a voluntary checkoff on state income tax forms. Colorado, for instance, had a 1980 nongame program with a budget of $650,000.

George Lowman, the consulting biologist, in 1975 spelled out the means of perpetuating wildlife ecosystems:

> Spruce-fir forests, cliffs, stream courses, bogs, and swamps should remain undisturbed. More forests should be permitted to attain climax vegetation. Certainly each area should be thoroughly studied before any harvesting is authorized. Maintaining environmental quality in general is absolutely necessary. Each of the national forests and grasslands should be protected from pollution, fire, human disturbance, reduction of stream flow, siltation, stream channelization, and other factors that might decrease or destroy the productivity of the land or its waters. Roads should be chained to limit access to remote areas.[11]

Broadened ecosystem thinking has dampened the debate of the 1960s and 1970s over benefits of clearcutting to wildlife—even game species. While it is true that clearcuts produce quail habitat, often where nonexistent before, and that an abundance of deer browse emerges on many clearcut areas, biologists note that these benefits are temporary, that before many years quail habitat and deer browse decline. Within ten years following planting, the pine canopy can be expected to close, and until it is thinned, this clearcut is of use to wildlife only as cover.

With increasingly short harvest rotations as projected in many clearcutting systems, "mast" (foods such as berries and nuts) becomes increasingly scarce for bears, turkeys, squirrels, and deer. Removal of mast trees and cover has threatened and destroyed prime squirrel and turkey habitat, and lack of mast may reduce the carrying capacity for deer after relatively few years. In the sequence of clearing and conversion to pine in hardwood areas or natural pine-hardwood areas, deer are not unlikely to destroy planted pines, leading to the demand for hunters to "bring the deer population into balance."

Recognizing the Mistakes

During the early 1970s, Harry R. Merriam, an Alaska field biologist, challenged the effects of Forest Service management. In a letter to the regional forester of March 16, 1972, he wrote:

> I believe I have now worked longer as a game biologist in Southeast Alaska than any other person (I started with the Department of Fish and Game in 1959), and probably am qualified to speak on game. I do not feel large block cutting is beneficial to any big game or to most small game species. It is too drastic an environmental change. In regards to deer, it removes too much winter range. Clearcuts in Alaska provide a great deal of summer food (when it's not needed), but it's all covered by snow in winter. Small cuts might improve some deer range. I'm presently recommending cuts not to exceed 100 acres, but do not expect much response.
>
> There are indications of change, but last week I reviewed a cutting plan for Red Bay, on Prince of Wales Island (extremely high recreation and wildlife values), which was as poor as any I've seen. Several cutting blocks approach 1,000 acres or more. Of primary concern to me is the aesthetic experience while hunting and fishing. No matter how much fish and game are in an area, the experience is changed when the forest is cut.

The regional office in Alaska rejected Merriam's challenge as invalid. In 1978, however, a technical study completed under the auspices of the Alaska Department of Fish and Game and Forestry Sciences Laboratory strongly asserted that only old-growth forests can provide both refuge from heavy snow and sufficient food to sustain deer through critical late winter and early spring seasons. These findings were discussed at a conference in Juneau attended by fifty wildlife biologists from Alaska, Canada, and the Northwest, who agreed: "The consensus of the conference was that current forest practices (clearcutting on a 100-year rotation) in Alaska and British Columbia are permanently eliminating the uneven-age old-growth forest on the northern coast and, further, that the result will most likely be a serious and long-term detrimental impact on deer habitat and consequently deer populations throughout this region."

The Alaska Chapter of the Wildlife Society (a professional organization of game biologists) went further, declaring: "There is strong evidence that clearcutting adversely affects marten, deer and eagles. It is becoming clear that under the existing practice of even-aged silviculture on a 100–120-year rotation, the climax forests of Southeast Alaska are, in fact, a non-renewable resource."

Alaska has also been a focal point of concern over anadromous salmon and trout. The complex reproductive process of these migratory species requires fresh water in high-quality streams, where they reproduce and spend the first part of their lives before beginning the great odyssey downstream to the ocean. They spend most of their lives in the sea before returning to the place of their origin, traveling determinedly upstream against all odds, guided by a powerful instinctive force that still defies full scientific explanation. But the great runs of salmon and sea-going trout have declined throughout the West as a consequence of industrial development, excessive siltation due to poor practices in road building, logging, mining and agriculture, and exploitation in the ocean and estuaries. An estimated 43,000 miles of anadromous fish-rearing, spawning, and migration streams in western national forests are home at one point or another for the eight species of Pacific salmonids, placing the Forest Service at the core of the issue of sustenance and survival.

The Alaska Department of Fish and Game has been especially critical. Following a clearcut on the Starrigavan watershed near Sitka, according to the Department, the winter survival rate of coho salmon dropped 80 percent. "Clearcutting drastically changes fish and wildlife habitat," the department asserted, "and causes permanent changes to the ecological system." Sport Fish Regional Supervisor Bob Armstrong said in 1980 that of 196 important coho salmon streams and 132 pink salmon streams in southeastern Alaska, only 36 percent were in protected areas; of 205 major chum salmon streams, only 44 percent may escape future damage.

On the other hand, Carl R. Sullivan, executive director of the American Fisheries Society, sees promise in artificial devices built with federal funds. Writing in *Fisheries Magazine* (September-October 1980), Sullivan described a $262,000 project at Anan Creek: a tunnel of concrete, steel, wood, and solid rock 105-feet long built to bypass falls often made impassible for salmon by flows of high or low water. "Anan Creek is Southeastern Alaska's best pink salmon stream and the addition of the bypass tunnel makes it far more productive and reliable."

As Sullivan viewed the scene: "Fisheries interests, led by the Alaska Department of Fish and Game, press hard for tight regulations of the timber harvest. Stringent controls, however, are not generally favored by timber interests so optimum development of all resources requires balanced decisions based on a thorough knowledge of both resources."[12]

Whether "optimum development" or "balance" is attainable may be subject to question in the light of congressionally mandated timber priorities in Alaska. Dr. Richard T. Myren, fishery biologist at the Auke Bay Laboratories (an installation of the National Marine Fisheries Service) in 1981 estimated a potential value of $350 million yearly for pink salmon in southeastern Alaska, but felt that logging abuses had depressed the runs to roughly one-sixth of what they might be. His forecast was gloomy: "Accelerated logging activity in Southeast Alaska will reach virtually every fish-producing watershed."

Salmon and steelhead have declined in California as well, to about one-third of their historic level, though even now they represent an annual renewable resource valued at $35 million. In 1979 the Forest Service Pacific Southwest Region proposed a Salmon and Steelhead Habitat Improvement Program in coordination with the State of California. The state plan involves major additions to fish-rearing habitat. Regulation of logging practices under the State Forest Practices Act is designed to reduce damage from that traditional source. The Forest Service, for its part, proposed a $20.5 million expenditure between 1981 and 1985 as a means to:

- Protect spawning and rearing streams in response to a water quality management program, including special attention to streamside zones.
- Remove barriers, such as log jams, which prevent migration to spawning grounds.
- Restore watersheds to prevent sedimentation and enhance water quality.
- Provide enough water flow to ensure support of fish life.

It was felt that through such efforts the salmon-steelhead resources in California could be restored to two-thirds of their historic level over the next two decades.

New Rules, Regulations and Policy

Fish and wildlife need are treated in detail in the 1979 rules and regulations issued by the Forest Service to implement the National Forest Management Act. As mentioned earlier, these were extensively modified in 1982 through the influence of Assistant Secretary Crowell. Nevertheless, the principles as originally stated continue to hold value for students, planners and managers of forestry resources, as made evident by these excerpts:

> The desired future condition of fish and wildlife, where technically possible, will be stated in terms both of animal population trends and of amount and quality of habitat.
>
> Management indicator species, vertebrate and/or invertebrate, will be identified for planning, and the reasons for their selection will be given. The species considered will include at least: endangered and threatened plant and animal species identified on State and Federal lists for the planning area; species with special habitat needs that may be influenced significantly by planned management programs; species commonly hunted, fished, or trapped; and additional plant or animal species selected because their

population changes are believed to indicate effects of management activities on other species of a major biological community, or on water quality. On the basis of available scientific information, the effects of changes in vegetation type, timber age classes, community composition, rotation age, and year-long suitability of habitat related to mobility of management indicator species will be estimated. . . .

Critical habitat for threatened and endangered species will be determined, and measures will be taken to prevent the destruction or adverse modification of such habitat. Objectives will be determined for threatened and endangered species that will provide for, where possible, their removal from listing as threatened and endangered through appropriate conservation measures, including the designation of special areas to meet the protection and management needs of such species.

Special attention will be given to land and vegetation for approximately 100 feet from the edges of all perennial streams, lakes, and other bodies of water and will correspond to at least the recognizable area dominated by the riparian vegetation. No management practices causing detrimental changes in water temperature or chemical composition, blockages of water courses, and deposits of sediment will be permitted within these areas which seriously and adversely affect water conditions or fish habitat. . . .

Diversity of plant and animal communities and tree species will be considered throughout the planning process. Inventories will include quantitative data making possible the evaluation of diversity in terms of its prior and present condition. For each planning alternative, the interdisciplinary team will consider how diversity will be affected by various mixes of resource outputs and uses, including proposed management practices. . . . Reductions in existing diversity of plant and animal communities and tree species will be prescribed only where needed to meet overall multiple-use objectives. Planned type conversion will by justified by an analysis showing biological, economic, social, and environmental design consequences, and the relation of such conversions to the process of natural change.[13]

A Policy on Fish and Wildlife announced by the Department of Agriculture on July 8, 1980, gave much the same directives:

Specific requirements of all management practices for National Forest System lands, to be met in accomplishing goals and objectives, will protect streams, streambanks, shorelines, lakes, wetlands, and other bodies of water. Management practices also will provide for and maintain diversity of plant and animal communities to meet overall multiple-use objectives. Practices will be monitored and evaluated to assure that they protect fish, wildlife, watersheds, soils, recreation, aesthetic values, and vegetation productivity.

And a statement issued almost concurrently, "Forest Service Policy, Goals and Objectives for Wildlife and Fish Habitat Management in the 1980's," pledges "coequal consideration to wildlife and fish habitat with other resources in Forest Service programs."[14]

The transfer of policy into practice doesn't necessarily follow quickly or easily. "The Forest Service will ensure that road construction on lands it administers fully considers the effects on fish and wildlife habitats," according to departmental directive. Yet in October 1980, the National Wildlife Federation demanded that the Forest Service halt road building in an area of the Pisgah National Forest in North Carolina, noted for its prime bear habitat, close access roads, and enunciate a policy closing roads. Officials of the national forest were then constructing high-standard roads to accommodate large logging projects, yet without any evident wildlife plan, management plan, or environmental assessment.

Earlier, the North Carolina Wildlife Resources Commission had issued a frightening report about the gloomy future of the black bear in the face of onrushing habitat destruction. The key to protecting the bear, the commission advised, lies in maintaining large roadless tracts in the mountain national forests of western North Carolina.

The Forest Service disagreed and denied the commission's recommendation. Of special concern was an area called the Twelve Mile Strip, close to the Great Smoky Mountains, known for the reproduction and rearing of black bears. It was already being roaded and logged, with plans underway for additional activity in nearby Harmon Den, a state-designed bear sanctuary. The Western North Carolina Sportsmens Association protested, citing a road construction cost of $27,000 to $30,000 per mile.

Officials of the national forest insisted that an interdisciplinary team had carefully planned the transportation system and a Forest Service biologist denied the potential of serious impact on bear and wild turkey populations. But Charles Hill, a state biologist with extensive experience in bear country, warned otherwise, foreseeing increased mortality due to poaching aided by better access, increased opportunity for disturbance because the new roads would provide a connecting link with Interstate 40 and the Great Smoky Mountains National Park, increased potential for disturbance and displacement in the Twelve Mile Strip, increased likelihood of more cub mortality due to abandonment resulting from disturbance, increased impact of bear population ranging outside and beyond the Twelve Mile Strip because of the mobility and extensive range of the bear, and disruption of turkey nesting due to increased public use (including hiking and unrestricted ORV). Hill noted also that the 80-year rotation cycle for timber harvest, rather than 100 or 110 years, would not allow sufficient regeneration of mast-bearing trees, causing eventual loss of wildlife food.

Critics both inside and outside feel that while the Forest Service may employ many more biologists than formerly, they do not always speak ecologically or freely. Few biologists have authority to make decisions. They furnish only advice, which sometimes is found wanting.

"Our biologists are receiving an excellent formal education. However, the education is irrelevant," observed Craig Rupp, Regional Forester of the Rocky Mountain Region, in a letter to Assistant Secretary M. Rupert Cutler in 1980.

There is some, but not much, variation, depending on the institution from which the biologist graduates. Basically, the biologist is well versed in the biological details of animals, and usually in depth about one species of animal or one class of animals. However, his knowledge of habitat and habitat requirements related to the population of "wildlife" in total is sadly lacking.

When asked to design a wildlife program to which other programs can be responsive, the biologist appears lost. When pressed to propose a wildlife habitat program for the purpose of benefitting wildlife, in terms of lateral and vertical diversity, increased numbers, improved distribution, and quality conditions rather than merely protecting the status quo, only a relatively small percentage appears to even want to respond.

Much of this is related to their professional training. Objective setting, decision making, and integrated planning appear foreign to the biologist. Universities are training wildlife management students to do animal research rather than management of habitat or wildlife species.

Approach to Wildlife in the Rocky Mountain Region

The Rocky Mountain Region covers 22 million acres on 15 national forests and two million acres on seven national grasslands, mostly in Colorado but also in portions of Kansas, Nebraska, South Dakota, and eastern Wyoming. Elevations range from 3,500 feet to 14,000 feet, from prairie to alpine tundra, and wildlife from prairie chicken to grizzly. Elk get the most attention, however, considering the Colorado harvest of 27,000 yearly, plus more than 100,000 deer.

During the 1960s the region had a few biologists collecting data on the deer and elk winter range, where use conflicted with livestock, and in the late 1960s moved toward placing a biologist as a staff assistant on every national forest. As an indication of increased emphasis, the fish and wildlife budget for the region rose from $742,000 in 1975 to a projected $3,600,000 in 1983.

In 1977 the region, in cooperation with state wildlife agencies, completed wildlife management plans for national forest lands in each of the five states, as called for by the 1974 Sikes Act. Each national forest was instructed to develop its own tentative wildlife objective, subject to change through the NFMA-RPA forest planning process and to draft a regional guide identifying deer and elk as "desired species" in all the five states.

Developing a management plan for wildlife in terms of numbers, units, targets, costs, and returns as though it were a production commodity, like timber or forage, is somewhat like asking a librarian or playground director to justify his budget before the city council against the competition of public safety, road repair, or sewage treatment. But that is the way resource professionals are expected to prepare and defend recommendations. In the Rocky Mountain Region, biologists have been asked to design timber sales to benefit wildlife. This involves determining species (game and nongame) that should be supported on each national forest and national grassland; average populations, by actual numbers or relative trends, as appropriate, for each desired species; the desired mix of ponderosa pine and gambel oak and the silviculture practice necessary to sustain optimum big game populations in southwest Colorado; and aspen age classes necessary to enhance wildlife (principally elk, deer, and ruffed grouse) and aesthetic values.

The original NFMA regulations require that diversity be provided and maintained for stability of ecosystems, and in quantifiable terms. Regional criteria, in turn, furnish guidance to national forest administrators for arrangement, size, and shape of various stands of trees and forest openings. Thus "vertical" diversity is likely to be achieved through uneven-aged, or selection, logging, and "horizontal" diversity and wildlife edges through even-aged, or clearcut, logging. In addition, the regional criteria direct leaving snags for cavity-nesting and roosting birds, and down-dead logs for many small mammals and ground-feeding birds.

Stephen P. Mealey, a Forest Service biologist, in 1980 introduced what he called "a linear programming resource allocation model" into management planning for the Arapaho and Roosevelt National Forests, attracting attention and interest throughout the region and elsewhere. His objectives specified ". . . proportions of forest successional stages, and numbers of residual trees per acre by size class, to maintain viable populations of all wildlife species, to maintain and improve habitat for mangement indicator species (MIS), and to provide for diversity of animal communities on a sustained yield basis."

Mealey listed his indicator animals in early forest succession as elk, mule deer, deer mouse, gray-headed junco, and ground squirrel; in late forest succession as goshawk, sharp-shinned hawk, Cooper's hawk, northern three-toed woodpecker, ruby crowned kinglet, red-backed vole, and tree squirrel. Accordingly, the Mealey prescription would apportion a 5,000-acre management unit as follows: 20 percent vertical diversity, or a minimum 1,000 acres, to provide habitat for one each of each accipiter (a family of certain hawks) species; 30 percent horizontal diversity, or 1,500 acres, to provide forage for 20 elk; plus 50 percent horizontal and/or vertical, depending on the indicator species to be featured.

Following are examples of application of this type of approach:

The Paonia Ranger District of the Gunnison National Forest instituted a habitat management plan in 1978 oriented in large measure on modification of aspen stands. This is based on the idea that pockets, or clones, of aspen growing from interconnected root systems, need cutting to stimulate regeneration and thus provide openings for elk, deer, and ruffed grouse. The plan calls for clearcutting overmature aspen clones so that "productive determinant age classes are uniformly distributed throughout the aspen management unit." In addition, a treatment involving burning of oakbrush is intended to improve availability and quality of feed for deer, elk, and cattle, and create an area suitable for introduction of ruffed grouse.

The Sopris Ranger District of the White River National Forest conducted several wildlife projects in 1981: burning 470 acres of oakbrush to improve big game winter range; burning and clearing 120 acres of aspen for regeneration; improving riparian habitat by building, in cooperation with the Colorado Division of Fish and Game, structures on the Crystal River and North Fork of the Fryingpan; creating snags for cavity-nesting birds; and completing two wildlife-designed timber sales covering 2,250 acres.

The Cebolla Ranger District of the Gunnison National Forest in 1980–1981 operated the Willow-Pines Wildlife Timber Sale, harvesting 25 percent of commercial timber volume in a unit of 3,100 acres. The objective is to stimulate diversity, since present stands are old, and the closed canopy has little or no understory. In sessions of the interdisciplinary planning team, the soils specialist questioned the validity of the sale in terms of fragile soils and cost/benefit ratio, but the sale was justified on the basis of wildlife benefits and measures to prevent erosion. In advance of logging, biologist Elaine Zeroth canvassed the sales area in order to mark for protection such special features as bird nests, wildlife trees, old growth stands, ponds, springs, streams, migration routes, calving areas, and denning areas. She marked for patch-cuts twelve small areas in old-growth spruce to be "sculptured to provide maximum edge and escape cover," and smaller patches of one-half acre to an acre next to dying aspen clones. She specified leaving three to five snags per acre for cavity-nesting and perching birds; leaving six old-growth patches of five to thirty acres for wildlife species dependent on old growth; closing access roads immediately after use; seeding skid trails, landings, and closed roads with clover-grass mixture to increase wildlife forage; piling slash for small animal cover and nesting; and avoiding logging road construction until after June 30 to avoid disturbing elk and bighorn sheep migration.

Whether manipulation and modification of the environment fulfill the requisites of an ecosystems approach may be questioned. It is true that the wildlife environment of Western Colorado already is being seriously modified by community growth, causing extensive loss of lowland winter game range, and by energy development with its spreading impacts. Through manipulation, wildlife managers *can* improve the habitat: prescribed burning, development of openings, and water impoundments for deer, elk, turkey, and quail; artificial dens, nests, and roosting structures for squirrels, wood ducks, eagles, and song birds.

On the other hand, conserving a rare plant or animal means conserving species and surroundings necessary to its survival, particularly when migrations are part of the life-cycle. "Populations and species, endangered or not, are all vital components of natural environmental systems," write Anne and Paul Ehrlich in their book titled *Extinction.* "Each kind of organism has its own role within its ecosystem and is to some degree—often a very great degree—essential to the continued healthy functioning of that system. It is impossible to separate protection of species from protection of natural ecosystems; they are two aspects of the same fundamental set of resources."

Plants and animals have come and gone naturally since the beginning of the earth, but nothing like the current wave of the last fifty years caused by "supercivilization." As part of their routine, personnel of the Rocky Mountain Region are asked to report unusual wildlife sightings: the nests of hawks, eagles, or owls; evidence of marten, wolverine, bighorn sheep, mountain goat, river otter, bobcat, lynx, black-footed ferret, or mountain lion. Nevertheless, as the Ehrlichs warn: "Great care must be taken whenever people intervene in nature. Moving animals and plants around may do a lot of good. It can also do a lot of harm. A great deal of care and biological knowledge are necessary in order to make transfers prudently."[15]

New Data to Help Solve Old Controversy

The stated goals of the Forest Service fish and wildlife habitat program, drafted in compliance with RPA and NFMA, are simple and direct:

1. To provide habitats for appropriate public use and enjoyment.
2. To assure through ecosystems management that species requiring special habitats do not become threatened or endangered because of man's activities.
3. To improve the status of threatened and endangered species to the point that they no longer need the protection of listing under the Endangered Species Act.

Admittedly, however, the weakest information provided to decision-makers has often been the possible consequences on fish and wildlife of various management alternatives. Accordingly, in 1981 the agency declared it would strive "to strengthen our knowledge of wildlife and fish ecology and to provide a systematic process for getting that knowledge into the decision process."

New research publications were designed to aid biologists working for national forests with their planning. As an example, a detailed study titled "California Wildlife and Their Habitats: Western Sierra Nevada" (issued by the Pacific Southwest Forest and Range Experiment Station in 1980), treats the habitat relationships of 355 wildlife species—26 amphibians, 27 reptiles, 208 birds, 94 mammals—showing how to consider protection of habitat in timber sale design, layout, and administration.[16]

The concentrated effort to accelerate the conversion of remaining old-growth forests of the Pacific Northwest into young, thinned, even-aged stands has raised questions about the effect on both game and nongame species. In 1973 a group of hunting organizations brought legal action to block the Department of Natural Resources of Washington State from intensive timber cutting on the Olympic Peninsula. Their complaint declared:

> These four elk herds all exist in the herds and the natural state that they do because of stands of virgin timber which have no access roads into them. These stands of timber provide safety and sanctuary for these elk in times when such are needed. Once these stands of virgin timber, relatively small in size though they are, are logged and have had access roads put into them, these elk herds will be reduced considerably in numbers and, speaking in terms of their natural states, will almost cease to exist on lands owned by the State of Washington.

As for nongame species, biologists warn that serious disturbance is apt to eliminate such species as spotted owls, pine martens, and goshawks. "Several of these predators and their prey are totally dependent on a specific forest type," according to Richard T. Reynolds, of Oregon State University, "namely old-growth or mature forests. At the present rate of harvest of mature forests, both predators and prey will be extirpated within our lifetime."

The spotted owl, in particular, has become a *cause célèbre*. The coniferous forests of Oregon support more than 20 hole-nesting bird species and hole-nesting mammals, yet the spotted owl, a close relative of the eastern barred owl and tawny owl of Europe, has been identified as the indicator of species that survive best in stands of old-growth Douglas-fir.[17] The Spotted Owl Management Plan, developed by an interagency team, calls for retention of habitat to accommodate 400 pairs of owls in their range between British Columbia and northern California (although up to 2,000 pair are believed to remain in Oregon alone). Biologists want a core of at least 300 acres of old growth for each pair, for a total of 120,000 acres in the state, principally on national forest and BLM lands. Timber spokesmen have complained this would deprive industry of more than $1 billion in raw materials. "If America needs more housing," they demand, "where will the wood come from?" They insist that Roosevelt elk, another species in question, do not need a thermal cover of dense old growth and that spotted owls are sometimes found in second growth. To which Jerry Franklin, a Forest Service research biologist, has responded with a basic wildlife precept: "The fact that a species can survive in other age classes of a forest does not mean necessarily that it can survive once the major reservoir of optimum habitat is exhausted."

Birds actually furnish the most efficient, least costly form of insect control in the forest. A single woodpecker has been estimated to consume 13,675 larvae of highly destructive wood-boring beetles per year. It is fair to generalize that the more numerous and varied the bird population, the broader the spectrum of natural insect control. John Smail, executive director of the Point Reyes Bird Observatory, a California research organization focusing on the ecology of nongame species, has reported on an analysis of nine breeding-bird censuses in coniferous forests in California, Colorado, and South Dakota. The analysis showed that 25 percent of the total number of birds using these forests are of species that nest in holes. These hole-nesters require older trees with some decayed portions in order to breed successfully (and feed large broods of young on destructive insects), although they forage on trees of various ages.

"Any forestry practice producing solid stands of trees of the same age reduces the diversity of bird species able to breed," according to Smail, "and this in turn reduces possible insect control."

In forests all across the United States wildlife species are hard pressed, yet this need not be the case in the national forests. Though timber and range have fixed targets, it is hard to explain the acre equivalent of wildlife habitat. Output numbers of animals with improvements, visitor days spent by birdwatchers, the wildlife-user benefits paid for by the taxpayer are becoming part of RPA calculations, but possibly the role of wildlife as an indicator of land health may become evident as a value in its own right.

11

"TO SECURE AN ENDURING RESOURCE OF WILDERNESS"

> This curious world which we inhabit is more wonderful than it is convenient; more beautiful than it is useful; it is more to be admired and enjoyed than used.
> —Henry David Thoreau

The Wilderness Debate

The first wilderness bill was introduced in Congress in 1956 by one Democratic senator, Hubert Humphrey of Minnesota, with eight co-sponsors and one Republican House member, John P. Saylor of Pennsylvania. Between June 1957 and May 1964, eighteen hearings were held on the wilderness proposal in both Washington and the West. After long discussion and debate, the bill was passed, and signed by President Lyndon B. Johnson on September 3, 1964. That historic act established a definition of wilderness in law and a National Wilderness Preservation System in fact. But it scarcely settled the sound and fury of controversy.[1]

Wilderness stirs the blood of partisans, pro and con. Almost twenty years after its passage, the Wilderness Act still challenges clear definition, or acceptance, of its role in the pattern of multiple use. It raises such questions as these:

- Is wilderness strictly a single use for the benefit of a few, in conflict with a combination of uses for the benefit of many?
- Do preservation and aesthetics interfere with the economy and prosperity of local communities in the vicinity of designated wilderness areas?
- Does wilderness retard the production of commodities essential to a growing nation?

These are important questions to personnel of the Forest Service,

174

from the Chief on down, considering that most designated wilderness in the "lower 48 states" and Hawaii lies in the National Forest System, and so do most areas proposed for addition to the wilderness system. The accompanying chart shows that as of January 1982, more than 25 million acres, located in 158 areas of the national forests, were also part of the National Wilderness Preservation System. And in 1980 the Carter administration, in the "high bound" Assessment and Program (discussed in Chapter 6), asked Congress to set a target of 42 million acres of national forest wilderness.

In many regions, national forests constitute the last reservoir of the unspoiled original America, offering, as the Wilderness Act provides, the chance "to secure for the American people of present and future generations the benefits of an enduring resource of wilderness."

But how much wilderness does it take? And is forestry the proper discipline to safeguard wilderness, when foresters are alleged by their critics to measure trees as timber and to see an old-growth grove simply as "decadent" and "unproductive?"

During the nineteenth century, men like Ralph Waldo Emerson and Henry David Thoreau evoked a wilderness conscience. "At the gates of the forest, the surprised man of the world is forced to leave his city estimates of great and small, wise and foolish," wrote Emerson in his essay on "Nature." "The knapsack of custom falls off his back with the first step he makes into these precincts. Here we find nature to be the circumstance which dwarfs every other circumstance, and judges like a god all men that come to her."

Emerson spearheaded a philosophy, while his friend Thoreau devoted himself to living as he wanted. He kept his life simple, flexible, and immediate, wary of possessions that might soon possess the possessor, observing, thinking, and doing for himself. Most citizens of Concord, Massachusetts, were baffled, but his view was plain: "If a man walks in the woods for love of them half of each day, he is in danger of being regarded as a loafer, but if he spends his whole day as a speculator, shearing off those woods and making the earth bald before her time, he is esteemed as an industrious and enterprising citizen."

Following establishment of Yellowstone, in 1872, as the first protected wilderness, other national parks were set aside, largely through the efforts of John Muir. "Thousands of tired, nerve-shaken, over-civilized people are beginning to find out that going to the mountains is going home," he wrote, "that wildness is a necessity; and that mountain parks and reservations are useful not only as fountains of timber and irrigating rivers, but as fountains of life."

Gifford Pinchot's concept of land use was different. In a letter to the author in 1968, Benton MacKaye, forester, regional planner, and a founder of the Wilderness Society, offered this recollection of Pinchot:

I was one of his cub foresters when he launched his F.S. in 1905. I knew him well. G.P., like his pal, T.R., was in person a nature lover. I

TABLE 11.1

NATIONAL WILDERNESS PRESERVATION SYSTEM

(As of 1/1/82)

Agency	Units	Federal Acres*	%
Forest Service, USDA	158	25,132,725	31.5
National Park Service, USDI	34(a)	35,334,482(b)	44.3
Fish and Wildlife Service, USDI	71	19,331,328	24.2
Bureau of Land Management, USDI	(4)(c)	12,206(d)	-
GRAND TOTAL	263	79,810,741	100.0
National Wilderness Preservation System (Lower 48 & Hawaii):			
Forest Service, USDA	144	19,770,826	84.4
National Park Service, USDI	26(a)	2,979,482(b)	12.7
Fish and Wildlife Service, USDI	50	655,026	2.8
Bureau of Land Management, USDI	(4)(c)	12,206(d)	0.1
TOTAL	220	23,417,540	100.0
National Wilderness Preservation System (Alaska):			
Forest Service, USDA	14	5,361,899	9.5
National Park Service, USDI	8	32,355,000	57.4
Fish and Wildlife Service, USDI	21	18,676,302	33.1
TOTAL	43	56,393,201	100.0

Footnotes:

*Some acreage estimated pending final map compilation.

(a) Does not double count Indian Peaks Wilderness with the Forest Service.

(b) Includes NPS portion (2,922 acres) of Indian Peaks Wilderness.

(c) Does not double count Santa Lucia, Weminuche, Wild Rogue Wildernesses with the Forest Service, and Oregon Islands Wilderness with the Fish and Wildlife Service.

(d) Includes BLM portion of Santa Lucia (1,722 acres), Weminuche (200 acres), Wild Rogue (10,160 acres) and Oregon Islands (113 acres) Wildernesses.

Source: Recreation Management Staff, Forest Service, USDA

doubt if anybody more than he thrilled at the feel of the "primeval environment." But to place it in the august class of "natural resource" would savor of blasphemy. The "rapture on the lonely shore" is all right in its place, but—! "Pleasure" and "business," he felt, should be carefully separated. And this despite a letter he wrote me extolling the Appalachian Trail.

But Pinchot was far more than a technical forester. He was the leader of a social movement. In his autobiography he told how the inspiration flashed into his mind that many separate questions concerning the nation's resources "fitted into and made up the one great central problem of the use of the earth for the good of man." Thus, said he, "conservation is the foresighted utilization, preservation and/or renewal of forests, waters, lands, and minerals, for the greatest good of the greatest number for the longest time."

When he put conservation into practice, Pinchot made it plain that it could not be limited only to protecting areas untouched. It was especially applicable to the ways in which resources devoted to production of commodities—whether timber, minerals, livestock, or water—were used and developed. Still, he did not disassociate wilderness from wise use. As shown in the chapter on state and private forestry later in this book, when Pinchot served as Commissioner of Forestry in Pennsylvania, he initiated a network of natural areas, designed for recreation and research, a precursor of Leopold's wilderness concept.

"To a greater extent than could be sensed when Gifford Pinchot was establishing conservation," wrote Howard Zahniser, a leader in the movement for the 1964 Wilderness Act while serving as executive director of the Wilderness Society, "this 'one gigantic problem' (as the Department of Agriculture and the Forest Service have so practically realized) includes the preservation of areas of wilderness. To the broad vision of Pinchot we owe much in the development of attitudes that now make possible the preservation of natural areas, at a time when the need is so deeply realized."

Zahniser did not see conflict between wilderness and multiple use. "The best apparent hope for success in the preservation of wilderness," he wrote, "is actually in application of the multiple use principle. To preserve some areas free from timber cutting will require adequate timber production on other areas. Preserving natural areas undeveloped will require adequate provision of developed areas with the facilities needed by the large numbers seeking outdoor recreation with conveniences."

Carhart, Leopold, Marshall—the Wilderness Pioneers

In its early years the Forest Service was engaged in instituting controls over logging, grazing, and mining, in reforesting denuded areas, in developing a system of fire protection and fighting holocausts like the fire of 1910 that swept across two million acres of Idaho and Montana, and in protecting the habitat of game animals. At the same time, it safeguarded millions of acres of wilderness, including areas that later became national parks.

In due course, as the national forests grew popular and accessible, the question arose of whether, and how, to develop these untamed places for recreation. It was during the period of 1919–1933 that the wilderness concept took root, emerging as a functional plan with actual results.

In those days a few advanced thinkers in the Forest Service found the depletion of wilderness alarming. One was Arthur Carhart, the first landscape architect hired by the agency, who was called the "beauty engineer." Carhart was working on a project in 1919 to choose a location for a cluster of summer homes along the shore of Trappers Lake in White River National Forest, Colorado. After completing surveys, he concluded there should be no development to mar this beautiful spot in the high Rockies. He convinced his superiors that the Trappers Lake area should remain roadless and that the many applications for homesite permits should not be honored. It was the first definitive application of the wilderness concept, a landmark in Forest Service history.

Later Carhart was sent to work on a recreation plan for the Superior National Forest in Minnesota, where once again there were proposals for roads and a great number of lakeshore homes. Carhart recognized, however, that the area could be "as priceless as Yellowstone, Yosemite, or the Grand Canyon—if it remained a water-trail wilderness." At first his was a minority viewpoint, but in time it prevailed. In 1936, the Superior Roadless Area was set aside, later to become the Boundary Waters Canoe Area Wilderness, renowned because it remains roadless.

Carhart was in touch about his ideas with Aldo Leopold, who was working for the Forest Service in the Southwest and developing his own original concepts.[2] Leopold warned of a "wilderness-recreation famine" and proposed a new detailed kind of management plan for the national forests. Logging would be restricted to the most productive, accessible, and economically viable forested regions and practiced on the sustained-yield principle, while remaining regions would be allocated to varying forms of recreation, game management, and wilderness.

Leopold was plainly ahead of his time. He gave to the American culture such expressions as "land ethic" and "ecological conscience," based on the responsibility of man to the rest of life on earth.

In A Sand County Almanac he wrote as follows (p. 239):

> All ethics so far evolved rest upon a single premise: that the individual is a member of a community of interdependent parts. His instincts prompt him to compete for his place in the community, but his ethics also prompt him to cooperate (perhaps in order that there be a place to compete for).
> The land ethic simply enlarges the boundaries of the community to include soils, waters, plants, and animals, or collectively: the land.

Leopold rejected the idea of questioning, of any plant or animal, "What good is it?" If the land mechanism as a whole is considered to be sound, or good, then so too, he insisted, must every part of it, whether readily understood or not.

Leopold, born in Burlington, Iowa, was a trained forester. After graduation from Yale in 1908, he joined the Forest Service and became absorbed in wildlife problems, which led him into the study of wilderness. In 1924, his superiors accepted his proposal to set aside an area in the Gila National Forest and mark it specifically for wilderness preservation. From that year forward, the Forest Service and Department of Agriculture have in one way or another set aside portions of national forests for such protection.

Leopold outlined a concept of wild areas for the Southwest to meet four goals: (1) prevent annihilation of rare plants and animals, like the grizzly; (2) guard against biotic disruption of areas still wild; (3) secure recognition, as wilderness, of low-altitude desert generally regarded as valueless for recreation because it offers no pines, lakes, or other conventional scenery, and (4) induce Mexico to cooperate in wilderness protection.

Since most Western forests were still undeveloped during the mid-twenties, there was ample opportunity to implement much of his outline. District Forester (the post now known as Regional Forester) Frank C. W. Pooler approved establishing the first wild land area in 1924 in the hazy Mogollon Mountains at the head of the Gila River in Gila National Forest, and Chief William Greeley encouraged other regions to follow a similar course with comparable areas. "The frontier has long ceased to be a barrier to civilization," Greeley wrote in 1927. "The question is rather how much of it should be kept to preserve our civilization." At his direction, Assistant Chief L. F. Kneipp undertook an inventory of available wilderness areas for use in formulating a service-wide policy.

In 1929, procedures were spelled out, under Regulation L-20, for designating "primitive areas," which were to be kept roadless and free of development and man-made structures. Within the following decade seventy-three primitive areas were established. Then, in 1939, Regulations U-1 and U-2 strengthened and refined facilities for protection: they provided that the Secretary of Agriculture, on recommendation of the Forest Service, could designate unbroken tracts of 100,000 acres or more as "wilderness areas" and others of 5,000 to 100,000 as "wild areas." Within their boundaries, commercial timber cutting, roads, hotels, stores, resorts, summer homes, camps, lodges, motorboats (except where such boating is well established), and airplane landings were prohibited. Grazing was accepted as a legitimate activity in wilderness. Mineral exploration and development, a long-lived sacred cow, continued to be permitted, reflecting the political power of that industry.

Leopold could hardly be classed as an "emotionalist" or "purist-preservationist." From the Southwest he was transferred to Madison,

Wisconsin, where he spent four years as associate director of the Forest Products Laboratory. He left the Forest Service to conduct a survey of game resources for the Sporting Arms and Manufacturers Institute (forerunner of the Wildlife Management Institute) and, subsequently, to become a professor at the University of Wisconsin from 1933 to his death in 1948. Leopold, in fact, held the first chair in game management anywhere in the country. He wrote extensively; *A Sand County Almanac*, published after his death, has become the classic statement in behalf of a lifestyle in harmony with the environment.

Robert Marshall was another Forest Service wilderness advocate of major influence. As a young New Yorker he walked all over the Adirondacks, the start of an intrepid hiking career across the continent. Following studies at the New York State College of Forestry and Harvard Forest School, Marshall joined the Forest Service in 1925 as a silviculturist at the Northern Rocky Mountain Experiment Station in Montana. Later he spent two years studying tree growth, climate, geography, and social conditions in arctic Alaska.

Marshall defined wilderness in words that were strongly reflected in the Wilderness Act, which came into law twenty-five years after his death. In *The People's Forests*, published in 1933, he wrote as follows (pp. 177–178):

> "Wilderness areas" are regions which contain no permanent inhabitants, possess no means of mechanical conveyance, and are sufficiently spacious for a person to spend at least a week of active travel in them without crossing his own tracks. The dominant attributes of such areas are: first, that visitors to them have to depend exclusively on their own efforts for survival; and second, that they preserve as nearly as possible the essential features of the primitive environment. This means that all roads, settlements, and power transportation are barred. But trails and temporary shelters, features such as were common long before the advent of the white race are entirely permissible.

As Chief of the Division of Recreation and Lands in the Forest Service, he reiterated this concept.

To arguments that wilderness designation "locks up" and withholds resources from multiple use, Marshall responded that a truly democratic society proves itself with respect for the rights of the few. How many wilderness areas, he was asked, does the country need? "How many Brahms' symphonies," he replied, "do we need?" For Marshall, wilderness was an essential part of the balance in recreation and public use.

Marshall's crowning achievement in the Forest Service shortly before his untimely death was the "U" regulations of 1939 already mentioned. In *The People's Forests*, he had proposed a national network of numerous primeval reservations to be located in all sections of the country, both to make them accessible and to avoid overuse. As a forester he urged saving specimens of each timber type, on the grounds that each has its charm and beauty and possible destruction of any species would distinctly lessen the aesthetic enjoyment of nature. His interests cut across bureaucratic boundaries of federal lands; when Marshall suggested recreational-use zoning, he employed categories such as "Superlative" for places like Yellowstone and Yosemite, "Primeval" and "Wilderness" for uninhabited forests, plus "Road-side," "Camp-site," "Outing," and "Residence."

Following World War II, Howard Zahniser took up where Marshall had left off. As early as 1951, at the Sierra Club's Second Biennial Wilderness Conference, he spoke on the question of "How Much Wilderness Can We Afford To Lose?" Zahniser saw test cases of civilization everywhere—"of the public interest with reference to wilderness preservation when in conflict with other enterprises"—cases so widespread that citizen conservationists were continually on the defensive, without time or energy to pursue a positive program to prevent recurrence of many controversies. Therefore, he called for an offensive to enlist public support and congressional action to establish something new and different: a national wilderness preservation system, based on legislation to be drafted through cooperation of federal land-management agencies and conservation organizations.

Zahniser, David Brower of the Sierra Club and other conservation leaders felt such a course was necessary because of pressures against national parks and national forests. They felt the wilderness that the Forest Service had protected administratively for years was disappearing, was being re-allocated for logging, mining, ski resort development, and other uses. And the forester, the defender of wilderness, now was considered a threat.[3]

The Pacific Northwest became a particular crossroads, a stronghold of undefiled nature and of an industry eager to unlock merchantable timber. In 1938 Bob Marshall had proposed that 795,000 acres between the North Cascades Primitive Area and Stevens Pass be studied for possible wilderness classification; but in 1940, one year after his death, the Forest Service set aside 352,000 acres as Glacier Peak Limited Area. Studies were dragged out or delayed, while logging continued up the valleys—eliminating virgin forests and wild rivers from potential wilderness. Finally, in 1960 the Forest Service designated a wilderness of 458,105 acres, a "starfish" design on tentacle ridges of rock and snow (or "wilderness on the rocks"), leaving the intervening valleys open to development. The issue became a virtual national crisis, debated in newspaper editorials across the country, reaching its climax in 1968 with legislation transferring 671,500 acres for establishment of a North Cascades National Park, and designating 505,524 acres as the Pasayeten Wilderness.

Another area of prime contention, the Three Sisters Primitive Area, had been established in 1938, embracing almost 250,000 acres astride the Cascade Divide. It included a portion of the Skyline Trail, beautiful lakes, fishing streams, and forested habitat of big game and birds. But the Secretary of Agriculture, on recommendation of the Forest Service, knocked out 50,000 acres when the area was reclassified as a wilderness in 1957. Similarly, road building and timber sales precluded enlargement of the Mount Jefferson Primitive Area to the size outdoor groups had anticipated. Centered on snowy Mount Jefferson, the second highest peak in Oregon, the area abounded in wild mountain flowers, unusual geology, and spectacular glacial scenery; it had been a favored objective of hiking clubs since 1900.

Meanwhile, in 1960, Congress enacted the Multiple-Use–Sustained-Yield Act, redefining the purposes of the national forests based on the "most judicious use of the land" for some or all of its renewable resources, referring to outdoor recreation, soil, timber, watershed, wildlife, and fishing. As mentioned in Chapter 5 one clause declared the establishment and maintenance of areas of wilderness consistent with purposes and provisions of the Act.

Although the Forest Service and citizen conservationists were having their difficulties, this was a way of recognizing wilderness management as part of forestry—a compatible, complementary function of other fitting uses of the land. "Management always implies use, but 'use' does not necessarily require the harvesting of a crop," as Samuel T. Dana and Evert W. Johnson wrote in *Forestry Education Today and Tomorrow*, the 1963 study sponsored by the Society of American Foresters. "It can provide also for recreation activities, conservation of water supply, scientific studies in natural areas and the enjoyment of scenic wonders."[4] In other words, forestry embraces management of wooded and related lands for a variety of goods and services required by society. And the well-rounded forester views the tree not simply as an item of production, but as part of a biotic community which performs a positive role as aesthetic backdrop for human activites.

Enter the Wilderness System

When Congress established the National Wilderness Preservation System in 1964, it reserved the right to designate components, or to withdraw them. It recognized wilderness as an area where the earth and its life community are untrammeled by man, and where man himself is a visitor who does not remain. It provided for use and enjoyment of wilderness, but in a manner that leaves wilderness unimpaired for future use and enjoyment as wilderness.

Only federal land that meets the following criteria can be included in the system (though several states have made designations of their own based on comparable criteria):

1. Generally appears to have been affected primarily by the forces of nature, with the imprint of man's work substantially unnoticeable.
2. Has outstanding opportunities for solitude or a primitive and unconfined type of recreation.
3. Has at least 5,000 acres of land or is sufficient in size—even when less than 5,000 acres—to make its preservation and use in an unimpaired condition a practical matter.
4. May also contain ecological, geological, or other features of scientific, educational, scenic, or historical value.

The Wilderness Act designated for inclusion in the Wilderness System all areas of the national forests previously classified as wilderness and wild areas, and also the Boundary Waters Canoe Area. Those fifty-four units, covering about 9.1 million acres, formed the core of the new Wilderness System. The law also directed the Secretary of Agriculture to review, within ten years, the status of all 34 primitive areas, covering almost 5.5 million acres, and to report to the President whether or not they were suitable for preservation. The President, in turn, was required to submit his recommendations to Congress. In evaluating each primitive area, a public hearing must be held. The Secretary of the Interior was charged with following similar procedures in roadless areas of the national parks and national wildlife refuges.

Other wildlands in unclassified status, or "de facto wilderness," were not required to be reviewed, but under broad interpretation could be considered for the Wilderness System and added to it. Millions of acres administered by the Bureau of Land Management appeared to be excluded from review. In 1964, however, the same year the Wilderness Act was passed, Congress adopted the Classification and Multiple Use Act (to remain in effect until 1970), authorizing BLM to classify lands in the public domain either for disposition or retention in manners that "best meet the present and future needs of the American people." The bureau was directed to manage retained lands for grazing, occupancy, industrial development, fish and wildlife, timber, outdoor recreation, mineral development, watershed, wilderness preservation, or preservation of public values "that would be lost if the land passed from federal ownership." In 1976 Congress brought BLM lands into the wilderness arena: the Federal Land Policy and Management Act of that year requires the agency to review, by 1991, all roadless areas of 5,000 acres or more and all roadless islands with wilderness characteristics under its jurisdiction.

The Forest Service was given its directives by Secretary of Agriculture Orville L. Freeman on June 1, 1966, when he ordered preparation of individual plans for each of the wilderness units already incorporated into the system and for others to follow. Based on terms of the law, he defined the basic management principle as follows: "National forest wilderness resources shall be managed to promote, perpetuate, and, where necessary, restore the wilderness character of the land and its specific values of solitude, physical and mental challenge, scientific study, inspiration, and primitive recreation."

The 1964 Act, as indicated earlier, specifically prohibited commercial enterprise and development, even including temporary roads, except as necessary to meet minimum requirements for administration of wilderness. But there were exemptions, or nonconforming uses, to be allowed. The most disturbing of these authorized the continuation of prospecting for minerals until December 31, 1983, a concession demonstrating anew the power of the mining industry in Washington. For claims filed until then, only the mineral rights can be patented; the land itself remains under government title. While machinery and motors can be used to develop claims, methods of access are subject to regulation to protect wilderness values.

There were two other principal exemptions: existing grazing, subject to regulations, and authorization for the President to approve prospecting for water resources, establishment and maintenance of reservoirs, power projects, transmission lines, and other facilities, should he find them in the best interest of the nation.

At the time the Wilderness Act was passed, it appeared that the National Wilderness Preservation System might ultimately encompass 50 to 60 million acres from these potential sources: about 22 million acres in national parks and monuments; 24 million acres in national wildlife refuges, and between 15 and 20 million acres in national forests. But that was before the inclusion of BLM and before the roadless-area review and evaluation revealed a much broader potential throughout the national forests.

The Forest Service implementation of the act was carefully thought through. It was constructive and far-seeing with reference to protection of classified wilderness, but limited in consideration of potential additions. The agency prepared field regulations and lengthy supporting guidelines, comprising no less than sixty double-spaced pages. In a speech on "National Forest Wilderness Under the 1964 Wilderness Act," delivered before the annual meeting of the Soil Conservation Society of America on August 22, 1965, Associate Chief Arthur W. Greeley listed three basic assumptions of the agency that were not specifically stated in the Act itself:

> 1. The intent of the Wilderness Act is to ensure an enduring resource of wilderness for the Nation. Hence the nonconforming uses which the Act authorizes need to be carried on in such a way as to minimize adverse impact on the wilderness character of the area.
> 2. Wilderness values must be given priority in the many kinds of decisions that must be made about how wilderness areas are to be used. At several points the act quite specifically indicates that the rules governing nonconforming uses must be "reasonable." But the administrators must also so

manage each area to "preserve its wilderness character." We do not see how to preserve the wilderness character except by finding reasonable ways to give priority to wilderness values.

3. Wilderness areas have different characteristics. They are not all alike. Some differences are the products of nature. Others are the result of human attitudes or of traditional patterns of use of a particular area. So, we do not expect to have one rigid set of directives that will cover all situations in all units of the National Wilderness Preservation System. Rather, within a generalized set of guidelines, the control will be a plan of management for each area which sets forth specific objectives and says how these objectives will be reached.

The Forest Service set a firm schedule for classification work on primitive areas, based on conducting field studies and field hearings within seven years, and kept to it throughout. The agency committed itself to impeccable compliance, except toward potential additions. The administrative handbook introduced the concept of availability, by which areas were to be considered "unavailable" if they contained economically minable minerals or timber. A long sequence of confrontation with citizen groups began with consideration of the very first proposed national forest addition to the Wilderness System: the San Rafael Wilderness in the brush-covered hills of the Coast Range in California, a part of Los Padres National Forest.

The agency's preliminary proposal for upgrading the San Rafael Primitive Area was published in August 1965, and field hearings were conducted at Santa Barbara that November. Public interest was very high. The California Citizens Committee for the San Rafael Wilderness was sparked by Dick Smith, reporter and photographer for the *Santa Barbara News-Press*, and Bob Easton, naturalist and author, both intimately familiar with the area. They and others cited examples of an unusual meadow life zone, highlighted by *potreros*, or grassy balds, dotted with rocky outcrops, components of a beautiful, unworldly scene. And there were other components to consider: prehistoric Chumash Indian cave art, among the best yet discovered in North America, and the presence of California condors, an endangered species of extremely limited habitat.

Because of these features, the Citizens Committee urged designation of 145,200 acres as wilderness. The Forest Service felt that 143,000 would be adequate to protect the wild environment of deer, bears, mountain lions, eagles, trout streams, and rare lilies. The difference was debated in Congress, relating not simply to San Rafael but to the public role in the decision-making process. The Forest Service prevailed, and from this point on, disagreements deepened into controversy and confrontation.

The Sierra Club, Wilderness Society, and other groups charged that unprotected wilderness was being roaded and committed to timber management before Congress had the chance to pass on proposals for classification. "There are many thousands of acres of presently undeveloped land throughout the National Forest System," wrote Acting Chief E. M. Bacon on July 13, 1967, in response to this viewpoint. "A large percentage of these have been at one time or another recommended for wilderness status by some organization or individual. To defer development of all such areas pending a Congressional determination would be to abdicate our responsibility for administering the national forests in the public interest."

This same point was made by Charles Connaughton, Regional Forester for Oregon and Washington, at the Northwest Wilderness Conference, conducted by citizen organizations in Seattle in February 1970. He maintained that the agency would only manage as wilderness those areas which it believed should be so classified, while management for purposes other than wilderness must be practiced in all other areas, regardless of any wilderness proposals by citizen groups. Conservationists lamented that in the preceding two decades at least ten million acres of wilderness in the Pacific Northwest had been lost due to road building and logging under the multiple-use program.

The scope of lands to be covered by the Wilderness Act was first expanded through legal action in Colorado in 1969: the case of East Meadow Creek, a part of White River National Forest.

East Meadow Creek lies adjacent to what was then the Gore Range–Eagle's Nest Primitive Area, approximately 80 miles west of Denver and just east of the renowned ski resort of Vail. It contains some of Colorado's most scenic and spectacular mountain country, with seventeen peaks over 13,000 feet, alpine lakes, and abundant wildlife, popular with many Coloradans for year-round recreation. The East Meadow Creek drainage provided one of the main access routes, its wilderness campsites having been used for years by hunters, fishermen, and hikers.

The Forest Service in 1968 agreed to a timber sale in East Meadow Creek; an old access road was believed to disqualify the area from consideration as wilderness. The announcement of the sale was answered by a suit entered by the town of Vail, the Eagle's Nest Wilderness Committee, the Sierra Club, and *Colorado Magazine*, asking that the sale be blocked. The judge ruled in their favor, holding that under terms of the Wilderness Act the President must be free to recommend the inclusion of contiguous areas, while removing "a great deal of this absolute discretion from the Secretary of Agriculture and the Forest Service." Based on this precedent, from then on the roadless areas contiguous to wilderness, wild, or primitive areas were to be protected until fully reviewed.

In late 1971 the Forest Service turned its attention to the roadless areas, "*de facto* wilderness" of unmeasured dimensions. Until then the agency had insisted it must first complete the study of primitive areas, which would take until 1974, before turning to the unclassified lands. But pressures for wilderness were clearly growing and the Service agreed to advance its schedule. Thus the program to be known as RARE—for Roadless Area Review and Evalution—and later as RARE I was born.

The results of the inventory were announced early in 1973. A total of 235 areas covering 11 million acres was nominated for further study for inclusion in the Wilderness System, and following public comment the number was increased later in the year to 274 areas covering 12.3 million acres. The review program would span 10 to 20 years; in the meantime, the areas involved would be managed to protect their wilderness values, while roadless areas released from review would be subject to provisions of the National Environmental Policy Act (including preparation and publication of environmental impact statements) in advance of roading, logging, or other developments.

The inventory of roadless areas revealed no less than 1,448 areas, encompassing 55.9 million acres. Most were eliminated based on "other resource values which might be lost or diminished if the area were classified as wilderness." It was the start of still another round of disputation. Yet the roadless inventory signaled that the early estimate of a wilderness system of 50 to 60 million acres was likely to prove low by far.

Bringing Wilderness to the East

One of the main questions left unanswered by RARE I was the feasibility of protecting wilderness in Eastern national forests. Citizen groups continually nominated one area or another, but the Forest Service insisted they did not qualify. As Chief Edward Cliff told the Sierra Club Wilderness Conference at San Francisco in 1967, "Personally, I hope very much that we will not see a lowering of quality standards to make acceptable some man-made intrusions or defects of other kinds simply for the sake of adding acreage."

Before passage of the 1964 Act, the Service had set aside as wild areas the 7,655-acre Linville Gorge and 13,400-acre Shining Rock tracts in North Carolina and the 5,400-acre Great Gulf in New Hampshire, but disclaimed validity of de facto Eastern wilderness on the grounds it had once been developed, before acquisition (under terms of the Weeks Act), and that "substantial imprints of man's works" still remained. Wilderness advocates, however, argued that large areas, benefiting from fertile soil and heavy rainfall, had regained their cover of mature trees—forests in the Southern mountains, in particular, providing the scenic, unspoiled beauty of wilderness. As a result of their efforts, Congress passed legislation commonly referred to as the Eastern Wilderness Act of 1975, immediately adding certain areas to the Wilderness System and designating others for further review. The following are among Eastern areas now set aside and protected:

Sipsey Wilderness, covering 12,646 acres of the Bankhead National Forest, Alabama, a chain of gorges threaded with streams and waterfalls, rich in wildlife, plants, trees, and history. The area contains eighty species of birds, ranging from northern whippoorwill to southern chuck-will's-widow (and more than half the species depend on the hardwood environment for survival).

Cohutta Wilderness, covering 34,102 acres mostly of the Chattahoochee National Forest, Georgia, and a little of the Cherokee National Forest, Tennessee, with rugged terrain offering outstanding hiking, backpacking, and camping; it is the largest wilderness in the Southern mountains outside of the Great Smoky Mountains National Park.

Otter Creek Wilderness, covering 20,000 acres of the Monongahela National Forest, West Virginia, with hardwood and hemlock forests penetrated only by trails, including some on grades of railroads built during the harsh logging era of 1905–15. The bowl-shaped basin drained by Otter Creek is one of the few remaining West Virginia mountain areas sufficiently remote to serve as breeding grounds for the black bear.

Other issues remained unsettled. Advocates and opponents of wilderness voiced dissatisfaction. Congress hurried to act on wilderness designation but reflected the cross-currents with piecemeal legislation that tended to overlap or conflict with comprehensive land management planning in the field. The big question of total national needs from the National Forest System, whether for wilderness or other uses of the roadless areas, was left unaddressed.

When Jimmy Carter became President in 1977 he faced the issue and tried to resolve it. In his environmental message to Congress, he declared: "When the Congress passed the Wilderness Act in 1964, it established a landmark of American conservation policy. The National Wilderness Preservation System established by this Act must be expanded promptly, before the most deserving of federal lands are opened to other uses and lost to wilderness forever."

This led to another roadless review, RARE II, initiated early in the Carter administration and designed to include analysis of social and economic impacts of management alternatives. The inventory generated massive public involvement and identified 1,920 areas covering approximately 65.7 million acres with wilderness potential.

During 1979–1980 Congress began considering RARE II results, processing 40 percent of the RARE II inventory, and adopting legislation for wilderness in nine states—allocating nonwilderness in those states for immediate management for purposes other than wilderness. Yet there was still wide discontent. California conservationists, for instance, strongly supported protection for the Siskiyous of northern California, home of the wolverine, peregrine falcon, pileated woodpecker, pine marten, mountain lion, chinook salmon, and steelhead trout, and spiritual grounds of the Yurok, Karok, and Tolowa Indians. As a result of RARE II, the Forest Service recommended only 41 percent of the Siskiyous (100,000 acres out of 240,000) for inclusion in the Wilderness System.

In 1979 the California Secretary of Natural Resources, Huey Johnson, brought suit to defend roadless areas in his state. And in January 1980, a federal judge ruled in his favor, declaring that the Forest Service had not dealt sufficiently with the adverse impact of opening roadless areas to development, had not seriously considered wilderness values, and had not given the public a fair chance to comment.

In testimony before a congressional hearing in San Francisco on April 15, 1982, Johnson declared:

> The State of California assisted in the review of roadless areas under the last administration. When the Forest Service failed to do a professional and fair job in evaluating wilderness, I insisted we go to court. California dramatically won the case and secured a federal court injunction against further developments in 43 roadless areas until Congress acted or the Forest Service did an honest job.

Only the week before, Johnson said, the California State Park Commission had dedicated as wilderness 250,000 acres of Anza Borrego State Park. Then he declared:

> Wilderness is an American phenomenon we in California cherish. The preservation of wilderness is basic to accepting the challenge to keep America great. Economically, wilderness and our new leading industry, tourism, are highly important to our state. But equally important is the concept of heritage. Values to be passed on to the next generation, a sense of our wilderness heritage is a vital part of our nation. In the words from an old European forestry book, "Sustained management . . . means thinking in terms of the needs of humanity and those of economics, both for the present and for the future. The individual is prepared to make certain sacrifices in order to benefit his children, his people or his country; he is ready to think in terms of the unborn."

The Forest Service has been accused variously of bias against wilderness and for wilderness. At various times and at various places both accusations doubtless have been valid. Hank Rate, a former forest ranger, helped organize the Cedar-Bassett Action Group in the early 1970s in order to campaign to have the Cedar Creek drainage included in the Absaroka-Beartooth Wilderness. As a former employee of the Custer National Forest, he had conducted the recreation study on the Beartooth and felt it was critical to sustain the vast unbroken wilderness adjacent to Yellowstone National Park. He felt that only wilderness classification would guarantee its protection and prevent some future change in plans by his old outfit. In a personal letter written on February 1, 1982, he had not changed his mind:

> Though multiple use is a byword, it is turned in such a manner that areas of greatest value for wildlife and watershed are roaded and logged anyway. Forests are considered useless until cut, since lands left in a wild

condition will die, burn up, be chewed up, harbor posey-loving hippies, and generally self-destruct. Wilderness is mostly for public relations. And when Congress takes away initiative from the agency, it creates resentment on the part of professionals toward local "involved citizens," intensifying problem and controversy.

This viewpoint is shared by Brock Evans, vice-president of the National Audubon Society and a former Northwest Representative of the Sierra Club. In a letter written January 6, 1982, to William A. Worf, long active in Forest Service wilderness policy and administration, Evans wrote:

The case of the Gila Wilderness—the first one established on the initiative of Aldo Leopold—is often cited by the agency as proof of its love of wilderness. After the area was established, it was cut in half, its boundaries shrunk over the years through one administration action after another, so that the existing area—even with the Blue Range—is not the same as it was when first set up. Some Forest Service attempts to open up the area did not succeed; but the fact remains that the Forest Service tried to undo its own work.

There are many other examples, but they have always illustrated in fundamental point: that the Forest Service is first and foremost a political agency, and it will support wilderness if the politics are there or if it perceives a threat to its own jurisdiction. But it is rare, at least in my experience, that it will do so if there is even a 50-50 split in public opinion—particularly where resources of any commercial kind, such as timber, are involved.

This belies any effort to use the "numbers game," citing all the millions of acres now in the wilderness system. Yes, there are millions of acres now in the system—but there are tens of millions of acres of roadless wildlands which are not, though they should be. Even granting that the Forest Service has many missions and responsibilities to provide timber and resources other than wilderness, that, in my mind, does not negate the fact that we could have had a truly magnificent wilderness system on national forest lands if only the Forest Service had not been so active in its opposition over the years.[5]

On the other hand, Steve E. Wright, director of admissions of the Sterling Institute, at Craftsbury Commons, Vermont, served two years on the staff of the Moose Creek Ranger District of the Nezpercé National Forest, under terms of the Intergovernmental Personnel Act (IPA). His assignment was based essentially on his professional background in resource management. On returning to Vermont, Wright was moved to express his feelings to Regional Forester Coston (in a letter of January 12, 1982): "From the volunteers up to the district ranger, there was an energy and commitment far beyond that required by job descriptions. These people believe in hard work and commitment to a wilderness ethic."

Perhaps the question should be asked whether personnel with "wilderness ethic" consciences or commitment occupy key positions in decision making. While timber production is a major function and timber specialists are held in high esteem, wilderness, despite the many millions of acres and crucial political issues it generates, has been one of several functions associated with recreation staff work. Wilderness management carries scant potential for advancement. When wilderness is measured, or quantified, in "recreation visitor days," it fares poorly, which doubtless has something to do with it. This standard scarcely treats with broader wilderness values, which might easily justify a separate role apart from recreation in the infrastructure. Professional foresters also feel that wilderness lacks the challenge of management, though it may actually demand even more skills in effective management than other forest uses.[6]

Nevertheless, in any assessment of accomplishment or systematic approach, the Forest Service must be ranked in the van of federal agencies, with the Bureau of Land Management, Fish and Wildlife Service, and National Park Service, all grappling with provisions of the Wilderness Act and how to apply them.

William A. Worf was intimately involved in shaping Forest Service wilderness policy under the Act. He was subject to some criticism, from both within and without, for cultivating and advocating the "pure philosophy"—that is, insisting that areas must be unspoiled to qualify as wilderness, and then be managed to remain so—but no one questioned his expertise. Worf was supervisor of the Bridger National Forest in Wyoming (which includes the 383,000-acre Bridger Wilderness in the Wind River Range) when the Wilderness Act was passed and was assigned to a four-member task force to draft national regulations and policy guidelines. Subsequently, he became Director of the Wilderness Program in Washington and Director of Recreation and Lands in the Northern Region until his retirement in early 1982. In an article appearing in the *Idaho Law Review* in 1980, Worf recalled the course of events to sustain his position on the purity concept.

The task force initially, wrote Worf, felt that Congress, through the Wilderness Act, was merely validating forty years of Forest Service wilderness management. But team members presently found many points on which they couldn't agree. The fifty-four "instant" wilderness units recognized by the act were all being administered differently, and management was subject to change with change in personnel. There was no consistent or distinctive philosophy. Declared Worf, "It became clear that Congress had not endorsed Forest Service wilderness management. It had mandated a new direction."[7]

In 1965, even while draft regulations and policy were being circulated nationwide for comment, new and different challenges accented the need for specific, consistent guidelines to prevent erosion of the wilderness resource. Miners sought to use helicopters and motorized equipment for prospecting. The National Aeronautics and Space Agency (NASA) also wanted to use helicopters and make a motorized installation for purposes so secret they defied description. Telephone companies wanted to install electronic repeaters. Cattlemen argued that bulldozers were necessary to maintain water ponds for stock, while water users pressed to install electronic devices to measure and broadcast snow depths. Fish and game agencies asked to use helicopters to plant game animals and stock fish. And even Forest Service administrators suggested that it was in the public interest of wilderness to cut young trees invading natural openings.

Congress opened the door to such proposals by listing specific exceptions to things that could not be done. The Forest Service recognized that there must be some differences between wilderness units, but Worf insisted it must be based on a consistent philosophy:

> Wilderness is recognized as a resource of the land—a fragile and essentially nonrenewable resource. Man's use of the area must always be in context with the idea that maintaining an enduring resource of wilderness for the future is our overriding mandate.
>
> The cumulative effect of nonconforming occupancies and mechanization is sometimes subtle, but nonetheless real. For that reason, the fact they can sometimes be hidden from visitors may reduce their impact somewhat, but it does not make them more compatible with wilderness.
>
> The wilderness which would evolve under the "practical" approach would by no means be unattractive to many people. It would produce high quality recreation. Outfitters would have permanent, neat, tidy and comfortable camps. They would use less horses per client day because equipment would be left in place. There would be more flexibility on camp location because we could develop and pipe water to presently unusable campsites. Management of outfitter stock would be made easier by drift fences and corrals. One disadvantage is that permanent camps may be viewed as resorts by public health services; however, we could permit necessary food handling and sewage facilities to meet health standards.
>
> In my personal view, our Wilderness System should be sufficiently distinctive and valuable that it would be as inviolate as our "crown jewels"—the national parks. The energy crunch has demonstrated that this is not yet so. The famous Overthrust Belt with its oil and gas potential covers parts of the Bob Marshall, Scapegoat and Great Bear Wildernesses and there are strong pressures to invade them in the name of energy. On the other hand, the possible petroleum potential under Glacier National Park is probably at least as great, but no one has suggested that we need it.

Measuring the Economic Values

It is simple enough to place a dollar value on tons of coal, barrels of oil, or board feet of timber, but analysis of costs and benefits have generally failed to assess the economic worth of wilderness, open space, wild and scenic rivers. Thus, the arguments have supported extraction and commercial development on national forests and other public lands.

Recreation has been measured and found to be a significant industry, but wilderness has not. Moreover, it holds meaning in terms other than recreation. In 1982, however, a research team at Colorado State University issued a detailed pioneering report showing how the dollar values of wilderness areas can be weighed against competing mining, logging, high-density recreation, and other commercial uses.

Working with funding provided by the CSU Experiment Station and the American Wilderness Alliance, the researchers employed standard, accepted survey procedures to measure the value of Colorado's wilderness to the general public of the state. Colorado provides an ideal laboratory for such a study. The state has been undergoing rapid growth, with marked industrialization, urbanization, and pollution. A total of 2.6 million acres has been set aside as wilderness, an amount equal to 4 percent of the state. But there are nearly 10 million acres of potential wilderness on federal lands, equal to 15 percent of the state. And at the time of this study, the early 1980s, the future of these lands was being seriously contested.

The Colorado State study, directed by Dr. Richard Walsh, an economics professor, sought to provide useful data by assigning "recreation values" and "preservation values"—based on the idea that scarce resources benefitting the public, like clean air or wilderness, are as valid as those traded in the marketplace. Dr. Walsh and associates cited the participation of Colorado residents in the "nongame checkoff program" on the state income tax as a willingness to pay for preservation of noncommodity resources. They might also have cited a more recent study by the U.S. Fish and Wildlife Service showing that nearly 100 million Americans spent almost $40 billion during 1980 on one or more forms of wildlife-related recreation, and the overwhelming majority were not even hunters or fishermen.

Returns from querying a cross-section of 218 Colorado families during the summer of 1980 showed the public values wilderness primarily for preservation of water quality, air quality, and wildlife habitat, and for the satisfaction of knowing that future generations will have access to such areas. These were considered more important than opportunities for recreation now. The CSU report placed a value of $1.9 billion on Colorado's currently designated 2.6 million acres of wilderness, while projecting a phenomenal rise in recreational use of wilderness over the next half century. Yet it added this cautionary note: "There may be long-run ecological values that are not included here. It is difficult for biologists to predict what these might be, let alone measure and incorporate them into an economic benefit estimate. The inability of economic analysis to place a dollar value on unknown ecological effects should be recognized in making decisions about future wilderness designation."[8]

Notwithstanding the rights or wrongs, or values, in wilderness, some agency personnel feel that classification achieved through a political process represents an unfair and unneeded restriction on their professional options. Yet RARE II advanced wilderness in Forest Service planning and programs; it broadened understanding of the old-growth and restored forests, and it identified types of land forms less than spectacular and yet worth saving.

Good things have been accomplished. "Wilderness on the National Forests in California" is the title of a brochure designed to help visitors plan enjoyable and safe trips and to remind them of the importance of "no-trace" camping. Wilderness management, in the face of heavy and growing use, is necessary to assure that resources will not be degraded and values lost. Permits are now required and group sizes limited for all the wilderness areas in California. The John Muir Wilderness, a composition of granite mountains, alpine lakes, and clear streams in the heart of the Sierra Nevada range, attracts 850,000 visitors annually—all on quota. In the San Gorgonio Wilderness, a part of the San Bernardino National Forest, where rough, rugged country rises above meadows and lakes to 11,499 feet, camping on the summit is limited to five permits per day.

The 1980 RPA document, using data gathered during a five-year period, recommended a wilderness system in the national forests that ultimately would cover between 33 and 42 million acres. And the report noted that setting aside even more land would not significantly affect national timber yields. But this idea has not been universally accepted or cheered.

In North Carolina, for example, the Forest Service in 1977 identified 24 areas covering 207,000 acres for study under RARE II. These included 19 areas on the Pisgah and Nantahala in forested western North Carolina, one on the Uwharrie, in the central Piedmont, and four on the Croatan, near the coast. Four wilderness areas already designated (Joyce Kilmer–Slickrock Creek, Linville Gorge, Shining Rock, and Ellicott Rock) total 36,000 acres.

The Forest Service held information sessions in two mountain communities, Franklin and Robbinsville, to receive public comment. The environmental impact statement prepared in connection with the proposal estimated only modest economic loss from wilderness designation, but this point was lost. Opponents turned out by the hundreds, proponents by the handful.

Businessmen, loggers, mill workers and their wives jammed the sessions. They interrupted the programs with boos and jeers. They brought tractor-trailers loaded with logs and lumber, bulldozers and other equipment, blocking roads and streets, displaying placards that read: "Stop the Sierra Club," "Timber is a crop—don't let it rot," "Save our jobs," "Western North Carolina needs more timber sales—not wilderness."

Reverend Rufus Morgan, a patriarch of the hills, was quoted in the media as saying, "I wish that all our people could forget the dollar sign long enough to get the inspiration that comes from climbing a mountain and seeing the world spread out before them." But the coalition of miners, loggers, businessmen, mayors, and county commissioners was extremely forceful. The owner of one logging firm, Tom Thrash, said the wilderness proposal was "a land grab by hikers and backpackers, mostly from Florida and the Midwest, for their exclusive use."

The wilderness proposals, finalized in 1979, were sharply reduced. President Carter in his submission to Congress increased them slightly (to 64,800 acres in wilderness, plus 23,800 acres for further study), but most of them were without prime tree cover.

In a foretaste of things to come, John Crowell, then general counsel and vice-president of the Louisiana Pacific Corporation, speaking before the annual meeting of the Society of American Foresters in October 1980, called RARE II "particularly ill-conceived" and entirely outside the planning powers mandated by law. Following the November presidential election, when Mr. Crowell was named Assistant Secretary of Agriculture, the Reagan administration pressed for "release" legislation, setting rigid deadlines for Congress to consider RARE II wilderness recommendations and releasing all lands not classified for logging and other uses. But this proposal was killed in committee.

"A federal court injunction has not stopped the Reagan Administration's rush to exploit our wilderness areas," declared California's Natural Resources Secretary Huey Johnson in his April 1982 Congressional testimony. "Over 85 small hydro projects are planned in California RARE II areas. Over 250 other developments are planned by the Forest Service in wilderness areas, many protected by the California RARE II lawsuit. These developments include roads, mines, geothermal leases, oil wells and timber sales."

The next chapter in this book details conflicts between mineral development and wilderness protection. It is fitting at this point, however, to note the introduction of the Wilderness Protection Act of 1982, sponsored by the Administration to assure that wilderness lands are "last to be exploited, if ever." Known as the Watt Bill, for its principal advocate, the Secretary of the Interior, it would close to January 1, in the year 2000 (if Congress does not act again), wilderness and wilderness study areas to energy, minerals, and commercial development, thus ending pressures contained in the Wilderness Act to consider mineral leasing until December 31, 1983. But it would also automatically open the system in the year 2000 and, in the meantime, allow the President, acting on his own, to open any area to mineral development by making a finding of "urgent national need." It also provides for release of RARE II and BLM study areas within a specified period of time at the discretion of the Secretaries of Interior and Agriculture, and for an inventory of "resource opportunities foregone"—that is, of the energy and mineral potential of wilderness and wilderness study areas.

Wilderness champions were quick to respond with bills of their own. The Udall-Conte Resolution, introduced by a coalition led by Representatives Morris Udall of Arizona and Silvio O. Conte of Massachusetts, expressed the "sense of the House" that:

(1) the Nation's wilderness areas should be the last areas where mineral leasing should take place and, therefore, the Secretary of the Interior should exercise his discretionary authority under applicable law to refrain from issuing mineral leases in wilderness areas and in areas under formal consideration for wilderness designation;

(2) if, in contravention of the sense of the House of Representatives expressed in paragraph (1), any Federal official proposes to issue a mineral lease in a designated wilderness area or an area under formal consideration for wilderness designation, he should provide sufficient prior notice to Congress to permit the Congress to take appropriate action to prohibit the issuance of such lease;

(3) the congressional consideration of additions to the Wilderness System should proceed expeditiously but without the imposition of arbitrary deadlines; and, pending congressional decision, lands under formal consideration for wilderness designation should be managed to preserve their suitability for inclusion in the Wilderness Systems; and

(4) the Federal land management agencies should continue to consider wilderness values, together with other multiple-use values, in making land management decisions and recommendations.

The battle over wilderness likely will continue; the less there is of it the sharper the issues become. There may be differences over where and how much wilderness should be saved, but there is little doubt that wilderness as a resource will have a significant place in national forest management.

People inside and outside the agency have a lot to do with identifying areas that deserve saving. Dick Smith, as an example, joined the staff of the *Santa Barbara News-Press* in 1948 and proceeded to make the back country of the Los Padres National Forest his special "beat." He studied the California condor and helped gain recognition for it as a species in need of special protection. Then he sparked the effort to designate the San Rafael Wilderness, the first one to be reviewed by Congress and set aside under the 1964 Wilderness Act. When he died suddenly in 1977, the mayor proclaimed Dick Smith Week and the whole city mourned.

In *Condor Journal*, published in 1978, Smith wrote as follows:

Motorized vehicles came snorting up toward the last wild places, shattering the stillness. For the first time it became possible to take a machine almost anywhere, even to level the mountains if man willed. There was also an increasing feeling of "what of it?" if wilderness and condor died. "These lands belong to the people." Man's will must be served. If the Pleistocene Age was obsolete, so maybe was the condor. But at the same time there was an ever-increasing feeling among many people that something had to

be done to save both the condor and wilderness, that they both belonged
to the people—the people for all time to come—people who ought to have
a chance to know wilderness and condor as they are now and always have
been.[9]

Smith had helped get a large roadless area adjacent to the San
Rafael Wilderness placed on the wilderness study list. It includes the
chaparral-covered slopes of Big Pine Mountain, eroded sandstone for-
mations, Indian cave paintings, and 6,541-foot-high Madulce Peak. Smith
was captivated by the remoteness of this country, the steepness of its
trails, elevation differences (from 3,300 to 6,660 feet), vast panoramas
from the highest peaks, the variety of wildlife and plant communities.
All of these, he felt, would instill in any visitor a sense of serenity,
solitude, beauty, and challenge. He saw discovery, adventure, and a
thoroughly American experience for those willing to walk or ride. As
he wrote in *Mountains and Trails of Santa Barbara County* (with Frank
Van Schaik):

> Many of the trails entering the wilderness have been used for centuries;
> remains of ancient Indian camping places may be found along them, in
> the deepest reaches of the forest. Many trails, while nearly overgrown, may
> still be traced and followed by a good woodsman.
> Traveling along one of these well-marked trails offers many pleasant
> surprises. In the farthest meadows there are remnants of early homesteads.
> Old and crumbling abodes, log cabins and frame houses long deserted, are
> often seen. Along the upper Sisquoc are the abandoned farms of a colony
> of Mormon settlers. Only an occasional buckboard or rusty plow remains
> to give a hint to the kind of life these people tried to carve from the
> narrow, fertile but remote valley.[10]

When it came to considering wilderness status, support was voiced
by all kinds of groups and virtually all the elected officials of Santa
Barbara County. And their support extended to naming the area the
Dick Smith Wilderness, which is what the legislation provided when
written for action by Congress in 1981.

Smith saw something special that others had missed. But then again
the Boundary Waters in northern Minnesota might be like any other
resort region if it hadn't been for Arthur Carhart. If ten foresters were
asked to select the first wilderness, nine of them would probably opt
for some spectacular mountain stronghold of rock and ice in the Rockies,
Cascades, or Sierra Nevada range, but the tenth, Aldo Leopold, was
working in the brushy Southwest and figured that wilderness is where
you find it.

PHOTOGRAPHS

Gifford Pinchot was a pioneer of photography in forestry. He believed that photographs should provide a factual permanent record and a basis for prediction of future land use.

He himself took more than 1,400 photographs. While conducting a Western survey for the Department of the Interior in 1897, he supplemented his notes with graphic illustrations of erosion, fire, and destructive redwood logging of the 1880s. He took photos of land uses in Borneo, China, and Russia during the course of his extensive overseas trip in 1902.

Once he became Chief Forester, Pinchot issued instructions on photography to field personnel. They were not to be content with mere scenery but were to concentrate on scientific photography of trees and forests; they were, in fact, to repeat certain pictures every ten years as the means of tracing growth and change. That system is still in use—the Northern Region of the Forest Service alone has 22 "picture points" where photographs are taken at ten-year intervals.

Logging in the Adirondacks. (*top*) This crew is shown at work in a typical river drive in the wildlands of upstate New York, probably in the early 1890s. (Photo by Gifford Pinchot)

Rooftop of the East. (*bottom*) Secretary of Agriculture James Wilson (in straw hat) and others are shown in 1901 erecting the monument atop Mt. Mitchell, North Carolina, highest point in the East. (U.S. Forest Service, photo by Gifford Pinchot)

Taking the test. This candidate for a ranger's job early in the century, wearing his best duds at the Fair Grounds, Leadville, Colorado, shows he can pack a horse. (Courtesy U.S. Forest Service, photo by H. A. Bliler)

When life was harder—but simpler. The early-day ranger had little, if any, professional training. He was a western man, who may have worked as a cowhand or sheriff, possibly with some political connection. This ranger is ready for anything. Note his two badges. (U.S. Forest Service photo)

Training the professionals. Students of the Yale Forestry School are shown at summer camp, a part of the Pinchot estate at Milford, Pennsylvania, in the Pocono Mountains. (Photo courtesy Pinchot Institute for Conservation Studies)

Off to the woods. The ranger and two forest guards are preparing to leave the Idlewild Ranger Station, of the Arapaho National Forest in Colorado, for a day's work. The man at the right is carrying his transit. (U.S. Forest Service photo)

A Portfolio by K. D. Swan

A legendary and accomplished field photographer, K. D. Swan joined the Forest Service at Missoula, Montana, in 1911, following forestry studies at Harvard. Photography was merely a hobby until 1913, when he became official photographer of the Northern Region. From then until retirement in the mid-forties he took pictures of every possible facet of Forest Service activity and was instrumental in establishing the Forest Service's storage system.

A few of his pictures are presented here. Swan developed and printed his own photographs in the laboratory he maintained at region headquarters in the Missoula Federal Building. He died in 1970 at the age of 83.

Ranger on the trail. At the base of the Chinese Wall in the Bob Marshall Wilderness of northern Montana, Ranger Horace Godfrey follows a quiet path in 1946. (U.S. Forest Service photo)

Homestead in the hills. (*top*) The rancher found a haven when he chose this location on the Yaak River below the Olson Ranger Station in the Kootenai National Forest, Montana. Swan took this picture in 1926. (U.S. Forest Service photo)

Fire aftermath. (*bottom*) Swan shows the results of the Teakettle Mountain fire along the North Fork Road, Flathead National Forest, Montana. His photos attracted wide attention for the moods they evoked. (U.S. Forest Service photo)

Camping in style. (*top*) A 1938 party is shown close to Ha-Nana Lake in the Hillgard area, Gallatin National Forest, Montana. (U.S. Forest Service photo)

Summer's end, 1937. (*bottom*) A band of sheep and their herder coming down from summer range on the Lolo National Forest, along the Montana-Idaho line, headed for the railroad. (U.S. Forest Service photo)

Loading supplies. Dick Johnson, a famous pilot in the Montana back country, carries a 25-man ration box at the Libby, Montana, airport in 1939, preparatory to a drop over a high, remote firefighting camp. (U.S. Forest Service photo)

An exhibit of transport machinery. These trucks and heavy equipment vehicles are shown at the Forest Service repair shop at Missoula in 1940. (U.S. Forest Service photo)

The ranger station. It served as an administrative outpost far from cities, often remote even from towns. This one, the Philipsburg Ranger Station, on the Deerlodge National Forest, was built in 1939 (pictured here one year later). It includes the administration building, ranger's house, warehouse-cook-bunkhouse, shop, and garage. (U.S. Forest Service photo)

Wilderness enthusiast. Although Swan chronicled everything about the national forests, he exulted in unspoiled landscape, as evidenced in one of his last official pictures, taken in 1946 in the heart of the Bob Marshall Wilderness, Montana, looking into Pearl Basin from the Continental Divide. (U.S. Forest Service photo)

Theodore Roosevelt confers on shipboard with his close confidant, Gifford Pinchot, during the trip of the Inland Waterways Commission down the Mississippi River in October 1907. Do top hats and morning coats seem strange dress for two outdoorsmen? (U.S. Forest Service photo)

Dwight Eisenhower seems pleased with his company, Forest Service Chief Richard A. McArdle, on his right, and Montana Governor Hugo Aronson, during the course of an inspection of the Forest Service Aerial Fire Depot at Missoula in October 1954. (U.S. Forest Service, photo by Ross Angle)

John F. Kennedy is seen with Chief Edward P. Cliff at the ceremony dedicating the Pinchot Institute for Conservation Studies at Milford, Pennsylvania, September 24, 1963. Secretary of Agriculture Orville L. Freeman is partially obscured behind the President. The Pinchot Institute is in the former home of the first Chief Forester, donated by his family to the government. (U.S. Forest Service, photo by Bluford Muir)

Lookouts Across the Land

"A young man who hoped for a career in forestry might begin on a trail crew, advance to manning a guard station, and then reach his first starring role as a lookout. Although he was not in fact very far along in the ranks of Forest Service personnel, he received a little more pay and, more importantly, a certain recognition. He had a peak of his own for the summer, and, what with all the reporting back and forth, everybody in the district knew him as an individual up there." —from *Lookouts, Firewatchers of the Cascades and Olympics,* by Ira Spring and Byron Fish (Seattle: The Mountaineers, 1981).

Fire lookout towers reached their peak in 1953, when there were more than 5,000 in national forests across the country. Their number has since steadily declined, yet their occupants have consistently spotted more fires than have people in aircraft. They are distinctive in architecture, and part of forestry legend and lore.

Chimney Rock Lookout Tower, San Juan National Forest, Colorado, 1942. (U.S. Forest Service photo)

Billy's Peak Lookout Tower, Shasta National Forest, California, 1953. (U.S. Forest Service, photo by Jack Rottier)

Scenic Spectacles That Draw Americans Outdoors

(*top*) Crooked Lake on the Superior National Forest, part of the vast water domain of northern Minnesota. The Boundary Waters Canoe Area, a significant portion of this national forest joined with Quetico Provincial Park of Canada, is the most celebrated canoe area in the world. (U.S. Forest Service, photo by Freeman Heim, 1962)

(*bottom*) Redfish Lake on the Sawtooth National Forest, near Sun Valley, Idaho, reflects the rugged crags of the Sawtooth Wilderness. The lake got its name from spawning salmon, which make their way upstream from the Pacific to lay their eggs and die. (U.S. Forest Service photo)

(*top*) The billowing plumes of steam escaping from Mount St. Helens on May 18, 1980, give fair warning of an imminent eruption of this volcanic peak within the boundaries of Gifford Pinchot National Forest, Washington. Mt. Rainier in background. (U.S. Geological Survey photo)

(*bottom*) The mountain has blown its top, as evidenced by this photo of the crater profile, taken from the north on June 28, 1980. Congress subsequently designated a Mount St. Helens National Volcanic Monument of 110,000 acres. (Photo by Walt Conner)

(*top*) Rafting the Middle Fork of the Salmon River, within the River of No Return Wilderness, Idaho, has soared in popularity. In 1950, according to ranger reports, only two tourists floated the Middle Fork. By 1980 that number had climbed to nearly 4,000. (U.S. Forest Service photo)

(*bottom*) Fishing for rainbow trout is a popular sport on the South Branch of the Potomac River where it flows through Monongahela National Forest, West Virginia. Mountain forests heavily logged early in the century now feature a range in recreation activities. (U.S. Forest Service, photo by Paul Steucke, 1966)

Members of the Mazamas, a well-known Portland outdoors organization, are shown making their way up the hogback towards the summit of Mount Hood (elevation 11,245 feet), the highest peak in Mount Hood National Forest, Oregon. (U.S. Forest Service, photo by Roland Emetez, 1963)

This lush waterfall in the Craggy Mountain Scenic Area of North Carolina, part of the Pisgah National Forest, was particularly entrancing to the late Supreme Court Justice William O. Douglas. Regional Forester James K. Vessey, with whom he visited the site, decreed that it would henceforth be called "Douglas Falls." (U.S. Forest Service, photo by Dan Todd, 1962)

A horse party rides along the Oregon side of the Snake River, where it flows through Hells Canyon, the deepest gorge on the North American continent. Low clouds actually hide the upper rims. After many proposals to construct a hydropower dam, Congress voted to safeguard this section of the Middle Snake in a free-flowing condition. Oregon and Idaho national forest river shores and mountain borders are administered as Hells Canyon National Recreation Area. (U.S. Forest Service, photo by James W. Hughes, 1970)

Maroon Lake and the peaks of the Maroon Bells in the background typify the glorious Rocky Mountain scenery within White River National Forest and other Colorado national forests. (U.S. Forest Service, photo by Bluford W. Muir, 1951)

(*top*) District Ranger Charles Cartwright, a native Virginian, found summer employment with the Forest Service encouraging and decided to make it a career. In 1980 he became a district ranger on the Okanogan National Forest, later on the Gifford Pinchot National Forest— a long way from Virginia to the State of Washington. (U.S. Forest Service photo)

(*left*) District Ranger Janet L. Wold, of the Umatilla National Forest, Oregon, began her career with the Forest Service in research in 1974. In 1982 she became the third woman district ranger. One year later there were six. (U.S. Forest Service photo)

(*top*) Charlotte Larson is the second woman ever to become a career Forest Service pilot. Larson, assigned to the Intermountain Region, Ogden, Utah, often flies as lead pilot in forest fire action, guiding fire-retardant bombers. (U.S. Forest Service, photo by Art Whitehead, 1980)

(*bottom left*) Deanne Shulman, smokejumper on the Payette National Forest, Idaho, is shown here testing rigging in a ground exercise, but she bails out fires along with the men. (U.S. Forest Service photo)

(*bottom right*) Joyce Allgaier-Ohlson, patrol person on the Targhee National Forest, Idaho, knows the back country. Her job is to help visitors with information about trails and natural features, encourage clean-up, and report on patterns of use and overuse. (U.S. Forest Service photo)

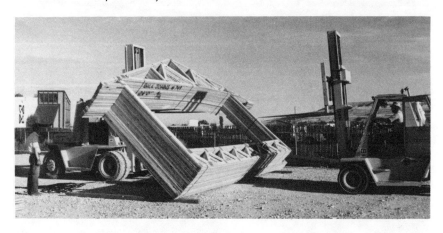

(*top*) A truss-frame (TSF) package being unloaded at a building site. TSF is part of the new look in light-frame construction. It was devised by Forest Service technicians at the Forest Products Laboratory, Madison, Wisconsin. (U.S. Forest Service photo)

(*bottom*) The house frame can be set up in 11 working hours. The system is designed to produce a stronger frame with less lumber (eliminating the need for basement posts and beams). (U.S. Forest Service photo)

(*right*) Mineral development has changed the face of Wyoming. This oil exploration equipment is seen at work at high elevation in the Bridger-Teton National Forest—where 1,500 leases cover 2.2 million acres. The national forest expects more than 500 miles of seismic lines to be completed each year during the 1980s. (U.S. Forest Service photo)

(*bottom*) Phosphate mining on the Caribou National Forest in southern Utah indicates the scope of mineral activity on the national forests. Pressures to prospect and extract have risen sharply since the "energy crisis" of the early 1970s. (U.S. Forest Service photo)

(*left*) Roadless logging is demonstrated by this Sikorsky Skycrane operating under Forest Service auspices in California. (U.S. Forest Service, photo by James W. Hughes)

(*bottom*) Cables from the helicopter are being attached to the log shown here during the experiment in helicopter logging. This system is considered promising, though many questions remain to be answered. (U.S. Forest Service, photo by James W. Hughes)

The Chief and a PR Man

(*right*) R. Max Peterson, Chief of the Forest Service, in August 1982. A Forest Service career professional, Peterson in 1979 became the first forest engineer to be named Chief. His midcareer training included studies at the School of Government at Harvard. Every Chief, without exception, has worked up through the ranks. (Photo by John Dennie)

(*bottom*) Clinton Leon Davis (at left), the unforgettable "Clint," known as the greatest PR man in the Forest Service since Gifford Pinchot, is shown with a clutch of pressmen on a 1940 bow and arrow hunt in the Chattahoochee National Forest, Georgia. Davis was an Atlanta-based writer on the outdoors before beginning his long career with the Forest Service. (Photo courtesy Harry L. Rossoll)

12

SEARCHING FOR ENERGY AND CONSERVING IT

> We can go into a forested area and drill and develop oil and gas resources without destroying the forest or wildlife habitat. It doesn't need to be maintained in its pure sense, without any intervention of man, just to have a healthy stand of trees. We can have the trees, the wildlife and the oil and gas production as well.
> —James G. Watt, Secretary of the Interior, April 1981.

> Sometimes I feel as though I'm not managing a forest anymore. I'm managing wood lots in between well sites.
> —Paul Brohn, district ranger, Allegheny National Forest, December 1980.

Wilderness Protection Versus Energy

Bob Marshall, the vigorous director of the Division of Recreation and Lands, died in November 1939, two months after announcement of the Forest Service U regulations, which he had helped draft. (Chief Ferdinand Silcox, who had supported Marshall in setting aside wilderness, died in the same year—a double blow.) The following year, the Forest Service honored Marshall with a living memorial, giving his name to nearly a million acres of rugged Montana country in the Lewis and Clark and Flathead National Forests.

It is classic country; the classic national forest wilderness, it might be called, a companion piece to the adjacent Glacier National Park and one that shines equally as a natural jewel of the northern Rockies. The serpentine backbone of the Continental Divide runs through the Bob Marshall Wilderness and at the escarpment called the Chinese Wall breaks away eastward in thousand-foot cliffs for fifteen miles. The Flathead Alps, an offshoot of the divide, challenge the best rock climbers, and the Bob Marshall Wilderness is one of the last great refuges of the grizzly bear south of Canada, as well as home of a large elk herd, deer, mountain goat, bighorn sheep, and mountain lion.

The Bob Marshall, or the "Bob," is connected with two other wilderness units established during the 1970s, the Great Bear and the Scapegoat. These cover about 1.5 million acres, an area roughly the size of Delaware—a fantastic environment of high mountain snowfields, rushing streams, and narrow valleys carpeted with wildflowers, long favored by hikers, horse people, hunters, and fishermen.

When these areas were set aside, that simple process seemed, on the face of it, to ensure lasting protection. But this was not to be, for even wilderness is open, by law, to mineral exploration and development. The Bob Marshall is part of what petroleum geologists like to call the Overthrust Belt, which the *Oil and Gas Journal* described in 1978 as the "hottest new area for drilling in the United States, offshore or on."

Thus Consolidated Georex Geophysics (CGG), a French exploration firm under contract to several oil companies, applied to conduct seismic exploration in the Bob Marshall complex. Their survey would require detonation of 165,600 pounds of explosives in 3,312 separate charges along 207 miles of line, including 161 miles inside the Bob Marshall. (Explosives create a shock wave that sends soundings into the earth's crust and are picked up and recorded by electronic devices, helping to determine locations where drilling for gas and oil is most likely to be successful.) Friends of the Bob Marshall sought to block the exploration, citing threats to wildlife from seismic activity, while defenders of wilderness in Montana and elsewhere claimed it would create a precedent for invading wilderness throughout the national forests. Besides, they argued, the Bob Marshall at best might contain gas to supply the United States for only two months, while its natural values would endure forever.

Recent years have brought several large gas discoveries in the Alberta portion of the Overthrust Belt and finds of both oil and gas in southwestern Wyoming and northeastern Utah. But exploration in the plains of Montana east of the Bob Marshall has produced only dry holes without significant petroleum discoveries. Critics insist that petroleum reserves remain almost completely hypothetical, that optimistic projections in the media have been overstated, and that drilling would exchange an unavoidable immediate loss for conjectural long-term gain.[1]

This, however, is not the way that industry views the scene. "The assumption that oil and gas developments are incompatible with wilderness values is incorrect," asserted Patricia L. Roberts, of the American Petroleum Institute, early in 1982, as quoted in the Washington, D.C. media. "Industry argues that that is not the case—that impacts from exploration and development are temporary and do not cause permanent impairment. We don't know what is in there until we explore."

The American Mining Congress complained publicly in the same year of rigid rules imposed by law and the Forest Service restraining exploration in wilderness: "We know there is a lot of mineralization in the wilderness. The United States Geological Survey has reported that 42 percent of the wilderness areas have mineral potential. That is a large area to be put off limits when the country needs access to resources."

Is there some basis for deciding at an early stage whether a deposit may prove insignificant relative to wilderness values—or whether it may, in fact, represent a major long-term source? Has engineering been advanced to a level of sensitivity to environmental and ecological values? Surface mining and current reclamation programs often appear to be wanting, but can subsurface extraction be performed in a way that is technologically compatible with the use and protection of other resources? Since no commercial firm undertakes a project that is not potentially profitable, or economically feasible, can there be some logical and acceptable measure of environmental feasibility as well?

At the request of the Northern and Pacific Southwest Regions of the Forest Service, George D. Davis undertook a study to answer the last of these questions. "My values lie toward wilderness but I, like most Americans, use a good deal more than my fair share of the world's energy resources," Davis conceded in the preface to his report completed in 1982. He suggested that "until and unless the state of the art improves dramatically, we not attempt to quantify wilderness values in a detailed, scientific way." He did, however, propose various evaluation criteria: natural ecological processes, natural landscape, air and water quality, options for future generations, spiritual values, therapeutic values, primitive and unconfined recreation, solitude, mental and physical challenge, cultural values, historic values, educational values, economic values, wilderness watershed, unique ecosystems, and proximity to population centers.[2]

Regional Forester Charles T. Coston may have had these criteria in mind in 1981 when he rejected the CGG application (while approving proposals to explore outside the wilderness and in other disputed areas). But Chief R. Max Peterson directed Coston to review his decision, citing a provision in the Wilderness Act specifically allowing prospecting in wilderness through 1983.

Subsequently, on May 21, 1981, the Interior Committee of the House of Representatives invoked a clause of the Federal Land Policy and Management Act of 1976 (FLPMA), which gave it authority to protect public lands in an emergency. The Committee declared that an emergency did exist in the Bob Marshall, Scapegoat, and Great Bear, and that extraordinary measures must be taken to preserve values that otherwise would be lost. It directed the Secretary of the Interior to withdraw immediately affected lands within the Flathead, Lewis and Clark, and Lolo National Forests from all forms of disposition pertaining to mineral leasing.

The Secretary responded by making the withdrawal, with clear reluctance, but the firm he formerly headed, the Mountain States Legal

Foundation, then went to court, contending that the committee lacked constitutional authority to block mineral leasing. Resolution of this conflict clearly would be a long time in coming. It showed at once, however, that national forest wilderness, while accorded considerable protection under law, is not as secure as wilderness in national parks, and that national forests are open territory to mining of all kinds.

National forests present logical and inviting exploration targets. About 45 million acres, or one-fourth, of the National Forest System are considered to contain potential for oil and gas development, with 20 million acres already under lease. In some cases, the actual use of national forest land began prior to their protection. The famous Sespe oil field, which lies under both private and national forest land in southern California, was first developed in 1887. By 1900, when the Los Padres National Forest was two years old, more than 35 wells already had been drilled. Approximately 6.5 million acres throughout the national forests are underlain by 50 billion tons of coal; 300,000 acres by phosphate; 300,000 acres by oil shale; 17 million acres in national forests of the Pacific states and Great Basin by geothermal energy sources. Idaho national forests hold significant cobalt, and others in one part of the country or another contain copper, molybdenum, uranium, chromium, platinum-group metals, barium, lead, nickel, gold, tungsten, and antimony.[3]

In recent times the Carter administration spurred development by issuing leases on a massive scale and by sponsoring the Energy Security Act, or Synfuels Bill, which took effect in late 1980. One section of the bill directs that mineral exploration and drilling in the national forests must be given priority regardless of planning for other uses, ordering the agency "to process applications for leases of National Forest System lands and for permits to explore, drill, and develop resources on land leased from the federal government, notwithstanding the current status of any plan being prepared under section 6 of the Forest and Rangeland Renewable Resources Planning Act of 1974."

But the history of mining, and the problems that go with it, are deep-rooted. The Mining Law of 1872 is the foundation of national policy. That ancient document was enacted at a time in history when the country had more land than people and could afford to provide portions as encouragement to "the hardy miner opening new frontiers." Population in the 1870s was about 40 million, with less than one million in the West. The prospector of that day was interested in gold, silver, other precious metals, copper, or lead. His $100 for assessment work might be considered a fairly significant and bona fide contribution by standards of that day.

The 1872 law gives any citizen or candidate for citizenship the unqualified right to enter federal lands—except for national parks and other areas specifically closed by law or administrative action—and extract its resources. He is welcome to stake his claim, or claims, without fees or leases, required only to file with BLM and keep his claim current and conform to state law.

If he makes a valuable discovery, the miner can purchase the land at a modest price ranging from $2.50 (in a placer deposit) to $5 (if the mineral occurs in a vein or lode) per acre, providing he has spent $500 on each claim for improvements and complied with other legal requirements. If he finds minerals, he is permitted to hold the claim and occupy the site indefinitely, as stated above, without "going to patent," as long as he performs the annual assessment work of $100 per claim.

The Mineral Leasing Act of 1920 changed things to some extent and for some resources. The 1872 law still applies to such "hard rock" type minerals as gold, uranium, lead, zinc, and molybdenum. Rights to these valid "locatable minerals" become sole possession of the applicant, with no direct receipts to the Treasury. "Leasable minerals," including coal, oil, gas, phosphate, and geothermal resources, entail payment by lessees of annual rental plus royalty. Title to minerals and land remains with the United States, which has full discretion whether to make the resources available, i.e., to issue leases. Various administrations, lawmakers, and consumer groups have urged a leasing system for the hard rock minerals as well, but the lobbying influences have been too strong against it.[4]

Critics contend that the decision to lease or not to lease is the vital management choice relating to oil and gas development versus wilderness protection. Once drilling begins, they say, and especially if a commercial discovery is made, pressures and timetables prevent further environmental assessment or adequate planning. The ultimate decision on national forest land rests not with the agency or with the Secretary of Agriculture, but with the Secretary of Interior, who has delegated the authority to the Bureau of Land Management. The 1920 Leasing Act, applying to lands reserved from the public domain, provides for recommendations by Agriculture to Interior, and the BLM, in fact, has delegated environmental assessment activity to the Forest Service. The 1947 Mineral Leasing Act for Acquired Lands (applying principally to eastern national forests) provides that the Secretary of Agriculture must consent to issuance of leases.

During the 1950s new claims were filed in the national forests by the thousands, not for valuable metals, but for sand, gravel, cinder, and building stone. Large areas were disturbed in the search for uranium. Evidence presented to the Senate showed that in one three-year period, claims rose in Arizona by 700 percent, and in Utah and New Mexico by 400 percent. During 1955, claims were being filed at the rate of 5,000 every month. One speculator patented his sand and gravel claim at the edge of a growing city for $2.50 an acre, then sold it soon after for ten times as much.[5]

The most notorious case of the period, known as Al Sarena, took place in the rich Douglas-fir country of The Rogue River National Forest in southern Oregon. It began in 1948 when a syndicate applied for patents on twenty-three old lode mining claims which had not been worked for years. The Forest Service objected to some of the claims, provoking a celebrated issue that continued for six years and involved the White House itself. In 1954, Al Sarena, Mines, Inc., won its patent and by 1967 had cut 6.5 million board feet of timber. In the Pacific Northwest this practice was known as "green gold mining," a reference to a not uncommon timber harvest worth $50,000 on a 20-acre mining claim.

The Forest Service estimated that during the 1950s only 15 percent of all mining claims going to patent were used for commercial mining operations. Jurisdiction in issuing patents, as already stated, lies not with the Forest Service, but with the Bureau of Land Management, successor to the General Land Office of homestead days. The Forest Service has only the right to challenge any claim on grounds of insufficient minerals. But the issue of mineral discovery doesn't arise until the patent application is filed or when the claim conflicts with national forest administration. The law and court decisions require only that the discovery be sufficient to warrant that an "ordinarily prudent man" might invest additional time and money in an effort to develop a commercial mine. When legal requirements are met, issuance of a patent is mandatory, and the new owner receives full title to the land.

In this manner, whether through law, administration, or interpretation, numerous individuals staked claims for building stone in Oak Creek Canyon, one of Arizona's most scenic areas, located in Coconino National Forest. Their clear intent was development of summer cabins and recreation business. And much natural beauty was heedlessly despoiled and lost to the public before a special act of Congress eliminated Oak Creek Canyon from purview of the mining laws.

Then the Multiple Use Mining Act, or Common Varieties Act, of 1955 closed the gap to some abuses by eliminating the discovery of common sand, stone, gravel, pumice, or cinders as a basis for mining claims. It prohibited the use of unpatented claims for anything other than mining. Another provision enabled the Forest Service to undertake a vast reexamination of pending claims filed over long periods of time, and thousands were found to be invalid or abandoned.

The marked shortcoming in the Common Varieties Act and other mining laws is the inadequate provision for determination of relative values. Whether in letter or spirit, mineral extraction take precedence; asserting an equal role for other uses, or for environmental protection, presents a continuing challenge.

Protection of eastern national forests is complicated further since in many areas the government purchased only the surface rights, with individual landowners retaining the right to mine the subsurface, or to dispose of it to others. Fortunately, where minerals are controlled by the government, they are subject to leasing laws.

Reviewing the Public Land Laws

Despite growing public concern and pressures, Congress in the 1960s dodged meaningful revision of the mining laws. Instead, it established the Public Land Law Review Commission (PLLRC) in 1964 to embark on a six-year study of the laws and procedures of all federal land agencies. With Wayne Aspinall, the powerful chairman of the House Interior Committee and a known partisan of the mining industry, as chairman of the Commission, retention of the status quo appeared likely.

The PLLRC report, *One Third of the Nation's Land*, was issued in 1970 and recommended some reforms—in payment of patent fees and in perfecting mining claims—but no outright modernization of the 1872 Mining Law. It conceded weaknesses, but looked on them more from the viewpoint of the extracting industry than from that of the public. It included the following statements:

> Public land mineral policy should encourage exploration, development, and production of minerals on the public lands. . . . Mineral exploration and development should have a preference over some or all other uses on much of our public lands. . . . We also urge the establishment of a program to determine the extent of mineralization of public land areas where mineral activities are presently excluded but mineralization appears to be likely.

In the case of the Wilderness Act, Representative Aspinall (who had kept it bottled in committee year after year) insisted on a provision to allow mining as his price for support of the bill. December 31, 1983, was fixed as the cutoff date for new mineral locations, serving to stimulate prospecting in areas that had been combed many times over. The Wilderness Act does not affect previously existing valid mining claims. They may continue to be developed and worked and go to patent, but the patent applies to the subsurface only, the title to the surface remaining with the government.

Secretary of Agriculture Orville Freeman in 1966 set forth significant regulations to restrain mining from running roughshod over wilderness. Prospecting, he insisted, must be conducted in a manner "compatible with preservation of the wilderness environment." Persons with valid claims must obtain permits for access. Except in unusual circumstances, only primitive transportation—horse, mule, or burro—can be used. Where heavy equipment is needed, helicopter transportation may be required to ensure the least damaging approach.[6]

With the advent of the "energy crisis" in the early 1970s, pressures for prospecting and extraction rose sharply, as did the potential for damage, temporary or permanent, to soils, vegetation, and water resources. In 1974 the Forest Service issued regulations over mining, citing the Organic Act of 1897, which authorizes the Secretary of Agriculture to regulate occupancy and use in protection of the lands involved. These regulations require mining operators to submit plans describing proposed activities, the nature of disturbance anticipated, and steps to be taken in protecting the resources. FLPMA in 1976 gave a further basis for control over mining, since it granted the Secretary of the Interior authority to protect public lands "against undue and unnecessary degradation."

Under the old mining law, the only requisite for development is discovery of a "valuable mineral deposit." Critics contend that NEPA, FLPMA, and other conservation measures now provide different standards of judgment: that determination of a deposit's value must weigh impact on other uses and environmental consequences, as well as profits versus cost. But this is difficult to accomplish. For one thing, the Forest Service shares responsibility with the Interior Department agencies over the land it administers. For another, laws are subject to abundantly diverse interpretation.

The Wilderness Act, for example, declares in Section 2(a) that areas "shall be administered for the use and enjoyment of the American people in such a manner as will leave them unimpaired for future use and enjoyment as wilderness." But Section 4(d) adds: "Nothing in this Act shall prevent within national forest wilderness areas any activity, including prospecting, for the purpose of gathering information about mineral or other resources, if such activity is carried on in a manner compatible with preservation of the wilderness environment."

But what exactly constitutes compatibility? When mining was proposed in the Boundary Waters Canoe Area, the Izaak Walton League went to court and blocked it (*Izaak Walton League* v. *St. Clair*, 1973), with the judge ruling:

> A Wilderness purpose plain and simply has to be inconsistent with and antagonistic to a purpose to allow any commercial activity such as mining within the Boundary Waters Canoe Area. There can be no question but that full mineral development and mining will destroy and negate the wilderness or most of it. Even any substantial exploratory operation such as core drilling will require a means of ingress and egress, a communications system of some kind, the establishment of various camp sites, the importation of food, clothing, etc., power lines and the modification to a greater or lesser extent of the environment.

Administrations preceding that of Ronald Reagan had generally avoided issuing leases in wilderness, feeling that they contained relatively scant resources and that primeval places should be the last explored. One month into the Reagan administration in 1981, however, Interior Secretary Watt announced that no federal land was sacred and that all available sources would be considered for development. He was supported by an opinion from the Interior Solicitor, the chief legal officer, on October 19, 1981: "We do not believe the Secretary can decline to issue mineral leases or permits solely on the basis of a desire to protect the wilderness character of an area."

By early 1982, more than 1,000 oil and gas leases were being sought in 200 wilderness areas throughout the country, and additional applications were being filed. All around Yellowstone National Park pressure was brought on national forest wilderness and nonwilderness. In January 1982, the Forest Service recommended that leases be issued for drilling within the Washakie Wilderness, east of the national park. The decision ultimately would rest with Secretary Watt. As in the case of the Bob Marshall, however, public sentiment was strongly against it. Rep. Dick Cheney, of Wyoming, gave this expression in February 1982: "There is a general feeling in my state that much as we would like the economic benefits from the energy resources in the Washakie, we'd like even more to save a few acres and declare them off-limits—Yellowstone, the Grand Tetons and the wilderness areas around the parks, which account for less than 8 percent of the state. Wilderness areas should be the last place we look for energy."[7]

Case Histories Across the Country

Allegheny National Forest

This national forest in Pennsylvania was established following acquisition in the 1920s and 1930s of a few superlative virgin tracts and other lands that had been extensively cut for black cherry. In one respect the Allegheny is not typical, since many wells predate establishment of the national forest. The government purchased only the surface rights, leaving virtually the entire forest open to mineral development. Petroleum is the resource here, of a type used chiefly for motor oil and other lubricants. With the price hikes of recent years, extraction has increased sharply, with 500 new oil and gas wells drilled in 1980 alone. On the face of it, the Forest Service holds little authority; state law prevails, guaranteeing holders of mineral rights or their lessees "reasonable access," including the right to build roads, drill wells, lay pipeline, and establish power lines.

Even so, large independent firms like Pennzoil and Quaker State have been cited by the state for water pollution violations, and smaller ones, like the Minard Run Oil Company, for causing excessive damage to trees and terrain. The *Philadelphia Inquirer* reported on December 14, 1980, serious concern by Forest Service and state officials over erosion and runoff, choking stream beds with mud, endangering fish and aquatic life. The discharge of contaminated waste water into tributaries of the Allegheny River threatened this source of drinking water for Pittsburgh and other downstream communities. The clearing of too many trees for well sites, access roads, utility lines, pipelines, and storage tanks caused further erosion. Officials were worried about the future of the beautiful Tionesta area, believed to contain the largest virgin climax forest in the northeastern United States, an important resource in forest research—in which forty oil and gas wells had already been drilled, with more planned.

At least it could be said that in the previous month, December 1979, the Forest Service had obtained an injunction against Minard Run. That firm was ordered to halt operations until it furnished a detailed plan of operations, including road and pipeline access and means to control erosion and sediment. "The court ruling is a warning to oil and gas developers that a legal remedy can be used to avoid damage to surface resources," commented the U.S. attorney, Joseph Strauss, to the media following the court decision. "The increase in drilling has been rapid; this resulting in asking, 'What are the rights of surface owners?' In the past the question wasn't asked, though such rights existed all along."

Possibly the Forest Service should have moved earlier. "We fully expect private oil and gas owners, who own 97 percent of oil and gas rights under this 503,000-acre national forest, to exercise their rights," declared Supervisor John Butt, "but this clarifies and strengthens the rights of the United States as a surface owner."

Perhaps the most significant statement regarding the Allegheny was made by John Anderson, a U.S. Fish and Wildlife Service biologist: "You cannot cut a road through the woods without getting some erosion problems. You cannot drill a well without spilling some oil on the ground."

Nantahala and Pisgah National Forests

These forests in North Carolina are richly endowed in natural resources: hardwood forests, clear streams, abundant wildlife, and scenic mountain ranges. Mining has not been significant—though it may become so. As this possibility was being considered, the Hendersonville *Times-News* of October 6, 1980, editorialized:

> Based on Pisgah's special relationship to Western North Carolina, its special role as the watershed to Tennessee, North Carolina, the Gulf of Mexico and the Atlantic basins, this newspaper believes it is inadvisable to drill for oil either to test or to fully develop the oil fields—if any.
>
> It is simply too risky.
>
> The economic dislocation to cities and towns alone would be horrendous. That's not a consideration at this point.
>
> The key, this newspaper believes, is in the ability of the forest to handle another exploitation and survive. If Pisgah is damaged, by fire, oil spill, erosion or logging, it can unbalance the entire ecology and environment downstream.
>
> The entire effort isn't worth it, no matter the benefit to our nation's industrial economy, if the forest and, subsequently, the downstream country, is damaged.
>
> No matter the assurance that it is safe, there should be no exploration for oil in Pisgah National Forest. The risk is too great because of Pisgah's role in providing water to a regional civilization.

The editorial followed the announcement by the forest supervisor, George A. Olson, of plans to lease almost the entire Nantahala and Pisgah for exploration. The only areas excepted were a strip bordering the Chattooga Wild and Scenic River, withdrawn for mineral entry by Congress, congressionally designated wilderness areas, study areas, experimental forests, and municipal watersheds—and Olson recommended that the Chief lease oil and gas rights in wilderness.

Seismic surveys would be conducted, but the Forest Service in its assessment foresaw "relatively little impact on the forest environment." The second phase, drilling and development of any resources actually discovered, would require access roads, drilling sites, pumps, tanks, sediment ponds, rights-of-way for pipelines, and possibly utility lines. "We have been asked time and again if granting leases to the oil companies would block off areas from other uses, such as recreation," declared a Forest Service spokesman, Karl Tameler, when the proposal to lease was announced. "The answer is, 'No.' All other multiple uses will continue. In fact, exploration should be beneficial to all users of the forests. If access roads were built by the oil companies, it could open areas to everyone."

Bridger-Teton National Forest

By the beginning of the 1980s, this national forest in Wyoming was almost completely leased for oil and gas—about 1,500 leases covering 2.2 million acres, everything of the national forest except for two established wilderness areas and the proposed Gros Ventre Wilderness.

Mineral development was changing the face of Wyoming. New dirt roads pressed into wildlife country, with huge trucks hauling water, diesel fuel, and supplies. According to Forest Service reports, they were penetrating steep, high country, prone to avalanches and massive earth slides. At 9,200 feet elevation in the Kettle Creek drainage, near the east entrance to Grand Teton National Park, Exxon was operating a deep drilling site in grizzly bear and elk country. "The protection needs of grizzlies and/or their habitat will be considered in all watershed management problems, and in all mining and special use administration activities which have the potential to affect grizzlies and/or their habitat," read the guidelines for grizzly bear management. Toward that end, precautions were taken at Kettle Creek, though it's impossible to give constant supervision to contractors and subcontractors, men in a hurry and untrained in environmental ways.

At their best, immediate drilling sites show only bits and pieces of a much larger picture. Change comes on so fast that the best intended government official or oil company executive, or even the critical observer, cannot grasp its full dimensions and side effects. Spurred by energy development, Wyoming in 1981 was the third fastest-growing state in the nation. Ranch country was being replaced with boomtowns, like Rock Springs and Evanston.

The same sort of pattern was taking shape elsewhere in the Overthrust Belt and Inter-Mountain West. The largest underground coal mine in the West was being developed in Manti-LaSal National Forest, near Price in central Utah. In announcing the opening of this mine, the Interior Department (under Secretary Cecil D. Andrus during the Carter administration) said the environment would be fully protected; yet small farming communities were destined to be transformed into bedroom communities, while deer and elk range and fisheries on the Manti-LaSal faced inevitable impact and constriction.

Most public attention on conflict between wilderness and mineral development focused on renowned areas like the Bob Marshall and the Washakie. In 1981, however, the Rocky Mountain Region identified pending leasing applications in a score of wilderness and wilderness study areas in Colorado (Flattops, Raggeds, West Elk, Mt. Sneffels, Lizard Head, Eagles Nest, Piedra, San Juan, Spanish Peaks, and Sangre de Cristo) and Wyoming (Washakie, Glacier [Whiskey Mountain], Fitzpatrick, and North Absaroka). The lease proposals were being processed either through forest plans or environmental impact statements, with deadlines for completion ranging from February 1982 to March 1983, with decisions on all areas to be made by April 1983.

A Closer Look at the Bridger-Teton. During 1981 the National Wildlife Federation listed the Bridger-Teton National Forest among the nation's ten major endangered wildlife habitats, specifically because of oil and gas development. The Forest Service, for its part, continually undertook to tighten controls, which led to criticism from one side that they were too tough and from the other that they were too soft.

During the 1980s the national forest expects 500 to 1,000 miles of seismic lines to be completed each year. This may be conservative; in 1982 more than 500 miles of line were laid across a single ranger district, the Big Piney. In testing with explosives, a series of holes are drilled, four to six inches in diameter, 10 to 200 feet deep, with detonating charges of 5 to 75 pounds of dynamite at the bottom of each hole. In some cases it takes large drilling rigs moving over-road from one site to another to make this system work. Sometimes small portable drill rigs are flown in by helicopter. Another approach utilizes a gravimeter, an instrument that provides a key to underlying structures by detecting variations in gravitational pull.

Impacts of seismic exploration include dust from trucks and other vehicles, craters and ruts caused by drilling rigs, wildlife disturbance from explosions and overflying helicopters, and possible disruption of historical and archaeological sites.

Sometimes the priority goes to "scenery over oil," as in the case of the pristine Cache Creek drainage, a part of the national forest at the edge of Jackson, Wyoming. More than 2,000 citizens of the area signed a petition during 1979–1980 urging denial of an application for exploration filed by a Kansas firm, the National Cooperative Refinery

Association (NCRA). Their concern was understandable, considering that they live in one of the most attractive, unspoiled communities of the West.[8]

Opponents cited a 1947 ruling by the Secretary of the Interior of that period, J. A. Krug, that because of prime scenic and wildlife values in the Jackson Hole area, no wells should be drilled on public lands unless the Geological Survey expressed confidence in finding oil or gas. Even though NCRA had held a lease on the Cache Creek site since 1969 and the Forest Service recommended approval of the application, the company withdrew in the face of overwhelming community opposition.

"It shows that when the people of Jackson Hole unite to fight a definite threat to our economic base we can have an effect," declared the tourist-minded executive director of the Jackson Chamber of Commerce. "I welcome this decision," Senator Malcolm Wallop said. "It is my hope that other lessees in the same area will see fit to do the same thing and that the Forest Service, with whatever help it needs from the congressional delegation, will see fit not to open the area for leasing again."

The issue involving the Palisades area was less resolvable. In the spring of 1980 the Forest Service issued an Environmental Analysis Report (EAR) proposing to allow exploratory drilling in the Palisades, a rugged 290,000-acre portion of the Snake River Range on the Targhee and Bridger-Teeton National Forests southwest of Jackson. Some areas of very steep slopes were reserved, or deferred, with stipulations prohibiting construction of roads or erection of drilling rigs. Because the Palisades had been designated through the RARE II process for further planning, the Idaho Environmental Council and Sierra Club appealed the proposal; they argued that when the oil companies were through, the wilderness character of the area would have been lost. But the Mountain States Legal Foundation initiated legal action from the opposing position, charging the Forest Service with violating FLPMA by not reporting the withdrawal (of the steep slopes) to Congress.

It was difficult for the Forest Service to steer a clear course. An "interim directive" issued in 1981 declared that further-planning areas like the Palisades could be recognized as wilderness given one or a combination of certain conditions. These include the following: government and industry agree that exploration has been completed and proven unsuccessful; industry does not express interest in exploration; federal and state governments reduce their estimates of the area's energy potential on the basis of scientific exploration data; exploration and development can continue with acceptable impact on the wilderness character; the area could be reclaimed to a sufficient level of wilderness character after depletion of the energy resource; leases and other private mineral rights can be acquired through purchase or exchange by the government, where development would seriously damage the wilderness resource and full evaluation of relative values has been made.

The difficulties were implicit in the ambivalence of law. "Congressional intent seems clear not to accord primacy to the protection of wilderness values," as the interim directive stated, "at the expense of reasonable mineral leasing and associated operations until January 1, 1984, to provide a period for discovery and recovery of mineral resources by private enterprise."

Gunnison National Forest

Located in Colorado, this national forest embraces ski slopes above the resort community of Crested Butte. On a tract of private land above 8,000 feet, AMAX (originally American Metals Climax) discovered and proposed a $1 billion program to develop what it said was likely the world's largest deposit of molybdenum, a metal used to harden steel. The deposit lay on the slopes of Mount Emmons, 12,392 feet high, known as "Red Lady" (because it takes on a red cast at sunrise). The High Country Citizens Alliance and other opponents, sparked by the mayor of Crested Butte, W (no period) Mitchell, complained that it would critically impact the town and its tourist industry. AMAX would hollow a portion of the mountain above Crested Butte removing 20,000 tons per day in order to extract a molybdenum concentration of less than one-half of one percent. This would mean the construction of a mill to refine and extract the mineral, with large quantities of water required for the process and for mixture with immense residues, or tailings, to be pumped into ponds.

On August 7, 1981, AMAX announced it would defer the project for at least two or three years because of a soft molybdenum market. Mayor Mitchell celebrated by helicoptering to the summit of Mt. Emmons and asserting that public sentiment was a factor in the company decision. "We demanded they do the job right, and that means higher costs," he said. "There are other deposits they will explore first."

Tongass National Forest

This forest embraces Alaska's Misty Fjords National Monument with some of the wildest and most spectacular scenery on the continent. When Congress passed the Alaska Lands Act of 1980 (formally, the Alaska National Interest Lands Conservation Act), it designated a Misty Fjords Wilderness of 2,136,000 acres, specifically excluding 149,000 acres of the national monument which President Carter had earlier proclaimed. The reason was to enable the U.S. Borax and Chemical Corporation (a subsidiary of the British conglomerate Rio Tinto Zinc) to pursue a molybdenum claim that might prove to be one of the largest known deposits in the world.

The Borax molybdenum claim had been a focal point in the bitter Alaska lands fight. The Wilderness Society called the exclusion "a major setback." The Society's publication, *Living Wilderness*, in its Spring 1981 issue charged that:

The company's mineral claims total no more than 2,500 acres, yet the act excludes nearly 150,000 acres in Misty Fjords from wilderness. The exclusion is 10 times the size that could possibly be necessary for mining operations, and it jeopardizes all three of the area's vital salmon-spawning rivers—the Wilson and Blossom to the north and the Keta to the south. The act guarantees that a permit to construct a road will be issued within 21 months, regardless of the necessity of the road or the adequacy of pre-construction baseline studies and impact analyses.[9]

The law gave Borax five years to prove its claim. The company expected to spend about $37 million on exploratory drilling by the end of 1981. It planned to construct the mine in 1984 and start operation late in 1987. The life of the mine at the Quartz Hill site is estimated at 70 years, with an ore processing rate of 60,000 tons per day. At early 1980s molybdenum price, Borax estimated the total gross value of the deposit at $25 billion.

During 1981, Forest Service personnel conducted frequent fish sampling surveys of the Blossom River. Environmentalists have voiced concern that disposal of tailings into the rich fishing grounds of southeast Alaska could harm that valuable resource. The Borax operation is expected to produce less than three pounds of molybdenite per ton of rock, while the remainder must be disposed, either on land or in adjacent fjords. Another concern has been the possibility of building a town near the hitherto isolated area of Quartz Hill for employees and their families. About 1,000 persons were expected to be required for construction, and 800 for the mine operation.

Don Finney, Ketchikan manager of the Quartz Hill project, was quoted in the *Anchorage Times* of September 6, 1981, on the possibility of environmental issues causing a delay in the project. Finney, chairman of the Alaska State Chamber of Commerce, had been a top executive with the Louisiana Pacific Corporation (one of the largest purchasers of national forest timber in Alaska) before joining Borax. Finney said that environmentalists "are criticizing everything we do," but he did not see them stopping or slowing down development: "The law mandated it. I really don't think a judge is going to give them a stay. And with the climate now in the Reagan Administration, I don't think we'll have any trouble with administrative rulings from the Forest Service."

Biomass—Wood Energy Potentials and Problems

For the 200 years prior to 1940 more than half of all the wood harvested from the nation's forests was used as fuel, so the ideas of biomass—all tree and woody shrub materials from root tips to leaf or needle tips—as an answer to oil supply depletion and of "wood energy opportunities" are not new. What is new and challenging? Between 1973 and 1981, fuelwood harvests on the national forests increased 10-fold.

The 1981 harvest, an estimated 5.3 million dry tons, is equal to the energy in 15 million barrels of home heating oil.

"A National Energy Program for Forestry," prepared by the Forest Service in 1980, reports that already more than 1.5 percent of the 80 quads of energy used annually in the United States, comes from wood and wood byproducts. (A quad is the measure of a quadrillion British Thermal Units, or BTUs.) By 1990, the Energy Program forecasts, an additional 5 quads could be derived from currently unused forest biomass, thus providing markets for surplus tree growth, logging residues, material removed in cultural and management activities, processing residues, and urban wastes. But this forecast is based on the assumption that one-half of the currently unused biomass could (1) be recovered economically, and (2) have favorable environmental impacts.

The Forest Service indicates three goals in this field:

1. Production annually of 6.4 quads of energy from forest biomass by 1990.
2. Conservation annually of 2.0 quads of energy by 1990 through improved efficiency in the production and use of wood products.
3. Development and implementation of environmental protection guidelines in connection with production and use of wood for energy.[10]

The Society of American Foresters has also taken a look at biomass potential. In a 1978 report by a Society Task Force on Forest Biomass as an Energy Source, a supply of 9.7 quads of energy is foreseen by the year 2000, while still providing for other needs from the nation's forests. And there could be an even larger supply, given the cultivation of biomass plantations. This seems to indicate the same approach to industrial tree farming that critics have questioned. The SAF report takes note of this point: "We caution, however, against excessive resource uses for energy or other products in the interest of short-term gain. Overuse could seriously reduce the nation's forest biomass growth and disrupt the forest such as wildlife, aesthetics, recreation, and general environmental amenities. . . . Further evalution of nutrient drain resulting from intensive harvests must be made."

A cord of wood (a measure for stacked wood four feet by four feet by eight feet) has the energy equivalent of 3.5 barrels of oil, which explains the substantial enthusiasm for burning wood, and the advent of various government incentives. The New England Pilot Fuelwood Project, as an example, was launched in 1979 through the Agricultural Conservation Program in an area of high demand (and high petroleum prices) to provide technical assistance to landowners wishing to harvest firewood. Various aspects of this cooperative pilot program were designed to encourage firewood cutting in overstocked stands to improve the quality of the stands, to discourage unguided cutting with possible erosion and destruction of wildlife habitat, and to reduce pressures for firewood on public and private forests.[11]

Cutting wood for home consumption is the simplest aspect of using wood as fuel (though growth in demand ultimately may cause a new set of problems). Under other programs, federal agencies have embarked on broad and ambitious efforts to encourage large-scale use of biomass as a source of oil-substitute fuels. But leaders of the forest products industry and environmentalists, in what might be called a curious alliance, share apprehension over how far this can or should go. Speaking at a Bio-Energy Conference at Atlanta in April 1980, T. Marshall Hahn, Jr., president of the Georgia Pacific Corporation, warned:

> Our greatest concern is that in the rush to replace oil and gas with any possible alternative, our government will subsidize an uneconomic use of wood as fuel. In doing this, it could disrupt existing forest products industries and create shortages of building and paper products. It could easily establish a wood-for-fuel industry forever dependent on subsidies to keep it afloat.
>
> Although wood is our greatest renewable resource, it is not an unlimited resource and it cannot solve our energy problems. If we were to burn the entire annual growth of our commercial forest, we could replace less than 10 percent of our energy requirements each year.

Ronald J. Slinn, vice president of the American Paper Institute, expressed similar concern in a speech before the American Institute of Chemical Engineers in August 1980:

> There is strong argument for what forest-based industries currently do—turn half of a tree into high value end products, such as lumber, plywood and fiber, and use the residues for a large proportion of the energy required to make those products. Burning the residues close to where they are generated makes economic sense. It would not be in the nation's best interest even to run the risk of sawlog trees being cut up into firewood.[12]

How this is apt to work in practice may be illustrated in these two case histories:

In Michigan, two power companies and the State Department of Natural Resources joined in 1980 to promote construction of an experimental 25-megawatt wood-fired power plant near Hersey, in lower Michigan. The boiler would be fueled with wood mostly from state forestland within a 75-mile radius.

The Committee for Rational Use of Our Forests was organized by citizens in the region to study the proposal. Following 18 months of review, this group declared that 50 percent of homes in the area already were using wood as a heating fuel, with 50 to 70 percent efficiency (much higher than the proposed plant); therefore, priority should be given to getting the greatest "energy mileage" out of fuel resources.

The plant would require 400,000 tons of wood yearly, enough fuelwood to provide heat and hot water for 20,000 homes; yet the

Committee for Rational Use claimed it would waste 80 percent of the heat generated. Likewise, the prescribed system of whole-tree chipping removes leaves, twigs and stems from the forest, which otherwise would replenish the soil.[13] "Go slow," the Committee urged, "before endangering small loggers, the budding firewood industry, wildlife habitat and tourism."

In Vermont, Burlington voters in 1978 approved a $65 million bond issue to enable the Burlington Electric Department (BED) to proceed with planning for a 50-megawatt wood-chip-fired plant and two other related power plants (one based on waste incineration, the other on hydroelectric power on the Winooski River). This operation was estimated to require 470,000 tons of wood annually. "Harvesting low quality trees will improve the quality of residual growing stock," according to a study by BED's consulting firm. It presumed the availability of Vermont state forests, Green Mountain National Forest, and 2.5 million acres of privately held land as sources of supply, stressing the potential of short rotation "energy plantations."

The Chittenden County Committee on Energy, and later the Burlington Friends of Safe Energy, however, questioned the assumption that residents would double energy demands by 1986, in the face of national and local trends to the contrary. These groups also asserted that for each 10-hour working day it would take 20 truck arrivals and departures at the proposed incinerator and wood-burning plant, to be sited in a floodplain favored by wildlife. Richard M. Brett, a well known naturalist of Woodstock, Vermont (formerly an investment banker), made this comment in a personal letter:

> The long-range consequences of heavy machinery to harvest wood in Vermont are: soil compaction; erosion and siltation; destruction of growing stock; depletion of soil; increase in flood potential; decrease in forest capital, and damage to the tourist trade. Hidden costs to the public include expense of building roads and bridges and controlling pollution. The final question is: If all direct and indirect costs are included, is this capital-intensive system of wholetree harvesting sound, even on a narrow economic basis?

Saving and Stretching Energy and Timber Supplies

Experience of the pulp and paper industry in meeting rising fuel costs of the 1970s provides evidence of the potential implicit in improved utilization and conservation. Pulp and paper constitute one of the country's most energy-intensive industries—ranking third, after primary metals and chemicals—in purchased fuels.

In 1972 the industry set a target of reducing consumption of fossil fuels and substituting residues of the manufacturing process, to be measured continually by an Energy Monitoring System of the American Paper Institute. Resources formerly classed and discarded as "waste" were put to work. By 1980 spent liquors left by chemical pulping provided more than 37 percent of the industry's total energy consumption; when the industry's use of bark and hogged fuel are included, the combined contribution of wood-based residues equaled 48.1 percent. For 1980 alone this was equal to saving 120 million barrels of oil.[14]

For the eight-year period of 1972 through 1980, production of paper and paperboard rose by 11 percent, while purchased energy consumption was reduced by 14 percent—including a 25 percent reduction in fuel oil use alone. The industry forecast continued gains toward energy self-sufficiency and reduction of fossil fuel consumption.

Resource scarcity has indicated opportunity for conservation at various points along the route of forest harvest and production. A 1980 report of the Research and Sawmill Improvement Program (SIP) of the Forest Service estimated that it is theoretically possible to save 186 to 257 million cubic feet of softwood logs, the equivalent of 1.3 to 1.8 billion board feet of lumber. This is based on reducing the amount of wood lost during processing by .13 to .18 inches per slice when producing nominal 2-inch stock (dressed to 1.5 inches). The Forest Service has been encouraging the use of available technology to achieve this goal. Improvements initiated at 680 mills, producing almost 25 percent of the national softwood output, has already increased production by about 8 percent from the same volume of logs.[15]

Meanwhile, on the national forests felled wood—often of usable standards—left by purchasing firms has long been a problem, creating fire hazards, delays in reforestation, and resulting in cost increases in management. Under the Wood Residue Utilization Act of 1980, purchasers can be required to remove residues which they do not purchase to points of prospective use in return for "residue removal credits" against what they actually purchase. The five-year demonstration program authorized by the Act may provide worthwhile lessons.[16]

Better utilization of forest resources is one of the principal targets of the Forest Products Laboratory at Madison, Wisconsin. In a paper titled "Forest Products Research in the Forest Service" presented before the annual meeting of the Forest Products Research Society, San Francisco, July 9, 1979, and published in *Forest Products Journal*, October 1979, R. L. Youngs, Director of the Laboratory, wrote (page 57): "It is now possible to increase per-acre harvest of wood fiber as much as 50 to 100 percent by removing logs, tops, and branches that are commonly left in the forest. It is also possible to improve efficiency in use of wood products in construction and in manufactured items. But to do these well requires research."

He listed research with the following objectives:

1. Design efficient and environmentally acceptable harvesting systems.
2. Economically use low-quality and dead trees and residues.
3. Extend the resource base and conserve energy by improving processing efficiency.

4. Devise more efficient and economical methods for wood construction and protection of wood in use.
5. Develop efficient, economical methods of using wood as a source of energy and chemicals.
6. Provide a base of economic feasibility data from which to evaluate new concepts in forest products utilization.
7. Increase capability to evaluate trends and project demands in forest products consumption and trade.

Successful projects initiated at the laboratory include the following:

• Truss-framed system (TSF) for light-frame construction (mostly in housing) uses less structural framing lumber than conventional building and eliminates the need for posts and beams in the basement. The new system combines a trussed floor system, trussed roof system, and conventional wall studs into a unitized frame. These specially engineered components, when connected with sheathing, create a wood-frame building with maximum strength and durability. (And it takes less time to erect than a conventional house.)

• Press-dry papermaking uses heat and pressure to bond the stiff fibers of 100 percent hardwood pulp into high-strength corrugated fiberboard boxes. Press drying, combining heat and pressure to bond pulp fibers together, uses less energy than conventional processes. More than one-half the timber harvested in the United States goes into paper manufacture, and of that volume one-half is used to produce corrugated box material. Other potentials of the new process are under study, including the use of recycled fiber.

• Best Opening Face (BOF) increases the yield of sawmills for a given batch of logs by 10 percent or more, especially with small logs. The system employs log scanners, computer-made sawing decisions, and high precision controlled network. Nationally and internationally, most major new or expanded sawmill operations incorporate the BOF concept to eliminate waste.

• Structural flakeboard utilizes volumes of forest materials considered uneconomic to process into lumber and plywood. A major Forest Service research program has shown it feasible, technically and economically, to manufacture structural sheathing products from both logging residues and low-quality trees. Under fire exposure, flakeboard panels exceeded endurance requirements for exterior walls of one-family and two-family dwellings. They also met Uniform Building Code recommendations for maximum load and allowable deflection values for floor and roof sheathing.

Other than increased use of forest or mill residues, the final opportunity for increasing the life of wood products is through recycling, a field of unlimited and untapped opportunity. Urban wastes are estimated at 70 million tons annually, of which three-quarters is waste paper. About 20 to 30 percent of the paper generated annually is now used for fuel or is recycled, but this is low by European standards.

Forest Service research has produced studies showing that recycling urban waste is an economically feasible source of fiber for both paper and fiberboard products. The U.S. currently produces about 67 million tons of paper, with about 20 percent, or 13 million tons, being made from recycled paper fiber. Not all paper (such as toilet paper) can be recycled, but 50 percent—or 30 million tons of paper fiber—can be. Since one ton of wood yields about one-half ton of paper (a 50-percent yield), 60 million tons of round wood could be saved by recycling.

The primary factors inhibiting further recycling are said to be the costs of collection and the difficulty of removing contaminants. Various communities, however, have established successful recycling centers. They earn money by selling newsprint, glass, and metals, while eliminating, or at least reducing, the need for landfills and the cost in land and transportation to keep the landfills in operation. Maine's successful returnable bottle law is said to have reduced the amount of waste going into the state's dumps by as much as 10 percent.[17]

Some Forest Service facilities in recent years have consciously endeavored to improve their energy efficiency and reduce consumption of oil, natural gas, and electricity. A new ranger station, completed in Montana in 1980, utilizes passive solar energy for more than half its heating needs. In August 1981, the solar-powered, wood-burning Forestry Sciences Laboratory and Berea Ranger Station (of the Daniel Boone National Forest) was dedicated at Berea, Kentucky. A total of 4,500 square feet of solar panels provide 90 percent of hot water, 40 percent of heat, 80 percent of cooling. Much of the building is illuminated by natural light from skylights, while two wood-burning boilers supplement the solar collectors. A hillside protects the building from winter winds.

Such conservation measures and use of alternative energy sources provide substantial potential for further reduction in the demand for oil, gas, and even domestic coal. Few projections of energy demand made in the 1970s correctly estimated the amount of energy conserved in 1980. Projections of demand tend to be overstated, setting in motion "self-fulfilling prophecies." But positive steps increasingly indicate potential for saving and stretching resources, and by so doing saving the forests from which they come.

13

RESEARCH AND PUTTING IT INTO ACTION

> The time is not far off when the United States will have to weigh its aspirations against the limits set by availability of materials, technological prospects, and environmental needs. The soundness of the decisions made then . . . will depend to some degree upon the capacity of forest and range research groups to tighten and perhaps refocus their operations to meet increased responsibilities more efficiently.
>
> —S. Blair Hutchison and M. B. Dickerman,
> in a paper presented before a symposium on
> forest and rangeland resource policies,
> May 30–June 3, 1977.

The Beginnings

Research and federal forestry began together in 1876, when Congress appropriated $2,000 for a study and report on the forest situation and the best means to preserve and renew forests. Dr. Franklin B. Hough, the consultant hired by the Department of Agriculture to conduct the study, subsequently issued three reports, all of which were widely read and circulated to increase public sentiment for conservation.

This effort marked a signal departure from previous research activities conducted by botanists, explorers, and others who had been preoccupied mainly with identification, nomenclature, and description of trees and shrubs of North America. Peter Kalm, for example, a pupil of Carolus Linnaeus, the master naturalist of Europe, had been sent to this country during colonial days by the Swedish Royal Academy. In 1750 he wrote in his *Travels into North America:* "I found that I was now come into a new world. Whenever I looked to the ground, I everywhere found such plants as I had not seen before. When I saw a tree, I was forced to stop and ask those who accompanied me how it was called. I was seized with terror at the thought of ranging so many new and unknown parts of natural history."

219

Expeditions of discovery conducted by parties led by pioneers like Lewis and Clark, John C. Fremont, and John Wesley Powell were scientific investigations as well, ranking in their time as vast as outer space may appear today. Picture General Fremont in 1842, with Kit Carson as his guide and a full military expedition, reaching the gap in the Rocky Mountains now called South Pass, below the Wind River Range in western Wyoming. Fremont was drawn to the higher peaks by their shiny ice caps and determined to scale the highest. He and five companions rode beneath a perpendicular wall of granite working their way to the summit, then dismounted and climbed on foot in thin moccasins made from buffalo skin. Ever dramatic, Fremont drove a ramrod into the rock and "unfurled the national flag to wave where never the flag waved before." He chose a spectacular site (Fremont Peak in the Bridger Wilderness), but was mistaken in assuming it to be the highest point in the Rocky Mountains.

Despite all the advance in knowledge, the unknowns still outweigh the knowns. The research program of the Forest Service is based on the idea that the better the knowledge the broader becomes the span of alternatives and technologies. From its beginnings in 1876, the research wing of the Forest Service has grown to become the largest agency in the world devoted to developing scientific knowledge about forest and range resources. Research projects are variously practical and theoretical, short-term and long-term in nature; they directly support programs of the National Forest System and those of state and private forestry, and go beyond.

The total research budget is more than $115 million annually, underwriting 4,000 specific studies simultaneously. Many studies use sophisticated computer and technology transfer systems. Research personnel are in contact with virtually every aspect of domestic and foreign forestry. They are grouped into 240 research work units at eight regional experiment stations, the Forest Products Laboratory at Madison, and 81 different centers of research (66 connected with cooperating universities), conducting work on 93 experimental forests and ranges and 146 natural areas. Through these facilities, the Forest Service is involved in basic research in such areas as forest genetics, fire, insects, diseases, tree physiology, soils, and water. Fields of practical as well as theoretical study include timber management (the most prominent and substantially financed), watershed, recreation, surface mining, range, wildlife and fish habitat, engineering, and products utilization. Many procedures now accepted as good practice have come from research laboratories and field stations.

Forest products research was initiated to solve a practical problem. In 1889, the Carriage Builders Association declared that the industry was running short of supplies of northern oak; the industry felt that southern oak, while plentiful, lacked the characteristics needed for carriage construction. The Division of Forestry (as it was then called) of the Department of Agriculture commissioned Professor John Johnson, of Washington University in St. Louis, to research the problem. He did, confirming that suitable material could be obtained from the South as well as the North.

This problem-solving kindled a relationship among Johnson, Bernhard Fernow, head of the Division, and Filibert Roth of Michigan, who joined the Division in 1892 (and later served in forestry education at the University of Michigan), leading to a government program of "timber physics" research and information. In the early 1900s, under Pinchot's leadership, forest products research moved into naval stores, chemistry, wood preservation, timber testing, and papermaking. Small research units proliferated in various parts of the country, ultimately to be coordinated through establishment in 1910 of the Forest Products Laboratory.

Research at forestry schools and state agricultural experiment stations started early in the century, concurrently with the blossoming of the Forest Service. The schools acquired tracts of timber for training and experimentation and, in many cases, made research a part of the school program. But "the great leap forward" in research came in 1915, when Earle H. Clapp succeeded in gaining autonomy for the Branch of Research, which he headed, from national forest administration. Clapp argued that research was being used as a dumping ground for incompetents unwanted elsewhere and, furthermore, that freedom from administrative pressures was critical to objectivity and credibility. Based on this principle, highly qualified researchers have turned up choice independent findings, but getting them applied in practice has proven something else again.

Research programs at cooperating educational institutions received a boost from the McSweeney-McNary Act of 1928 and its amendments and supplements, leading to the McIntire-Stennis Act of 1962. These laws enable the Department of Agriculture to support cooperative activity with private organizations, universities, forestry schools, and state experiment stations. In recent times, however, new pressures on forest and range resources have resulted in legislation changing the framework for planning, developing, evaluating, and executing forestry programs—including research.

RPA and NFMA have laid down concepts of periodic resource assessment, based on best available data, and of development of sustained programs for use and protection. The Food and Agriculture Act of 1977 reinforced the tie between agriculture and forestry, and the concept of research programs aimed at effective use of renewable resources for consumer good, energy, and supporting trade and industry. Three laws adopted by Congress in 1978—the Renewable Resources Research Act, the Cooperative Forestry Assistance Act, and the Renewable Resources Extension Act—added legislative encouragement to research and to using fully its results.

The effectiveness of research, as well as its contents and direction, have been subject to question. Over a two-year period, 1977–1978,

Forest Service programs were examined in four regional conferences and a national conference in Washington, D.C., with participants representing government, professional, environmental, conservation, consumer, and industry clientele.

"Critics in recent years have leaned very heavily on federal and state agricultural research," declared a background paper presented by S. Blair Hutchison and M. B. Dickerman, Forest Service research specialists, at a 1977 symposium conducted as part of this review by the Renewable Natural Resources Foundation at Airlie House, Warrenton, Virginia.

> Perhaps the most widely quoted criticism is one by the National Academy of Sciences "that much of agricultural research is outmoded, pedestrian and inefficient." [1972 Committee on Research Advisory to the Department of Agriculture.] Some forest and range research no doubt deserves such criticism. Some is excellent. At any rate the statement does raise a fair question that should be explored. . . . What it amounts to is that if the United States is not to wake up one morning in 1995, or in some other future year, with a raw materials—natural resource crisis on its hands, the renewable resource professionals, including scientists, will have to provide leadership heretofore lacking.
>
> Unfortunately, the transmission of knowledge about ecology, wildlife, timber management, wood products, and other nonagricultural renewable resource subjects has, in contrast, been much less effective than transmitting research results to the farmer through county extension agents. Forest Service research, for example, has been criticized because, allegedly, its findings are not widely applied. Even if discounted 50 percent such statements are damning.

The two-year regional and national planning effort, conducted in cooperation with the National Association of State Universities and Land Grant Colleges, resulted in the National Program of Research for Forests and Associated Rangelands and in four regional program documents. The results have been said to tie directly to RPA by laying out key research needs on the basis of the present and projected forest research situation.[1]

Will such research reflect balanced use and the ecosystem approach? Or perhaps be tilted, as in the past, toward commodity production? In the field of forest products, one objective is to design efficient and environmentally acceptable harvesting systems. Among lines of investigation is equipment designed to harvest trees efficiently under difficult slope and soil conditions. Other phases of research address the question of whether trees *should* be harvested under difficult slope and soil conditions. But the two research programs seem unrelated and almost extraneous to each other.

The Forest Environment Research Group is one of seven research staff groups located in Washington, with responsibility for providing guidance and direction in six designated research areas: fish and wildlife;

range; urban forestry; outdoor recreation; forest hydrology; and rehabilitation of disturbed mine lands. While it has the leadership role for
approximately 50 projects nationwide and participates in another 25
projects led by other research staff groups, forest environment research
receives a low priority and low funding.[2]

Research in the Natural Areas

The Forest Service has been a leading influence in the research
natural areas (RNA) program, coordinated by the Federal Committee
on Ecological Reserves (under the auspices of the National Science
Foundation and Council on Environmental Quality). The purpose of the
RNA network is to preserve a representative cross-section of all significant
natural ecosystems as baseline areas against which other areas subject
to man-caused changes can be measured.

In addition, research natural areas serve as gene pools for rare and
endangered species of plants and animals. They provide the means for
studying environmental changes caused by air pollution, weather modification, and alterations in groundwater levels, and their effects on
natural vegetation. Research natural areas are used frequently by colleges
and universities both for research and field education in forestry, biology,
botany, soils, wildlife, and hydrology.

The Forest Service began setting aside research natural areas in
1927, continually designating new ones, particularly during the 1970s,
with a total of more than 140 established by the beginning of the 1980s.
They range in size from ten or twelve acres to several thousand acres.
In 1980, as an example, the 7,000-acre O'Hara Creek Natural Area was
established near Lowell, Idaho, on the Nezperce National Forest as the
eleventh natural area in the Northern Region. Thus O'Hara Creek, south
of the Lochsa River, became the first large, relatively undisturbed stream
in a research natural area in Idaho, containing a large number of aquatic
and terrestrial plant communities.

The tracts in the National Forest System are units of a larger federal
system of more than 440 natural areas (administered by the Fish and
Wildlife Service, National Park Service, Bureau of Land Management,
Department of Defense, and Tennesee Valley Authority), representing
a cross-section of vegetation types, soil types, fish and animal habitats,
land forms, and mineral deposits. In turn, they are also part of a
worldwide system under the auspices of the Intenational Union for the
Conservation of Nature (IUCN).

The administering agencies do not encourage public use of natural
areas, though some peripheral nature trails and interpretive signs have
been established. Picnicking, camping, swimming, hiking, and gathering
rocks, plants, nuts, and berries are generally discouraged and often
prohibited. Hunting, fishing, and trapping are not encouraged, but are

permitted subject to state regulation, except on restricted lands like the national parks.

In 1966, the Forest Service began significant research studies of the Boundary Waters Canoe Area (BWCA) in northern Minnesota as a means of better understanding and managing wilderness. The virgin forests of the Canoe Area were recognized as being among its truly great resources; these last remnants of the old north woods offer the only chance to travel for days through forests undisturbed by modern man. About fifty different studies were undertaken in conjunction with the Fish and Wildlife Service, Macalester College, the Minnesota Conservation Department, University of Minnesota, and Quetico-Superior Wilderness Research Center. During two summers of work, botanists collected data on nearly 200 species of trees, shrubs, herbs, mosses, and ferns from 106 upland natural plant community examples.[3]

But this work only demonstrated how much remained to be done. In the Spring 1969 issue of *Naturalist Magazine*, Charles T. Cushwa, leader of the Northwoods Wilderness Recreation Project, sponsored by the North Central Forest Experiment Station, reported:

> Forest Service scientists still must describe plant communities in the lowlands and disturbed parts of the BWCA as they did the upland natural plant communities. The forest history studies must continue in an effort to determine *when* different plants "come in" on an area following disturbance. Our future efforts are being expanded to include studies on mammals and birds living in the BWCA to learn more about their relationship to the various plant communities.

Still another aspect of the research involved social and economic studies of the effects of visitors. With the number of visitors increasing steadily, researchers focused attention on problems such as water pollution, deterioration of heavily used campsites and portages, increasing pressures on the fisheries resource, and conflicting interests among users themselves. Mr. Cushwa observed: "Overuse in some places is not only degrading the environment but is probably decreasing the quality of the outdoors experience for some visitors. We are now planning studies to determine optimum capacity and distribution of visitors so the public can continue to enjoy the area—so that we can have our wilderness and use it too."[4]

In this case, research contributed to national forest management. It furnished a basis of data for limiting use and types of use of the Boundary Waters Canoe Area. Administrators tend to complain that research lacks relevancy to their needs and that researchers are too independent of the National Forest System. On the other hand, studies produced by Forest Service scientists have brought out ecological reasoning not always welcome to or applied by national forest managers.

The Input of Ecological Reasoning

In 1973, for instance, the Forest Service requested the Environmental Protection Agency to lift its ban on the use of DDT in the Northwest. The agency proposed to employ this chemical in order to subdue the tussock moth, a native insect of the Northwest, which already gave indications of collapsing during the third year of infestation. Administrators insisted that the insects would cause tremendous damage to the Douglas-fir forests before dying off, and that DDT would prevent their spread.

Managers had available to them a research report, "Major Outbreaks of the Douglas-fir Tussock Moth in Oregon and California," specifically intended to guide foresters in their handling of the infestation. It did not support the argument that DDT is necessarily effective against tussock moths, nor did it contend that chemicals of any kind play a major part in population collapse:

> To prevent this defoliation and possible added tree mortality before population collapse, control has often been applied, but always during the declining population phase. Thus, the actual effectiveness of control applied at that stage of an outbreak can be difficult to demonstrate. For example, limited comparisons in California of two chemically treated with two untreated areas show no significant differences in total tree mortality.
>
> The most severe tree damage occurred in the second year. Additional loss of foliage before population collapse in the third year was usually of minor importance in terms of total impact.
>
> Justification of large spray programs has frequently been based on the assumption that outbreaks will spread extensively beyond a population center or "hot spot" unless controlled. This, of course, assumes that the outbreaks are often the result of spread from some point of origin rather than the natural buildup of a resident population. Historically, there is no sound evidence to indicate that tussock moth outbreaks ever expand much beyond the boundaries of the initial infestation.

On January 4, 1974, the Forest Service reapplied for permission to use DDT, citing the threat of imminent economic loss to the region—based on the loss of timber, dangers of fire and watershed disruption, and diminishing use of defoliated areas for recreation. Permission was granted; more than 400,000 acres were sprayed at considerable financial expense, and probably at a considerable ecological cost as well.

During the period of controversy over clearcutting, the research program of the Forest Service produced a variety of studies to give serious pause. The best known of these was conducted by F. Herbert Bormann of Yale and Gene E. Likens of Cornell at Hubbard Brook Experimental Forest, a unit of the White Mountain National Forest. It was a challenging study dealing with nutrient cycles of an ecosystem. "Our results," they reported in the *Scientific American* of October 1970, indicate that the capacity of the ecosystem to retain nutrients is dependent on the maintenance of nutrient cycling within the system. When the cycle is broken, as by the destruction of vegetation, the loss of nutrients is greatly accelerated. This effect is related both to the cessation of nutrient uptake by plants and to the larger quantities of drainage water passing through the system.

The conclusion they reached is simple, yet fundamental to using the earth's resources:

> Failures in environmental management often result from such factors as failure to appreciate the complexity of nature, the assumption that it is possible to manage one part of nature alone and the belief that somehow nature will absorb all types of manipulation. Good management of the use of land—good from the viewpoint of society at large—requires that managerial practices be imposed only after a careful analysis and evaluation of all the ramifications. A focus for this type of analysis and evaluation is the ecosystems concept.

Bormann and Likens did not criticize clearcutting *in toto*. They raised questions about possible detrimental effects on sites vulnerable to nutrient loss through leaching and erosion. Even so, they stirred criticism and attack from various quarters of the forest community. As independent researchers of standing, they could take it and continue their studies. In advance of its 75th anniversary in 1980, the Forest Service signaled its intention to recognize individuals making special contributions to forest progress. Robert S. Pierce, project leader in charge of watershed management research at the Durham, New Hampshire, laboratory, nominated Bormann and Likens, and they were indeed honored for adding significantly to knowledge of forests and forest ecosystems.

"I think, in terms of forest ecology, the Hubbard Brook study has opened a new dimension," said Bormann in 1982.

> We have enlarged the perspectives of what ecology takes into account. This project is an attempt to think wholely, and there really aren't a lot of them around.
>
> In the past, most biological ecologists focused only on the biological fraction of the system. They'd go and look only at forests, only at the populations that composed them. But we took the forest and linked it into biogeochemical cycles, things like hydrology, what came in through the rain, what left in evaporation, and what nutrients were removed from the system. Our studies have more effectively linked the forest ecosystem to the larger biogeochemical cycles. It's important in our attempts to understand man's impact on the forest.[5]

Hubbard Brook has become a well-used facility. Over the years it has provided the basis for a new research technique designed to draw information from the forest, a system Bormann calls "experimental ecosystem ecology."[6]

"The key to Hubbard Brook," in Pierce's view, "is the ability of a large group of people to share their information. If a person is studying insects, for example, he or she doesn't have to look at the weather or the soils or the vegetation. They can draw from the background of information that has been obtained. There is a mutual exchange of information, to the benefit of all."[7]

Jerry F. Franklin, principal plant ecologist at the Forestry Sciences Laboratory, Corvallis, Oregon, and Dean S. DeBell, research forester of the Crown Zellerbach Corporation, also studied clearcutting. In a report on "Effects of Various Harvesting Methods on Forest Regeneration," presented at a symposium at Oregon State University, August 1, 1972, they said that rarely do biologic considerations necessitate large-scale clearcuts:

> Do any types, species, or sites have characteristics that necessitate the use of clearcuttings? Biologically, no types or species appear to require large clearcuttings for successful regeneration—by "large" we mean clearcuttings that exceed 10 acres. Extensive clearcutting becomes increasingly unde- pendable as environmental conditions become more and more limiting. Consequently, using such a technique on severe sites is unwise, particularly if no dependable method of artificial regeneration is available.[8]

The Pacific Northwest Forest and Range Experiment Station reported successful trials of natural regeneration following shelterwood cutting in portions of the high Cascades in Oregon, where regeneration, both natural and artificial, had failed after clearcutting. Workshops conducted by Timber Management Research in West Virginia in 1975 and California in 1976 endeavored to clarify options in solving a multitude of silvicultural problems and to keep research ahead of practice.

One of the most practical studies, conducted in 1971 on "Forest Land Stratification," provided improved methods for classifying national forest lands for their best use. This study, involving six western national forests, showed that acreage suitable and available for timber growing to be 22 percent lower than previously estimated—that many areas were not high-volume timber sources, but were more valuable for wildlife, aesthetics, and watershed.

Robert W. Harris, then Director of the Intermountain Forest and Range Experiment Station, described the study as a step toward refinement of methods and criteria for quality planning and management of national forest land. It showed that improved information gathering for truly coordinated resource planning is needed, and that determining soil characteristics and steepness of slope are vital first requisites in decision making on logging. The authors of the report, J. H. Wikstrom and S. Blair Hutchison, put it this way:

> It could be said in short that the problem is inadequate financing in basic data collection, analysis, and planning. The problem is more than

one of financing for timber inventories. The timber resource cannot be described meaningfully unless it can be described within the context of the land, ecology, and land-use situation. Thus, there must be balanced financing for soils-hydrologic surveys, ecological habitat surveys, and resource inventories.[9]

Making such findings available in digestible and useable form has been a challenge in itself. A 1972 investigation by the General Accounting Office claimed that the Forest Service was not getting the products of research into use rapidly or efficiently, which led to the introduction of the technology-transfer process as part of the emerging pattern of the 1980s. The use of technology transfer likely would not be limited to research, but would extend Service-wide as a means of making useful data available, though actually putting it to work is something else.[10]

"Research application and technology transfer are becoming more demanding and specialized," R. L. Youngs, Director of the Forest Products Laboratory, declared in a paper presented at the annual meeting of the Forest Products Research Society, held in San Francisco in July 1979.

Research must increasingly be measured in terms of whether it really helped solve the problems used to justify it.

Those of us in the federal research structure, supported primarily by the taxpayers, are increasingly aware that the public expects us to do more useful work, more economically and more efficiently than ever before. And, through all of this, the standards and expectations of science are continually rising to new levels of sophistication and excellence. What a fantastic time to be in research!

Researching Cause and Effects of Acid Rain

The warning signals of Forest Service research bear heeding at times by society in general, as well as by the immediate forest community. Areas subject to acid precipitation, or "acid rain," have been widening steadily over the past three decades. Evidence strongly indicates man-made emissions of acid-forming materials—sulfur dioxide and nitrogen oxides emitted from the smokestacks of utilities and smelters and from automobile exhausts—as the primary causes. Fish and other aquatic life have declined sharply in affected areas. Acid rain is suspected of endangering forest growth, degrading soils, and affecting human health.

Scientific warnings were expressed at a four-day International Symposium on Acid Precipitation and the Forest Ecosystem, conducted in May 1975 at Ohio State University, with 300 participants from 12 countries. The symposium was jointly sponsored by the Atmospheric Sciences Program of Ohio State and the Northeastern Forest Experiment Station, with key figures including Gene E. Likens, professor of ecology at Cornell; Garth K. Voigt, professor of forest soils at Yale; Leon S. Dochinger, principal plant pathologist at the Experiment Station at

Delaware, Ohio; and Thomas A. Seliga, director of the Ohio State program.

Dochinger and others later showed that rainwater, particularly in the eastern United States, has become increasingly acidic and dangerous to forest growth. He warned that most pollutants from the burning of fossil fuels are being transported long distances so that no place is immune. In a paper presented before the Third National Conference on the Interagency/Environment Research and Development Program, held in Washington in 1978, Dochinger and two collaborators (Likens and Norman R. Glass, of the Environmental Protection Agency at Corvallis, Oregon) warned further that increased use of coal, as provided in President Carter's National Energy Plan, " . . . will cause an increase in atmospheric emissions of some or all of the precursors of acid precipitation, even if best available control technology is implemented on new sources and old sources are retrofit."[11]

Subsequently the National Academy of Sciences reported that evidence of acid rain's "serious hazard to human health and the biosphere" was so clear that continued emissions of sulfur and nitrogen oxides at current levels would be "extremely risky from a long-term economic standpoint as well as from the standpoint of biosphere protection."

Such expressions of concern led to passage of the Acid Precipitation Act of 1980, which established a ten-year program (the National Acid Rain Precipitation Assessment Program, or NADP) to be conducted by an inter-agency task force. Under NADP, assigned goals are to identify causes and sources of acid precipitation; evaluate environmental, social, and economic effects; and propose a program of action to limit or eliminate sources and remedy or ameliorate harmful effects.

The Forest Service plays a key role in monitoring acidity levels and studying ecological and biological effects, much of it cooperatively with universities and other public agencies. But when and where does the evidence provided by research become strong enough to compel action? In October 1981, the National Wildlife Federation reported on a study of its own. The federation said it had determined that fifteen of twenty-six states east of the Mississippi River were "extremely vulnerable" to the harmful effects of acid rain, while another ten were "moderately vulnerable," and only one, Florida, could be considered as "slightly vulnerable."

These ratings were said by the federation to be based on a study of each state's rainfall acidity, geology, soils, and water chemistry to estimate the potential for acid rain damage to fisheries, soils, crop foliage, and marble and limestone structures within the states. The study evaluated each state's visibility impairment—as indicated by sulfate and nitrogen oxide concentrations found in the air—and auto paint damage resulting from acid rain corrosion.

"Our study should lay to rest once and for all the claim that acid rain damage is confined to a few hundred lakes in upstate New York," declared the executive vice-president of the Federation, Jay D. Hair. "President Reagan's recommendation to Congress that acid rain merely be studied further represents a disastrous waste of time and research money. The studies have been done." Possibly Hair did not have all the evidence, yet his warning that, "It can't be long before every state is affected," suggests an urgency to respond to the warning signs of research with positive action.

14

COOPERATIVE STATE AND PRIVATE FORESTRY

> If the National Association of State Foresters agrees to a funding scheme which exclusively favors "demonstrated" economic efficiency, it will encourage a departure from the stewardship intent of the law which authorizes the Cooperative Programs, and contribute significantly to more procrastination of investments upon which good stewardship depends.
>
> —Tom Borden, Colorado State Forester, in a white paper to all state foresters, February 9, 1981.

> A town is saved not more by the righteous men in it than by the woods and swamps that surround it.
>
> —Henry David Thoreau.

Purpose and Beginnings

National forests have become the focal point of public attention, and of controversy, in issues of policy and practice, but they comprise only one part of a much larger picture. Forest and rangelands in private and nonfederal public ownerships constitute by far the major part of the nation's 1.7 billion acres of forest, range, and related watershed lands.

The scope and significance of these nonfederal holdings are not truly reflected in Forest Service activities. Cooperative programs through the State and Private Forestry (S&PF) wing of the agency received appropriation support in 1981 of $73 million, as compared with $1.258 billion for national forest management, and $115 million for research.

Nevertheless, these cooperative programs cover broad fields of rural forestry assistance, forestry incentives, insect and disease control, urban forestry assistance, rural fire prevention and control, management assistance, planning assistance, and technology implementation. These are all justified (some perhaps with more validity than others) on the basis of direct and indirect benefits to society in the form of income, employment, tax revenues, and social values from the production of timber and other commodities, the protection of watersheds, and other services derived from well-managed lands.

Of the nation's commercial forest land, the Forest Service administers approximately 17 percent, and other federal agencies an additional 4 percent. State, county, and municipal governments hold only 6 percent, as compared with 14 percent by forest industries and approximately 58 percent (of 488 million acres) by non—forest industry private owners (including coal companies, railroads, and land development companies).

Cooperative activities began almost on the day when Pinchot joined the Department of Agriculture in 1898. The federal lands were then administered by the Department of the Interior, and Pinchot had nothing to manage, for the moment, but his own ideas. These he put to work forthwith with Circular No. 21 of the Division of Forestry.

It was an invitation to lumber companies, offering technical advice on how best to exercise their own interests. "Mr. Pinchot is planning to introduce better methods of handling forest lands in public and private ownerships, the private owners paying the expenses of Department agents who give instructions," wrote Secretary of Agriculture James Wilson. The program was well received, with applications from all parts of the country keeping the little Division of Forestry busy. It was the mainstay of the Division until transfer of the forest reserves in 1905 gave Pinchot and associates millions of acres to manage directly.

Cooperative activity received another boost with passage of the Weeks Act of 1911. Though remembered mainly for bringing national forests to the East, it also authorized expenditure of $200,000 in federal matching funds (equivalent to about $1.9 million in 1981 dollars) for states with forest protection agencies that met government standards. The Weeks Law enabled the federal government to acquire land for the protection of watersheds of navigable streams, but these matching funds were made available to support protection of nonfederal lands for the same purpose. The uniform federal standards and matching funds became significant influences in stimulating state forestry.

Although Pinchot and his followers pressed for federal regulation over private lands, the Clarke-McNary Act of 1924 set a course of cooperation that ultimately prevailed. Regulation as an issue would be succeeded by efforts at more effective controls by the states.[1]

Industrial Forestry Patterns and Power

Early in this century, industrial forestry was based on readily and endlessly available supplies of wood. Timber scouts went first, sending back news of towering virgin forests. Then followed the sawmills and logging railroads. The flatlands of Mississippi and Louisiana were easy to exploit, and by the 1920s the land there had been cut barren. The big mills had departed, leaving in their wake fierce fires fed by resinous slash scattered on the ground. The magnificent pine forest was replaced by a raw jungle of scrub oak, rattan, and catbriers. Logging firms were little more than migrants who did their work and moved on when it was done.

In "Appalachian Comeback," published in *Trees, The Yearbook of Agriculture* for 1949, M.A. Mattoon depicted the scene:

"Handsome timber in increasing amounts fell to the axe, but there always seemed to be more. Sawmill towns sprang up in their temporary ugliness, thrived, and vanished as the cutting moved on. When Europe burst into the horror of warfare in 1914, demands on the forest mounted, and postwar reconstruction saw no let-up. So the large sawmills, accompanied by many little sawmills, marched across the face of the remaining Appalachian wilderness."

Equipment grew larger and more sophisticated. The geared Shay locomotive, "Model-T of the woods," was ideal for hauling heavy loads over the steepest mountain grades almost anywhere in America. There was the steam-powered Clyde skidder and also the sensational McGiffert log loader. They were effective, but fires were started by sparks from wood-burning trains and skidders. During the dry season the tops and branches scattered over the ground flashed into flame, and the sparks flew from one mountainside to another, with no way to control them. By the 1930s, devastation had spread across the land. In the Depression years vast holdings were left to county and state ownership for nonpayment of taxes.

Suddenly the industry which had leapfrogged from one virgin forest to another found itself pressed against the last timber frontier, the Pacific Northwest. Favorable tax legislation in 1943, coupled with war-borne timber shortages, brought the industry to undertake timberland ownership as a continuing source of raw material. But other factors influenced this decision.

National forests served as testing ground and demonstration areas, where many forestry practices were first tried. Cooperative programs with states and private landowners then extended application of protection and management principles beyond the boundaries of national forests.

Provisions of the Weeks Law of 1911 and the Clarke-McNary Act of 1924 furnished the framework for cooperative programs. These laws were later supplemented by the Cooperative Farm Forestry Act of 1937; Lea Act of 1940 (white pine blister rust control act); Forest Pest Control Act of 1947, and the Cooperative Forest Management Act of 1950. With the promise of improved protection, particularly for seedling and sapling stands, landowning companies were encouraged to harvest timber in ways favorable to natural restocking. They began to show an interest in growing timber as well. For example, when Champion Paper Company came to western North Carolina, company officials not only installed their own system of fire protection, but were conscious of the wisdom of cutting selectively, leaving some trees as a green protection against fire and as a means of regenerating spruce, rather than leaving regeneration to a new cycle of succession starting with fire cherry.

Enlarged opportunities for the forestry profession introduced systematic programs in the place of hit-or-miss operations. Since its organization in 1920, the National Association of State Foresters has promoted cooperation among private landowners and government agencies for improved technical management. In 1966, Bernard L. Orell, a vice-president of the Weyerhaeuser Company and a former forestry educator, estimated that nearly 10,000 professional foresters were involved in industrial forest management as compared with fewer than 1,000 only fifteen years earlier. "This to me is an unparalleled achievement in resource conservation," he declared. "It is a tremendous base from which to carry forward more intensive management, even better forest fire protection and control methods, and to refine through research utilization of the timber supply and the products which result." Of course, in the fifteen-year period, forestry education changed considerably. Until World War II, students had been heavily dosed with the early conservation idealism, along with silviculture, management, and forest administration. Most graduates went to work for state and federal agencies. With the new postwar industrial demand, schools shifted emphasis to training technologists, engineers, and economists.

Industrial forestry approaches to resource husbandry vary widely. Some programs are run by foresters, others by cost accountants responsive to corporate directors far from the land; in the latter cases, decision makers may never have been in the woods. Some firms cooperate wholly with fish and game agencies and public groups like the Nature Conservancy to protect habitats and unusual ecosystems, while others rely on advertising and public relations to justify drainage of wetlands, conversion of mixed native stands to plantations of pine, and the use of toxic herbicides and pesticides. Their systems are not geared to the caution expressed in 1980 in a personal letter by Dr. Carl Reidel, a prominent educator and former president of the American Forestry Association: "Large-scale, capital-intensive mechanization in the woods and mills will be unwise investments in an era of energy shortages and an uncertain future of inflation and tightening environmental controls. We will have to discover ways to decentralize and diversify forest industry, favoring multiple products and small plants that are not dependent on large quantities of uniform materials over long periods."

Forest products firms control huge areas and exercise considerable political influence. Most of the Maine woods is owned by half a dozen corporations. In the South, companies like International Paper, Bowaters, and Weyerhaeuser controlled about 5 million acres in 1945 (by ownership, leases, and cutting rights), but in the ensuing twenty years expanded their control to more than 25 million acres. And that was before

TABLE 14.1

Private Forest Landowners
Number Of Owners By Size Class Owned

SIZE CLASS (ACRES)	NO. OF OWNERS (THOUSANDS)	PERCENT
1-9	5,528	71.3
10-49	1,164	15.0
50-99	464	5.9
100-499	538	7.0
1,000-9,999	21	.5
10,000 +	2	--
TOTAL	7,757	100.0

Weyerhaeuser acquired the Dierks Company, with 1.8 million acres of mixed stand astride the Arkansas-Oklahoma border, and proceeded to convert it to plantation management. International Paper, said to be the largest landholder following the federal government, owns nearly 8 million acres in the United States and Canada, with an added 15 million acres in Canada under government license. As for influence and potency, while William Ruckelshaus served his first appointment as Administrator of the Environmental Protection Agency, Weyerhaeuser officials criticized him sharply for his efforts to restrict the use of chemicals, but after he left the government they hired him as a senior vice-president.

Nonindustry Forests Are Different

Private owners of small woodlots hold an estimated 58 percent of the 488 million acres of commercial forestland. Tables 14.1 and 14.2, based on Forest Service data, picture private ownership by size classes. For a long time it was assumed that owners of small woodlots, through one means of inducement or another, would provide a steadily larger share of the nation's timber, principally by meeting management needs—in terms of tree species and age classes—of the mills. A large part of the Cooperative Forest Management Program of the Forest Service was directed toward that end.

TABLE 14.2

Private Forest Landowners
Acres Owned By Size Class Owned[1]

SIZE CLASS (ACRES)	ACRES OWNED (MILLION)	PERCENT
1-9	11.0	3.3
10-49	28.1	8.4
50-99	32.9	9.9
100-499	102.6	30.8
1,000-9,999	47.6	14.3
10,000 +	84.0[1]	25.2
TOTAL	333.1	100.0

1/ Includes 55.0 million acres of industry owned land.

A minority of foresters, inside and outside the Forest Service, urged another course. In his book, *Woodland Ecology: Environmental Forestry for the Small Landowner,* published in 1975, Leon Minckler noted the criticism of small woodlots as being "unstocked or poorly" as compared with plantations subject to "intensive management" on large industrial holdings. He reasoned that on private forests, averaging 70 acres in size, use of heavy machinery is not feasible; instead, professional skill is required, particularly to protect old hardwoods and so-called "cull" trees vital in providing food and denning and nesting places for turkey, squirrel, raccoons, deer, owls, and scores of bird species.[2]

L. Keville Larson, a practicing consulting forester of Mobile, Alabama, expressed the viewpoint in the mid-1970s that private woodlands could furnish the best quality timber in natural stands, while serving other purposes as well. He cited figures of the Southeastern Forest Experiment Station showing that miscellaneous private ownerships, consisting primarily of natural stands, have been building their growing stock by 40 percent per year; industry ownership, however, with a much higher percentage of plantation, has less volume per acre and has been building growing stock by only 9 percent per year.

"Although with an understanding of silviculture, natural stands can be managed for wildlife, recreation, watershed, or any other single or multiple use objective," Mr. Larson declared, "natural stands will continue to be of great importance because the most common objective will be return to the owner.

"Selective management is much more intensive (than even-aged). It is practical forestry based on the judgment of men in the woods, following the basic guidelines of partial cutting, insuring full stocking, increasing the growth rate, improving the stand, considering each acre, and selecting the treatment that fits the forest condition and ownership objective."

"Improving Outputs from Nonindustrial Private Forests," a 1979 report prepared by a task force of the Society of American Foresters, largely supports these views:

> Many programs—ranging from distribution of literature to management plans—strongly emphasize the "needs" of forestland. Foresters are perhaps too concerned about "sustained yield" and regular outputs; these are valid concepts but more appropriate for large public and industrial holdings than for the small holdings with short owner tenures.
>
> Not enough attention is given to whether specific actions are economic on a specific owner's tract, fit his objectives, or match availability of his investment and operating capital. All too few assistance programs present or discuss management options, relative investment requirements, immediate and long-term revenues, and other vital interests. . . . Heavy equipment required in some situations multiplies the problems. Because of the unsightliness and possible lasting damage of logging, many owners are reluctant to sell timber."[3]

Who are the private owners? The composition has changed. Farmers have switched to town and industrial jobs, and old farm and country estates have passed to heirs who are often scattered, or into other hands. Many of the owners are teachers, doctors, lawyers, and citizens who enjoy the outdoors. They want a forest environment and an investment: an opportunity to make money selling timber in harmony with environmental values. Growing trees with high-quality timber, the small landholder can expect a regular income at short intervals without heavy investment in machinery.

Cooperating with the States

Cooperative programs have gone through internal structural changes. In 1966 the Forest Service established two area offices to serve 33 eastern states, where most private forestland is located. This action was based on the idea that S&PF is not a land management group but works through state foresters. As part of the 1966 reorganization, liaison with wood-using industries was transferred from research to S&PF, with a number of programs for technical assistance to loggers and processors emerging since then.

Then, in 1982, the Southern Area office, headquartered in Atlanta, was merged with the Southern Region; under direction of the regional forester it would continue to coordinate its work in 13 states, Puerto Rico, and the Virgin Islands. The Northeastern Area office at Broomall, Pennsylvania (a Philadelphia suburb), coordinates work in 20 north-central and northeastern states. (Only 7 percent of the land in the Northeast is in federal ownership, but large tracts are in state forests and parks.)

State forestry programs have been growing on various fronts, including timber production, environmental protection, wildlife enhancement, and recreation. In the 1940s, a renewed effort toward regulation had led many states to adopt "seed tree laws"—so named because they required that a specific number of desirable trees be left during timber harvesting for regeneration. But these "laws" were generally considered inadequate and unenforceable. The 1970s, however, saw an upsurge of stricter legislation.[4]

The Oregon Forest Practices Act of 1971, for example, directs the State Forestry Board to develop minimum standards in reforestation, road construction and maintenance, harvesting, application of chemicals, and slash disposal. It appears to be effective, with a high level of compliance. The California Forest Practices Act, adopted in 1973, is comprehensive and far-reaching.[5] The California State Forestry Board is empowered to regulate the size and shape of areas on which even-aged management is practiced. Logging operators must obtain permits and every timber harvest must include a plan prepared by a professional forester, with the plan providing for regeneration, prevention of erosion, and protection of streams and unique areas.

A number of states now almost completely support programs to furnish technical guidance and assistance to landowners, wood processors, and loggers, based on initial support from federal funds. Several states have begun their own cost-sharing programs financed by special taxes and state appropriations.

Recent national forestry laws have authorized intensive planning at state and regional levels. Of particular note are the Forest and Rangeland Renewable Resources Planning Act (RPA) of 1974 and the Cooperative Forestry Assistance and Cooperative Forestry Extension Acts of 1978.

The Cooperative Forestry Assistance Act reaffirms the cooperative principle and consolidates previous programs. It authorizes the Secretary of Agriculture to cooperate with state foresters to provide assistance for advancement of forest resource management; encouragement of timber production; prevention and control of insects and diseases affecting trees and forests; prevention and control of rural fires; efficient utilization of wood and wood residues, including recycling of wood fiber; improvement and maintenance of fish and wildlife habitat; and planning and conduct of urban forestry programs.

One section of the act authorizes the Secretary to provide financial and technical assistance to state foresters to strengthen their organizations. It encourages natural resource planning, with opportunity for the governor, legislature, and citizens to participate. It identifies issues, establishes goals and objectives, considers alternatives and effects, and increases the state forester's ability to participate in other planning efforts.

Some of these programs have lacked budget support. The entire expenditure for S&PF was reduced by 10 percent from 1981 to 1982 and faced a proposed cut of 25 percent for 1983. Organization Management Assistance (OMA), which provides aid in budget planning, training, personnel management, information and education, and computer programming, received $1 million in 1981, $775,000 in 1982, and faced a proposed reduction to zero in 1983, until state foresters stirred a strong protest.

The State Forest Resources Planning Program (SFRP) is a data source for the RPA. This led Tom Borden, Colorado State Forester, to express his apprehension, through a white paper sent to all state foresters on February 9, 1981, to the effect that funding would be based on increased wood production—at the very time when their agencies were broadening to encompass balanced use.

The Indiana Forest Resource Plan, issued in February 1981, reflected the first comprehensive analysis of the status of timber, fish and wildlife, outdoor recreation, water and related resources on four million acres of forestland. It consists of an assessment of resources, planning issues, and alternative programs for public comment. It shows some of the problems, too. For example, though Indiana leads all states in the volume of high-grade fine hardwoods, the rate of harvest exceeds the rate of growth, due to conversion of forestland to cropland.[6]

Jerry J. Presley, Missouri State Forester, in a paper presented at the Wildlife Management on Private Lands Symposium May 4–7, 1981, at Milwaukee, also discussed the potentials of federal aid:

> The USDA, working with the states and within federal jurisdiction, can greatly influence state and private forestry programs through directives, policies and regulations. For example, most states are in the process of preparing a state forest resource plan as well as a plan that provides the basis for federal funding of state programs.
>
> If wildlife management is to ever receive strong consideration in state and private programs, it would seem imperative that wildlife habitat must be fully addressed in any state RPA document. It is well within the jurisdiction of the USDA to require states to include wildlife habitat considerations in state RPA plans. States hold the key, however, to bringing wildlife and forestry together. State forestry departments and state fish and wildlife agencies must first get together to decide what they want to happen.

In Missouri, resource foresters work with biologists (both in the state Department of Conservation) and with forest landowners to enhance forest amenities, wildlife included. As Mr. Presley told the Symposium:

> Our resource foresters estimate that 80 percent of the people that request forest land management assistance are furnished with some degree of wildlife management assistance and advice. Last year total requests covered roughly 240,000 acres; so the yearly potential, in terms of landowner interest, is high. We have many standard wildlife management recommendations that are made to owners concerning such things as logging deck areas, timber stand improvement, old fields, planting, pine conversion areas, wildlife water holes, balancing size classes during cutting operations, protecting areas from grazing and fire, and den trees.
>
> Most of our foresters have been trained by our biologists to be able to readily identify forest stand conditions that offer the best opportunities for improving wildlife habitat. Also, our wildlife biologists are available to provide more detailed wildlife management recommendations when they are needed.[7]

In Penn's Woods (and Pinchot's)

The interest and influence of the Forest Service is not always evident, but these are pervasive virtually everywhere that forestry is practiced and forests grow. Public foresters, however, are not all in Forest Service employment. Many work for the states and counties, a few for cities and towns. Generally they earn less money than federal counterparts, but they are where they choose to be, close to the resource and to home, with less paperwork and with more opportunity for local decision making. Many of their activities and actions have significance reaching beyond their jurisdictions.

During the early 1980s the forested Pocono Mountains of northeast Pennsylvania became the setting of the first nongame project of the Pennsylvania Game Commission. In cooperation with the Pennsylvania Audubon Council, the Game Commission was endeavoring to reintroduce the osprey, a great bird of prey whose population had been decimated largely through the effects of DDT in the food chain. Though the osprey once ranged over most of Pennsylvania, the Poconos were chosen because of the sanctuary furnished by the woodlands.

Delaware State Forest covers almost 77,000 acres in several sections, a considerable portion of the Poconos. The primary industry of the region is recreation, with eighty percent of all of Pennsylvania's resort facilities "keeping the Poconos green." Besides contributing to tourism, 3,440 acres of the state forest are part of municipal watersheds. Approximately 56,000 acres are classed as commercial forest, capable of producing a usable timber crop, and are managed, as directed by state law and the administrative code of the Department of Environmental Resources, "to provide the maximum sustained yield of high quality timber products." On these 56,000 acres, the goal of management is to improve the quality and growth rate of the timber, sustain the production of timber by balancing the cut with the growth, and improve the forest for other resource uses.

Wildlife constitutes one of those other uses. The bear population in the Poconos is among the highest in the East. Turkey numbers have increased along with mast production from maturing oak forests. Squirrels also have benefitted from mast availability. Snowshoe hare and ruffed grouse are not abundant, but are present nonetheless, along with such furbearers as beaver, muskrat, mink, and opossum. Through a cooperative agreement signed in 1970, the Pennsylvania Game Commission and Department of Environmental Resources are committed to a program of optimum habitat conditions for a balanced wildlife program.

Gifford Pinchot was familiar with the Delaware State Forest. Not only was it handy to Grey Towers, the Pinchot estate at Milford, but in 1920—ten years after his dismissal by President Taft in Washington—he was named Commissioner of Forestry for Pennsylvania and proceeded forthwith with a major reorganization in his own image. "Keeping the public well and correctly informed about forestry" at once became a principal function of the Department of Forestry, as one might expect. Thus Pinchot left his legacy with two forestry institutions, one federal and one state.

Pennsylvania's state forests now cover almost two million acres. Pinchot fought to obtain a state bond issue of $25 million for the purchase of about 3.5 million acres of nonproductive forest land. He insisted that this would insure "future citizens of the state against timber want and . . . bequeath them a timber heritage necessary for their prosperity." Despite his typically energetic publicity efforts, Pinchot as a commissioner and later a governor (1923–1927, 1931–1935) was unable to get a constitutional amendment authorizing the bond issue. He did, however, spur an accelerated land acquisition program while he ran the Department of Forestry, under which 77,544 acres were purchased by the state at an average cost of $2 per acre.

Pennsylvania, despite its heavy industry and large population centers, is still 58 percent forested, with approximately 17 million acres of forestland. But, except in isolated tracts, there is scant resemblance to the great colonial stands of Penn's Woods. In 1860 Pennsylvania led the nation in lumber production, after which large cut-over areas were left abandoned and forest fires ran uncontrolled. By 1900 Pennsylvania was importing lumber from other states to fill its needs.

In 1897, when little of value remained, the legislature voted to establish a system of forest reserves in the headwaters of the main rivers of the Commonwealth: the Delaware, Susquehanna, and Ohio. The Delaware State Forest became one of the first of the 19 state forests. (The earliest purchase, of 1,521 acres, was made in 1898 for a price of 15 cents per acre.)

During his tenure, Pinchot saw more than timber in trees. While he had shown little concern for recreation as Chief Forester in Washington, in one of his popular Pennsylvania lectures he stated: "The woods are the people's playgrounds and were intended by Nature as such. While the forest is growing lumber it is also furnishing for the people who live in the cities and towns the refuge that attracts when play time comes." He approved funds for campsites and supported the establishment of seven state forest parks, nine state forest monuments, and two special scenic areas. He envisioned these and others not only as wholesome environment for outdoor activity, but as ideal outdoor laboratories for study of forest types.

In this way he laid the basis for a kind of state forest wilderness system for both recreation and research. Today, thirteen Wild Areas, covering 106,735 acres, are roadless and undeveloped, open to non-motorized public access for hiking, hunting, fishing, and the pursuit of peace and solitude. An additional forty-four Natural Areas, covering 57,710 acres, are considered of unique scenic, historic, geologic, or ecological value, to be maintained in natural condition, usually without direct human intervention.

Pennsylvania has undertaken extensive planning for the protection and wise use of its forest resources. The planning unit called the North Central High Mountain Area (NCHMA) covers approximately 2 million acres of land, one of the largest forested regions between the East Coast megalopolis and the Chicago-Detroit-Cleveland urban complex of the Midwest. It consists of 1,093,000 acres of state forest and state park lands, 173,924 acres of state game lands, and 733,076 acres owned by private individuals, sportsmen's organizations, and the forest products industry. With such solid blocks of public land, it constitutes a unique resource and an opportunity for rational long-term planning.

Resource-based recreation in the NCHMA is furnished by twelve state parks, seven Wild Areas, fifteen Natural Areas, four major hiking trails totaling 173 miles, primitive camping, choice canoeing and rafting, numerous short hiking trails, three cross-country ski trails, two offroad vehicle (ORV) trails, and snowmobiling (which is allowed on designated state forest roads). Wildlife resources (including Pennsylvania's only elk herd) provide major hunting and fishing and add significantly to the local economy.

Though logging once devastated the area, almost denuding the tree cover between 1870 and 1920, protection and restraint have provided the way for a new forest industry to operate compatibly with recreation, wildlife, and watershed. The value of the industry (including 76 sawmills producing 88 million board feet of lumber and 5 pulp mills utilizing 50,000 tons of chips and 219,000 cords of pulpwood annually) has been estimated at more than $60 million.

Recognizing rising demands and finite capacity, the Department of Environmental Resources defines its management objective: "To maintain the present remote and natural character of the North Central High Mountain Area in harmony with public needs and the economic and social well-being of the region." The details are designed to conserve a life-style as well as the resources.

"Cooperative Forestry programs continue to show a gradual movement away from what has been considered traditional forestry and toward urban and community forestry, whereby environmental concerns often outweigh the production of wood fiber." So notes the 1980 annual report of the Pennsylvania Bureau of Forestry. The report adds:

> The fiber resource and dollar value are often less important to the owner than some of the other woodland benefits. Many landowners rate wildlife, aesthetics, or just the security of owning land as their primary objective and fiber production a secondary asset. This does not indicate a waning interest in forestry, but just the opposite. There is a definite, steadily increasing interest in forestry and the environment. The number of landowner requests for forestry assistance is increasing. Last year, Bureau foresters assisted over 3,000 Pennsylvania landowners and prepared forest resource management plans on over 26,000 acres. These plans were written to be compatible with the landowners' objectives and multiple use of their woodland including: timber, water, wildlife, recreation, soils and aesthetics.[8]

Planning is a major factor in management of the nearly 2,000,000 acres of state forests. A system called the Forest Resource Plan is in effect for 1970 to 1985, designed so that it can be revised or updated at any time within the 15-year management period. A critical review reveals a clarity of purpose and expression. The Forest Resource Plan is intended to establish policies and objectives for all forest resources, coordinate their use and development, and eliminate conflict between uses. Coupled with the need for increased production of basic forest resources is the implicit recognition of social change: increased mobility and leisure, concern over the environment, renewed interest in the outdoors, and pursuit of new forms of outdoor recreation.

Besides the basic legislative mandates dealing with economic and social values of state forests, a 1971 amendment to Pennsylvania's Constitution adds new meaning to the development of the Forest Resource Plan: "The people have a right to clean air, pure water, and to the preservation of the natural, scenic, historic and esthetic values of the environment. Pennsylvania's public natural resources are the common property of all the people, including generations yet to come. As trustees of these resources, the Commonwealth shall conserve and maintain them for the benefit of all the people."

Each Forest Resource Plan is divided into five major sections: the report, giving detailed background and overview of the state forest; timber management section; watershed management section; mineral development section, and recreation section. On the Delaware State Forest, where recreation, wildlife, and watershed values are highly significant, the Forest Resource Plan led to acquisition in 1980 of a tract of 1,743 acres, containing a 14-acre lake and two miles of frontage on the Delaware River (complementing the Upper Delaware and Delaware Water Gap National Recreation Areas).

In terms of timber, the state forest is divided for administrative and control purposes into 60 compartments, ranging in size from 470 to 2,800 acres, identifiable on the ground by physical features. Uneven-aged management is applied to areas of commercial forest where aesthetics and recreation are primary values. Even-aged management is applied elsewhere, but with restraints and modifications, assuring that logging remains compatible with other forest uses.

The use of firewood is a growing influence in the Poconos and throughout Pennsylvania (as well as much of the nation). Thus far, the demand has been satisified with low-quality materials. The Bureau of Forestry 1980 Annual Report provides the following indicator of a certain future trend:

> Both public and private forest lands are feeling the impact of people seeking fuelwood for home heating and for industry. Supplies of readily available and inexpensive wood will dwindle, and it will be necessary to establish reasonably stable sources of wood for fuel which will not deplete supplies of high-quality timber. The Bureau is faced with the challenge of providing technical assistance to landowners to manage their woodlands properly so as to prevent indiscriminate cutting of high-value timber for fuelwood. It is also an excellent opportunity for the Bureau to make the public aware of other forestry programs and services that are provided to enable landowners to better manage their valuable forest resources.

Forestry in Urban America

William Penn wanted one acre of forest left wild for every five that were cleared. Henry David Thoreau proposed a 500-acre woodlot behind every village or town. But things haven't quite gone that way.

Most cities were born in the forest, or along a river valley with riparian trees and plants. In the beginning trees were cleared to make way for settlement, but then they were replaced in time with plantings along city streets, in public parks, around homes and public buildings. Thomas Jefferson personally planted Lombardy poplars along Pennsylvania Avenue in Washington.

But recent times have seen a marked trend in the opposite direction. The 1980 Bureau of Census figures reveal that 74 percent of the nation's population lives in and around 20,000 principal cities and communities, the focal points of growth, where each year three million acres of land are consumed and converted into concrete, or asphalt, for highways, shopping centers, parking lots, subdivisions, and convention centers. The estimated annual loss of forestland alone, according to the National Agricultural Lands Survey, is between 800,000 and 1,000,000 acres. The loss of forest leads to soil erosion, depletion of water storage in natural aquifers, degradation of water quality, and inevitable change in the quality of life and the quality of people divorced from nature.

Street trees have been valued at $25 billion, not including backyards, parks, cemeteries, or naturally wooded areas within urban boundaries. They make a city attractive and habitable, and enhance property values, as every realtor recognizes. Yet less than 1 percent of municipal budgets is spent on tree care (compared with 13 percent for police; 9.5 percent for fire protection; 4.8 percent for refuse collection). Most programs that do exist are inadequate; city arborists have little influence, dealing with tree maintenance rather than expansion of urban forests, and about 50 percent of all cities have no systematic tree programs.

Organizations like the United States Conference of Mayors and Association of Local Government pay little attention to the values of urban forests. Their focus is directed to economic development (though it benefits directly from community enhancement), crime, taxes, transportation, and waste collection. Old buildings have been recognized to some extent, but Golden Gate Park in San Francisco, Central Park in New York, and a thousand other last touches of forest are being filled, or threatened, with academies of science, galleries of art, skating rinks, solid waste disposal sites, and freeways.

There are still a few examples of doing things otherwise. Overton Park, in Memphis, Tennessee, with seventy-five varieties of trees, is one of the great urban forests in the world today. Though targeted for destruction by an Interstate highway, the citizens saved it after a struggle that lasted more than 15 years and was debated in Congress and the Supreme Court. Philadelphia has about 300,000 street trees, and is adding more than 5,000 yearly. Fairmount Park, the pride of the city, covers 8,400 acres; between 1972 and 1978 the Fairmount Park System acquired more than 500 prime acres of land. Fairmount Park has an estimated four million trees, reaching into the urban heart of the city along tree-lined Benjamin Franklin Parkway, a 1.5-mile-long greenway joining the main body of the park with the center city. And the famous downtown squares—Rittenhouse, Washington, Franklin, and Logan—are all well treed, in keeping with William Penn's plan for a "Green Countrie Towne."

Conditions may yet change for the better elsewhere as concepts of urban forestry take hold. Various people in the Forest Service expressed interest in urban forestry over the years, but not until 1972, with amendment of the Cooperative Forestry Assistance Act, was a technical assistance program authorized. The 1978 Cooperative Forestry Assistance Act, replacing the 1950 Cooperative Forest Management Act, contained a specific section authorizing matching funding for urban forestry technical assistance. To this date, however, no administration has ever proposed line item spending for urban forestry assistance. It has always been added to the Forest Service budget through Congressional action.

As a result, state forestry agencies now can utilize federal funding to provide technical and financial assistance to communities, educational institutions, organizations, and individuals for local forest planning, establishment, protection and management.

Florida has been one of the leaders. The Division of Forestry has had the objective: "to provide the services of a professional forester to assist and advise urban communities in planning, and to assist in the establishment and management of trees and plant associations to enhance the beauty and livability of the urban environment." Toward that end, Division foresters have supplied data under "Developments of Regional Impact" to local and regional planners on the impacts of development on forest resources and unique natural systems. They have helped with protection ordinances saving cypress stands, sabal palm hammocks, tropical hardwood hammocks, and mangrove forests, and assisted cities in establishing nurseries to provide future street tree-planting stock.[9]

In Maryland, the urban forestry coordinator of the state Forest Service has worked with land developers to reduce erosion, control storm water, save valuable vegetation, and utilize timber that must be cut for wood products. The developer who recognizes that a landscape with trees increases property values always benefits, along with the community. The state consultant shows how to design with natural features, explaining what best to take and leave, and what to plant.[10]

Forestry schools have offered few courses in urban forestry. Resource professionals have not shown much desire to work in cities (whether in dealing with urban forestry, urban national parks, or urban wildlife), possibly because it involves confronting tough political and social problems. Yet there is a continuing challenge to provide forests for sustained use by the urban dweller—to identify and fulfill man's social, physical, and psychological needs through green space and green belts.

Advances are likely to be made during the 1980s, even in the face of decreased funding for urban parks and amenities. A National Urban Forestry Conference held in Washington in November 1978, under the auspices of the Forest Service and the College of Environmental Science and Forestry of the State University of New York, brought together interested groups and individuals. Subsequently, a National Urban and Community Forestry Leaders Council was organized, with coordination provided by the American Forestry Association. A second national meeting was held in 1982, as part of the annual meeting of the American Forestry Association.

It is true that growing conditions for city trees have steadily degenerated because of air pollution, drought, heat, erosion, disease, and concentrated use of the land; the loss of trees invariably speeds other kinds of deterioration. On the other hand, urban forestry advocates can show how trees in the city help to reduce noise pollution, deadening sounds of traffic and construction; reduce air pollution, by trapping and holding dust and dirt particles; and filter water, giving nature time to purify it.

The sight and smells of blossoms and foliage produce pleasant feelings. Trees block wind, cutting down on heating bills in winter, and provide cooling shade and moisture in summer. In fact, scientists say that a single, isolated tree can have the cooling effect of five average room air conditioners running 20 hours per day. That gives some idea of what a forest can do for urban life.[11]

15

INTEGRATING
PEST MANAGEMENT

Today there are more insect species than ever before. Over 200 of these pests have developed resistance to chemicals. Costs of pest control have increased strikingly. And pesticides have polluted the biosphere.

—Robert van den Bosch, *Environment Magazine,*
April 1970.

The Chemical War

Chemical contamination of the environment is one of the most critical issues facing civilization during the 1980s. Poisons in the air, water, and food products are believed to endanger plant and animal life and public health, including the genetics of the human species.

"Chemicals are invaluable to society, and most are believed relatively safe under normal conditions of use," noted the eleventh annual report of the Council on Environmental Quality (CEQ), published in December 1980. "But there are many chemicals whose environmental and health effects are unknown and some that have proven harmful." Particular concern has been shown over the effects of synthetic chemicals used against unwanted insects and plants, which tend to be classified as "pests" and "weeds."

The Forest Service has been a focal point of this concern, although the agency makes the point that only a small percentage of total chemical use occurs on forest lands, that far more herbicides are applied annually to the country's lawns than to its forests. Nevertheless, insecticides have been used occasionally on very large areas of commercial and noncommercial forests, the treatments, however, to cope with a very few pests. According to a report titled "Integrated Pest Management," written by Dale R. Bottrell and issued by the Council on Environmental Quality in 1979, between 1954 and 1974 treatment of four insect species—western spruce budworm, easten spruce budworm, gypsy moth, and Douglas-fir tussock moth—accounted for 96 percent of the acreage sprayed for insect control by the Forest Service.

The Forest Pest Control Act of 1947 influenced the direction of forest insect and disease pest control programs. The Act provided subsidy for cooperative control on private land by cost sharing with federal funds. It authorized the Forest Service to conduct surveys of destructive insect and disease pests and to execute control programs as demanded. According to Bottrell: "The Act stimulated control programs that relied exclusively on chemical destruction of insect pests and, to a lesser degree, plant pathogens, with little or no emphasis on incorporating a variety of control methods. Chemical control was emphasized because the programs were structured to deal with emergencies that required only short-term remedies."

Certain new directions have been taken, if not wholly at least in part, based on the principle of integrated pest management, or IPM. A "USDA-Combined Forest Research and Pest Development Program" was initiated in 1975 to develop and implement technology to cope with the gypsy moth, Douglas-fir tussock moth, and southern pine beetle. It entailed collaboration of universities, four agencies in the Department of Agriculture, state organizations, and private forestry. Its major objective was "to find, evaluate, and implement environmentally safe management systems for the three insect pests and to provide the knowledge necessary to prevent or suppress future outbreaks."[1]

In an effort to spur the search for alternatives to toxic chemicals for pest control, President Carter's 1979 Environmental Message to Congress included particular reference to integrated pest management, which he hoped would prove effective and acceptable in agriculture and forestry. The President described IPM as "a systems approach to reduce pest damage to tolerable levels through a variety of techniques, including predators and parasites, genetically resistant hosts, natural environmental modifications and, when necessary and appropriate, chemical pesticides." At the same time, the President called on agencies to modify and alter pest control practices and apply IPM strategies. He created the Interagency Pest Management Coordinating Committee, chaired by CEQ, to encourage the effort.

In its 1980 report, CEQ noted that: "Encouraging adoption of integrated pest management is an important policy challenge. Many farmers now use crop production management systems that establish biologically stable agricultural ecosystems which include soil, plants, pests (at tolerable levels), and natural predators. Both research and practical experience have shown that stable ecosystems can lead to fewer pest problems and to more reliable farm income."[2]

The Forest Service responded to the President's 1979 message by adding a new chapter on IPM to its basic operating manual. Thus Forest Service policy, as written, would encourage IPM practices—natural controls, selective harvesting, resistant hosts, maintenance of diversity, and removal of damaged trees—in order to reduce excessive use of chemicals, while making forest and rangeland management more efficient. National Forest Management Act regulations described IPM as "a process in which all aspects of a pest-host system are studied and weighed to provide the resource manager with information for decision making. Integrated pest management is, therefore, a part of forest or resource management."

But regulations and policy declarations do not necessarily find their way directly into practice. Environmental and economic analyses are required to demonstrate that pesticides can achieve specific management objectives, and so they do. An average 498,113 acres (or 0.2+ percent) of National Forest System lands were treated annually between 1976 and 1980 with assorted pesticides, herbicides, fungicides, and rodenticides as part of vegetation management, insect and disease suppression and prevention, and animal damage control.

Old habits die hard and forestry has long relied on toxic chemicals to protect a timber crop. A substantial body of personnel has been engaged toward this end and funds channeled to chemical manufacturers and purveyors. "Forest entomologists have directed a major portion of their time and talents to finding ways to put the insects down," wrote William E. Waters, Chief of the Forest Service Insect Research Branch (and later professor of entomology and forestry at the University of California, Berkeley), in 1969. He made a strong case for the use of DDT, especially via large-scale aerial spraying, striking back at its critics, and expressing total distrust of other approaches.[3]

Insects have plainly been targeted as the enemies, and timber upheld as the prize to be defended from them. In *Eastern Forest Insects*, a book published by the Forest Service in 1972, the author, Whiteford L. Baker, estimated that during an average year losses caused by insects were enough to construct 1.3 million homes, and that insects caused more damage to woodlands than forest fires. In advance of the 1974 aerial spray of DDT over 400,000 acres in the Pacific Northwest, designed to remove the threat of the tussock moth, the Forest Service warned of imminent losses if nature was left to follow its own course. These losses, however, appeared to be measured primarily in terms of timber values. The Forest Service warned that livestock would have to be moved from sprayed areas, at the expense of livestock owners. While hunters would have to be alerted to the dangers of DDT residue in game animals, there was no mention of resultant loss in sport or hunting economics.[4]

The most publicized single incident, contested (after the fact) over the longest period of time, involved not timbered land, but rangeland. At issue was the method of use of a herbicide called Kuron, first manufactured by the Dow Chemical Company in 1954. Kuron is equivalent to Silvex. It has evoked disturbing comparisons with Agent Orange, used as a defoliant by U.S. forces in Vietnam.

In September 1965, an aerial spray program was initiated in the Kellner Canyon–Russell Gulch area in the Pinal Mountains, a part of

Tonto National Forest at the south and west edges of Globe, Arizona. The principal objectives were to thin out the brush, or chaparral, in order to increase the water flow and increase forage for livestock. The area selected, covering about 1,900 acres, was sprayed with potent herbicides: 2,4-D, 2,4,5-T, and 2,4,5-TP Silvex in 1965, 1966, 1968, and 1969. The June 8–11, 1969, project led to complaints by residents of the area of damage to vegetation, their domestic animals, and their own health.

The Forest Service and the Agriculture Department investigated. They reported that two task forces turned up no serious victims of property damage from the chemicals. The first attributed plant damage to root rot, woodpeckers, and sapsuckers. The second reported that one human illness, eye irritation and skin rash, "might be related to the spraying." However, Dr. Arthur Galston, a Yale University plant physiologist, who came to investigate, warned Globe residents that as little as 1 part per billion of any phenoxy herbicide makes water unpalatable and possibly dangerous. (The Arizona Department of Health had measured 3.4 parts Silvex per billion in Globe's water supply.) There may, in fact, said Dr. Galston, be no safe levels, in light of what is known about dioxin, the by-product of the chemicals involved.

Five families brought legal action against Dow Chemical and the U.S. Government, contending that they started to have health problems almost immediately after the spraying and suffered permanent injuries. Mrs. Billee Shoecraft wrote a book, *Sue the Bastards*, before her death of cancer in 1977.

Despite slow progress in the legal case, the Globe incident became widely cited in medical literature on phenoxy herbicides. It also became a focal point among consumer groups trying to outlaw the use of phenoxies by the government. The developments at Globe were watched closely by Vietnam veterans pursuing their own legal action, charging Agent Orange had caused them various ailments, including cancer.

In March 1981, just as the case was about to come to trial—fully 12 years after the incident—Dow and the federal government agreed to settle with cash payments to the five families. The settlement was negotiated by a federal judge who imposed a condition that the amount of cash be kept secret.

As a result of the compromise, the suits against Dow and the Forest Service were dismissed. The issue of responsibility and the question of right or wrong remained unanswered. But Jane Kay, a staff writer of the Arizona *Daily Star*, had already turned up a piece of highly illuminating testimony that extends beyond this incident to the value, purpose, and cost efficiency of chemicals in resource manipulation.

In its issue of September 22, 1980, the *Daily Star* carried an interview by Jane Kay with William Fleishman, retired from the Forest Service. As range and wildlife officer of the Tonto National Forest, he had overseen the aerial spraying in the brushy foothills around Globe. He personally was not aware of adverse effects and denied the spray had drifted as the families had charged. He had been around the chemicals involved without evidence of ill effect. "But that doesn't solve the question," he said. "It's a poor way to spend the government's money."

Clearly the operation was not scientifically based. As Fleishman stated: "We were changing herbicides, always trying to get the lowest rate, trying this and trying that. The program was absolutely new. No one was doing it."

Of the years of burning and spraying chaparral and seeding of grass to produce more forage, Fleishman told Ms. Kay:

> We were deluded. It looked real good when it started. As time went on, it became obvious that the treatment was not good. It produced a lot of grass. But there was no use putting the cows back on because those grasses needed every bit of energy to compete with rapidly growing shrubs.
>
> The environmental people started to ask questions that should have been asked before. There are many unanswered questions when you use chemicals. We didn't know the answers, and probably they still don't. We were actually turning a wide ecosystem of trees, forbs and grasses into one of scrub oak and weeping love grass.
>
> It is totally uneconomic. Let's assume you could do these glorious things: Take a hill, spray it with herbicide and have no chemical washing into people's water supplies. Say you've done it. You're in low-producing land, poor land. You get enough feed for between three and four cows a section—you can't justify it from a grazing standpoint.
>
> Considering the cost of chemicals, helicopter and seed to produce forage for three cows? The bank would just laugh you out of the room.
>
> It took me a long time to get it through my head. If you've got money to spend, don't spend it in the chaparral.

The Forest Service has changed direction to some extent. The accompanying charts show patterns of use on national forest lands of pesticides in general (covering the years 1974 through 1981) and of herbicides 2,4-D from 1975 through 1980 and 2,4,5-T from 1971 through 1981. These data will speak for themselves.

Research has continually shown that there are other techniques than chemicals for dealing with problems. The 1980 Resources Planning Act program emphasizes IPM:

> There would be nationwide emphasis on integrated pest management strategies that are cost-effective and that rely less on chemicals to control major insects and diseases. Integrated strategies would be initiated for spruce budworm and gypsy moth in the Northeast, western spruce budworm in the Pacific Northwest and Southwest, bark beetles in the South and West, and Douglas-fir tussock moth and dwarf mistletoe in the West.
>
> Prevention strategies against insect and disease attack would be initiated on National Forest System lands, especially silvicultural control of mountain

TABLE 15.1

Application of phenoxy herbicides on national forest lands

2,4-D Use						
FY	1975	1976	1977	1978	1979	1980
Pounds[1]	270,826	236,636	213,040	280,949	341,893	216,882
Acres	130,820	106,385	110,617	126,485	139,274	128,255

[1] Includes small amounts of other herbicides (e.g., picloram, but not 2,4,5-T) applied in combination with 2,4-D.

2,4,5-T Use									
FY	1971	1972	1973	1974	1975	1976	1977	1978	1979	1980
Pounds[1]	57,402	40,501	53,220	26,022	95,540	102,720	14,597	6938	1210[2]	220[3]
Acres	25,777	32,100	26,424	37,436	49,357	50,463	7,105	4254	405[2]	55[3]

[1] Includes small amounts of other herbicides (e.g., 2,4-D) applied in combination with 2,4,5-T.

[2] Involved two projects: one of 390 acres on the Lassen National Forest, California, for conifer release and one of 15 acres for poison plant control in Utah.

[3] Represents the final phase of a rangeland research project to control the noxious weed, tall larkspur (Delphinium barbeyi) in Utah.

pine beetle and dwarf mistletoe in the West. Other control methods—mechanical, cultural, manual and biological—would be implemented for other major insects and diseases.[5]

In late 1981 the Pacific Southwest Region initiated an environmental impact study covering vegetation management and use of herbicides in seventeen national forests of California, to replace an earlier one dating from 1974. It was to deal with all types of vegetation management, including hand, animal, machine, and chemical, with justification for each. In the same year, 1981, the Pacific Northwest Region completed an environmental impact statement in which it selected as the "preferred alternative" a plan calling for reduction in chemical use. Despite industry protest against deemphasis on spraying, the use of herbicides was programmed to decrease from use on 79 percent to 47.4 percent of lands requiring treatment, while manual release was to increase from 17 percent to 48.6 percent.

Nevertheless, a report, "Pesticide Use on National Forest System Lands," for fiscal 1980 stresses what it calls the beneficial effects of

FIGURE 15.1

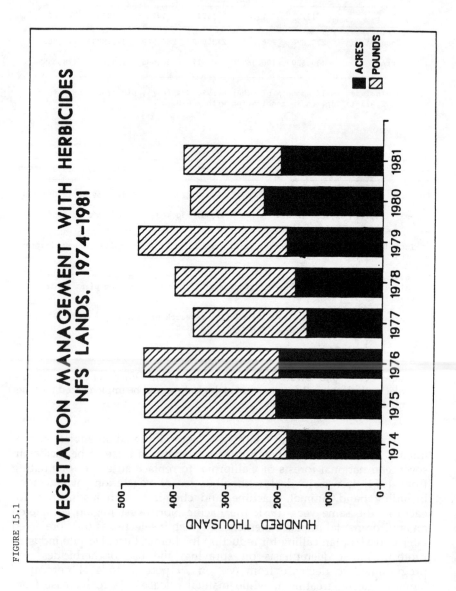

VEGETATION MANAGEMENT WITH HERBICIDES
NFS LANDS. 1974–1981

FIGURE 15.2

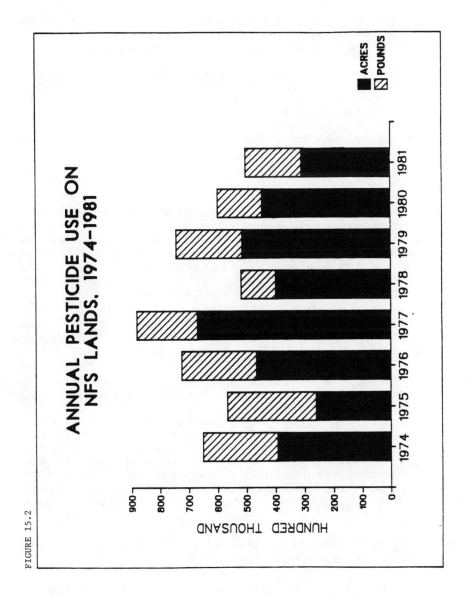

ANNUAL PESTICIDE USE ON
NFS LANDS, 1974–1981

ACRES
POUNDS

HUNDRED THOUSAND

chemicals in three general categories: vegetation management, insect and disease prevention and suppression, and animal damage control. In reference to vegetation management, the report explains:

> Vegetation management is the manipulation of the kinds, amounts, quality, and condition of the vegetation resource. Forest vegetation management programs range from complete protection of the vegetation resource to intensive management to favor a particular plant species. It is estimated that approximately 39 percent of commercial forest land is dominated by undesirable vegetation. Significant gains in forest productivity could accrue from increased vegetation management activities on these lands. These activities can be grouped into these eight major categories: noxious weed control, range improvement and maintenance, site preparation for timber production, conifer release, thinning, rights-of-way maintenance, fire protection, and general weed control.

And of rangeland projects:

> Some range improvements accomplished with range betterment funds include vegetation management with herbicides which reduce the growth and spread of undesirable species increasing forage availability and productivity and benefitting livestock and wildlife. Of the nearly 100.5 million acres of range associated with NFS lands, an average of 12,586 acres are treated annually with pesticides to control such plants as rabbitbrush, sagebrush, juniper, chapparal, manzanita and scrub oak.[6]

In a letter to the author of March 8, 1982, James L. Stewart, Director of Forest Pest Management of the Forest Service, stressed the relationship of chemical use to commodity production goals:

> We suggest you do not just look at total pesticide usage to determine trends of use on national forests land. Pesticides are a tool used in the management of the various resources. Therefore, pesticide use should be evaluated in conjunction with the program objectives for the various resources. For example, in "A Recommended Renewable Resource Program" reforestation is to increase until 1985 to remove the reforestation backlog. Timber harvesting is also scheduled to increase. If timber harvesting, reforestation, and reduction of reforestation backlog are increased, site preparation and release treatments must also increase, especially when considering the amount of competing vegetation that occupies the acres in the backlog.

Regional Forester Steve Yurich (of the Eastern Region) expressed the same view in a letter of April 2, 1981:

> The Forest Service is committed to an integrated pest management approach. We still need considerably more information from research, however, to make nonchemical treatments effective in many situations. For example, our Region has a real need to use herbicides to release pine and

spruce seedlings from early invasion by competing hardwoods. This need will continue in the 1980s.

Concerns no doubt will continue to be raised by some segments of the public. Thus we expect to explain the reasons for herbicide use and demonstrate its safety while we seek efficient and cost-effective alternatives.

Another school of thought, however, clashes with the forestry viewpoint, insisting that insects and weeds perform constructive ecological roles at no cost. Weeds are seen as helping prevent erosion and flooding, while an abundance of insects produces an abundance of birds, which control the insects. Many plants are totally dependent on insects for pollination and continuation of their species. "Weeds" and other plants would get out of hand if leaf-eating insects did not keep them in check, while countless insects in old trees are said to induce decay, enriching soil for seedlings and saplings to follow.

Weed trees have low market value or none at all. That is what makes them undesirable. In the Pacific Northwest, much brush control has been aimed at eradicating red alder to make way for Douglas fir, the desirable crop tree of the region. Yet alder has its own supporters, who note that its decomposing leaves compare favorably with famous nitrogen-fixers, such as clover, alfalfa, horse beans, and lupine, as valuable organic fertilizer. The alder, with attractive gray and white trunk and dome-shaped crown, produces luxuriant foliage in streamside groves. In time it yields to Douglas fir anyway, which thrives in the rich humus layer and buildup of nitrogen from the alder leaves.

Adherents of this school would agree with the late Dr. Robert van den Bosch, a University of California entomologist, that poisons are likely to knock out natural enemies of pestiferous insects while the pests themselves develop resistant strains and the costs of control spiral. "The plain truth," declared Francis Lawson in 1971, on retiring as director of the Department of Agriculture Biological Control of Insects Research Laboratory at Columbia, Missouri, "is that many important insects can no longer be controlled by *any* insecticide, and one or two pests have reached this level of resistance over very large areas."[7]

Long War Against the Gypsy Moth

The gypsy moth moved in and made itself at home in the East. Government experts have attributed its surge in recent years to abandonment of DDT in favor of what they call "more acceptable but less effective insecticides." The war against the moth has been directed from the Department of Agriculture and its various bureaus, with considerable media material warning of the evil doings of this creature and the need of funding to exterminate it or control it at "acceptable levels."

A USDA publication of June 1969 entitled "The Gypsy Moth" begins by warning: "The gypsy moth, a serious pest of trees in Northeastern United States, continues to threaten the nation's hardwood forests." A

March 1970, USDA Picture Story for the Media, titled "Gypsy Moth: Forest Killer," bears a suggested headline: "Gypsy Moths Threaten Our Woodlands, Commercial Hardwood, and Gardens."

Over the years, experts of USDA and the Forest Service have found the gypsy moth steadily spreading. Consequently, they have issued yearly warnings to homeowners and landowners through a variety of news releases, booklets, and books. So, too, has Union Carbide, manufacturer of a poison called Carbaryl, or Sevin, whose literature notes: "This voracious pest can quickly defoliate your shade, fruit, forest and ornamental trees."

The war against the gypsy moth has never been universally supported. When the Forest Service in 1971 officially estimated that the gypsy moth was causing an economic loss of $10 per acre, it was challenged by the Environmental Defense Fund. "Man Vs. Gypsy Moth," an article in the *Connecticut Conservation Reporter* of February-March 1971, urged that there be no statewide, publicly financed gypsy moth control program, and particularly that Sevin not be used, since it kills nontarget insects, including pollinating species and natural enemies of the moth. "We Can Live with the Gypsy Moth," headed an article by Irston Barnes, in the *Washington Post* of August 1, 1971, and "The War Against the Dreaded Gypsies" headed one by Frank Graham in *Audubon Magazine* of May 1972. Massachusetts and Connecticut actually gave up trying to eradicate the moth. New Jersey adopted a policy of not recommending spraying for anything less than 500 egg masses per acre.

While Union Carbide and government experts have called Sevin completely safe and "nonpersistent," the label includes the following cautions: "Keep out of reach of children and animals. . . . Harmful if inhaled or swallowed. . . . Avoid contamination of food, feed, water supplies, streams and ponds." Critics insist, however, that Sevin is nonselective, killing beneficial insects, threatening birds and other wildlife. It is, admittedly, toxic to honeybees, the invaluable pollinators, and is reported to have caused a high incidence of birth defects and malformations in test animals.

In 1975 a Forest Service research ecologist, Robert Campbell, endeavored to take a fresh look. In an Information Bulletin titled "The Gypsy Moth and its Natural Enemies," Campbell noted that when conditions are favorable, insect populations tend to build to epidemic proportions. This makes things favorable for their parasites and predators, which also increase in number. Through overpopulation, their food supply is reduced, while the increase in enemies wipes out the epidemic and reduces insects to what might be called normal numbers. Campbell wrote:

> Insect-pest management has been identified so closely with chemical pesticides in recent years that we tend to ignore the fact that these pesticides are only a part of pest-population management. We should know our enemy beforehand in enough detail so that we can use pesticides only in bona

fide emergencies; and other methods—including doing nothing—would become more prominent in the management scheme.[8]

Campbell's views are generally supported by the experience and observations of Daniel Smiley, who, since 1930, has recorded natural history on his family estate bordering Mohonk Lake near New Paltz, New York. Because of its no-spray policy, the Mohonk lands have been the site of research projects conducted by both USDA and New York State. In June 1965, Smiley recorded this observation and comment:

> Thousands of acres of woodland in Ulster County are without leaves, and look nearly as brown as in early April. Where homes are located in these areas, or people are outdoors for recreation, large fuzzy caterpillars drop down their necks. It is a real problem, in which biology, economics and emotion become involved.
>
> Technicians and certain public employees looked at the "facts" and arrived at the conclusion that spraying—to the extent that money is available—is the answer.
>
> Ecologists and naturalists study the same "facts" and conclude that most of the proposed spraying is neither desirable nor in the long-range public interest.
>
> In my opinion there is no one satisfactory solution to the gypsy moth problem, but by considering the relationship of some of the facts to each other we may see how to live with the problem with the least hurt.

After another look in 1971, Smiley reported:

> I am very pleased with what happened to the gypsy moth at Mohonk this summer. There was a crash in population without spraying. I observed many control factors in operation. There was considerable parasitism by a number of insects. Disease probably took its toll also, although harder to recognize in the field. I saw chipmunks eating live caterpillars. Birds and gray squirrels consume pupa. The best evidence of the population downswing is that there were relatively few eggs laid and the masses were very small in size.

During 1981 the gypsy moth at Mohonk showed a fast build, followed by quick decline. In "1981: The Year of the Gypsy Moth," a supplement to his earlier study, Smiley indicated that egg masses are poor indicators, since many in 1981 were small and placed in disadvantageous locations. Defoliated trees, noticeable in June, subsequently put out new leaves and began making food again. Some defoliated trees would die in the next few years, he reasoned, but they were weakened from other causes, before the moth buildup of 1980 and 1981 and would have died anyway. He forecast they likely would be replaced by different species, in a more varied and healthier forest, less subject to moth damage. Smiley wrote:

If it were possible to eradicate the gypsy moth from North America, and if this could be done at reasonable cost and without injury to other living organisms, I would probably favor the effort. But there is no evidence that it could be done. Even the proponents of a vigorous spraying program no longer speak of eradication, they speak only of damage control. The gypsy moth is no longer an "invader," it has become a resident. Like the dandelion and many other introduced species, it is here to stay. We shall have to accommodate to one another.[9]

Nevertheless, the war against the gypsy moth has continued. A news release issued by USDA on September 27, 1972, bears this heading: "Gypsy Moths Defoliate More Than One Million Acres in 1972." And a decade later, in May 1981, "News from the Northeast Forest Experiment Station" opened with: "For the first time, the gypsy moth is causing an outbreak of allergic reactions." This was followed by an Issue Briefing Paper from USDA on June 4, 1981, titled "Why Fight the Gypsy Moth?" It provided all the reasons to continue financing the war: the moths are chewing more trees; they are stripping the forests, and they are spreading—while the only persons who want to reduce the spray are those who "haven't experienced the devastation the moth can cause."

The year of 1981 was a tough one for the Pocono forests. At a July meeting, the Pike County Commissioners received a report on their $246,000 investment to spray 27,500 acres. They were told that it had not been very successful in controlling the gypsy moth infestation. The Commissioners voted to endorse the return to DDT. The chairman claimed that Pike County would have solved its problem with DDT (which it had last used in 1965), and another commissioner distributed and quoted from an editorial commentary published in *Barron's*, a business journal: "Thanks to the efforts of environmental extremists, both in government and out (not to mention ecological nuts), the climate has turned more and more hospitable not only to gypsy moths, but also to grasshoppers, locusts, chinch bugs, beetles, ticks, and other insect hordes."

Nipping the Spruce Budworm With Federal $$$

"It has become a common practice for private pesticide users to seek taxpayers' money to subsidize spray programs on their privately owned lands," editorialized the April 1981 issue of *Maine Environment*, published by the Natural Resources Council of Maine (NRCM).

First it was the spruce budworm program they wanted publicly funded; now it's the gypsy moth. These private landowners recite a litany of what they claim are "public benefits" resulting from these spray programs. Unfortunately, theirs is a rather one-sided analysis of what constitutes "public benefits."

Especially curious is their claim that spraying is the only "economical means" of pest control. If it's so "economical," how come they need a public subsidy to pay for it?

We have seen very little evidence to justify any kind of broad-scale spraying for the gypsy moth. Sure, gypsy moths are a nuisance, and they do kill a few trees, but there also are environmentally acceptable methods for controlling their effects without resorting to spraying. That is why NRCM opposes all spraying of chemicals for gypsy moths.

The spruce budworm presents a somewhat more complex problem. There are, at present, no alternatives to chemical sprays which can prevent the loss of thousands of acres of trees to this voracious insect. At the same time, it is clear that spraying does not work in the long term in countering the spruce budworm menace. NRCM has been willing to accept a spray program which includes a planned reduction in the dependence on chemicals and requires that landowners alter their forestry practices to reduce the forest's vulnerability to future budworm damage. Such steps were taken in the 1980 budworm spray program.

In 1980 the Forest Service reexamined its role in activities in Maine relating to the spruce budworm. From 1972 through 1979 almost $35 million had been spent on budworm suppression in that state. Of that total the federal government contributed 42 percent (through cooperative forestry assistance programs), the state government 7 percent, and forest landowners, principally large timber companies, 51 percent. During 1980 a total of $4.8 million was spent on chemical insecticides, mostly Carbaryl, and all of it paid for by forest landowners. It was a period of public concern over the effects of spraying on the bald eagle, peregrine falcon, the endangered Eastern panther, and two rare forms of Arctic char. In addition, attention was being directed to the needs of Maine blueberry growers, and the effects of Carbaryl on the 2,500 beehives (with 50,000 bees per hive) that they normally import in May and June for use in pollination. The Forest Service made no direct contribution to the 1980 pest control program except to fund research projects, including a 20,000-acre demonstration (in five separate tracts) of *Bacillus thuringiensis*, or BT, a biological material being developed for insect control.

In February 1981, the Forest Service issued an environmental impact statement on a cooperative five-year Maine Spruce Budworm Management Program covering eight million acres, all privately owned. The Service chose an IPM alternative involving reducing the use of chemicals, increasing the use of biologicals, and emphasizing improved silviculture. This approach called for better forest surveys to concentrate protection on the most productive stands, while marginal timber would be left unprotected. Harvesting, thinning, and new marketing techniques would be focused on killed or threatened stands.

The Natural Resources Council of Maine noted that federal subsidies would continue and commented critically:

Unfortunately, the Forest Service took a giant step backward. They now propose to provide 36 percent of the funds to buy and apply chemical pesticides for the budworm. Even more disconcerting is the omission of requirements for landowners receiving this subsidy to practice better forestry.

The Forest Service is also offering up to 50 percent funding for gypsy moth chemical controls funneled through local government spray projects.

NRCM cannot support this use of taxpayers' dollars for any chemical pesticide application. Public funds should be used only to finance research into alternatives to chemical spray. The landowners are the ones who should be paying the full cost of chemical controls. After all, they are the ones who keep telling us that chemical pesticides are the only "economical" control measure.[10]

The Forest Service felt this criticism unwarranted and countered with this refutation:

> The percentage of Forest Service funding in Maine is correct. But these percentages relate to mixed ownership maximum allowable cost-sharing with the State of Maine through cooperative cost-sharing ratios established for each state by the Forest Service. These maximums frequently are not realized. In 1981, due to unavailability for Forest Pest Management funds, federal cost-sharing with Maine was 27, not 36 percent, and this was for many activities, not just purchase of pesticides. In 1982 the Forest Service offered cost-sharing at 12.5 percent, not to exceed a suppression cost of $14.30 per acre. We could cost-share up to 50 percent for gypsy moth control.[11]

Bacillus thuringiensis, the promising biological control of the gypsy moth, is slower acting than chemical control, but does not kill natural enemies of the moth.[12] As Dan Smiley writes of BT: "In order to be lethal, some of the spray must be ingested by the gypsy moth caterpillar, which then dies within several days. . . . Its great advantage is that, being derived from a disease of caterpillars, it does not affect creatures other than insects. In particular, it is harmless to fish (pesticides often end up in water) and to warm-blooded animals."

Opposition to chemical poisons has led to the organization of activist groups like the Northwest Coalition for Alternatives to Pesticides (NCAP) and to confrontations and sit-ins. Efforts were made during 1980 to prevent use of 2,4-D by occupying scheduled spray sites on the Umpqua and Siskiyou National Forests.[13] Occupational safety was made the issue on the Willamette National Forest during August 1979, when 80 temporary workers and crew bosses, ordered to cut roadside brush which had been sprayed with 2,4-D, refused to expose themselves. The Northwest Forest Workers Association, formed to enhance the position of reforestation personnel, pressed two key issues: (1) recognition that pesticides constitute an occupational hazard, and (2) a shift from chemical spraying to manipulation by hand to aid local economic development.[14]

Conflicts between insects and timber are not readily resolved. Along the drainage of the North Fork of the Flathead River, comprising portions of Glacier National Park and Flathead National Forest, both in northwest

Montana, and across the border in Canada, the latest epidemic of mountain pine beetle was first detected in 1972. By 1980 the infestation covered 175,000 acres, and was estimated to have killed 12 million trees in the national park, six million in the national forest. Mortality due to beetle damage was considered the equivalent of more than half the total annual cut conducted on the national forest.

The mountain pine beetle has probably been a normal, functional part of lodgepole pine ecosystems for almost as long as lodgepole pine has existed, and that is well recognized and accepted. Many infestations, however, followed extensive fires during the last century. Still, the beetle became a problem only when its effects began to conflict with management objectives.

Beetle damage is said to disrupt management of timber, wildlife, watershed, grazing, aesthetics, and fire. Since lodgepole pine largely inhabit areas with dry summers and frequent lightning strikes, beetle-killed trees become volatile fuel with high probability of burning. For these reasons, the Forest Service in 1980 and 1981 planned extensive salvage logging of dead trees on the North Fork drainage while they were still harvestable, and an accelerated harvest of susceptible green trees in the path of the beetle infestation.

Critics sought to block this project, arguing on wildlife and ecological grounds. Infestations, they reasoned, likely result in greater diversity, consequently a greater range of habitats favorable to more species; and the endangered grizzly could survive the mountain pine beetle better than the intrusion of logging road building. And fires in lodgepole tend to prepare soil for seed germination and cause cones to release their seeds. As things turned out, the Forest Service had difficulty in getting the timber harvested because of poor market conditions.

Foresters advance silviculture as an alternative to the use of chemicals in pest management. The economics may or may not be in its favor. Weighing the objectives and effects in the course of decision making is not an easy job. When that "system approach to reduce pest damage to tolerable levels through a variety of techniques"—as enunciated in the Integrated Pest Management program—ultimately takes hold, the answers are apt to be easier to reach and more widely accepted.

16

GLOBAL HORIZONS

This nation has few well-qualified, internationally recognized experts on tropical forests, and inadequate language capabilities hinder placement of U.S. nationals in international forestry organizations. In addition, U.S. institutions and agencies often do not encourage or reward staff participation in international programs, and this country has not been as aggressive as others in pursuing positions for its experts in international organizations or in providing appropriate training and career opportunities in federal programs.

—Excerpt from *The World's Tropical Forests: A Policy, Strategy and Program for the United States*, report to the President, May 1980.

An area one-half the size of California is being deforested yearly, principally in the tropical belt of the Amazon Basin of South America, Southeast Asia, and West Africa. The effects of this process are felt throughout the world, in ways that are still not manifest or understood.

"The fate of tropical forests will be the major factor that determines the biological wealth of Earth in the future," according to Anne and Paul Ehrlich in their book, *Extinction*. "Those extraordinarily vulnerable ecosystems are the greatest single reservoir of biotic diversity on this planet. A reasonable assumption is that about two-thirds of the species of the tropics occur in the rain forests. If this is correct, then something on the order of two-fifths to one-half of *all* species on Earth occur in the rain forests, which occupy only 6 percent of the Earth's land surface."[1]

There may seem to be little if any way to reverse course, in the face of increasing demand for industrial wood products and fuelwood inevitably outrunning supply. Experts forecast gloomily that tropical forests may be destroyed by the end of the century, that almost a million species may become extinct as a result of this deforestation, and that these losses are likely to upset the world's climate. At the same time, much of the impetus for deforestation derives directly from government programs and from international aid subsidies. The long overdue recognition of the urgency of international cooperation in sustaining a productive, inhabitable earth may yet provide the foundation for a change in direction.

To Theodore Roosevelt and Gifford Pinchot, conservation on a world scale, spurred by the forestry movement, was the key to peace among nations. With that idea in mind, Roosevelt in 1905 convoked a North American Conservation Conference at the White House, attended by representatives of Canada, Mexico, and Newfoundland. Plans for a world conference were developed but later rescinded by the Taft Administration.

Pinchot never gave up. Throughout human history, he insisted, the exhaustion of resources and need of new supplies have been the prevalent causes of war. "Conservation is clearly a world necessity," he declared before the Eighth American Scientific Congress in 1940, "not only for enduring prosperity, but also for permanent peace." Even as he neared the end of his life in 1946, he was trying to set the wheels in motion for a post–World War II international conservation congress.

Over the years the Forest Service was called on to furnish specialists for advisory missions abroad. It cooperated with world bodies like the International Union of Forest Research Organizations and the Food and Agriculture Organization (FAO) of the United Nations, helped train foreign nationals who came to the United States, and provided technical assistance to the Peace Corps in forestry, range, and watershed projects. The Forest Products Laboratory, under contract with the federal foreign-aid administration, during the 1970s undertook a three-year project on "Utilization of Secondary Tropical Hardwoods," demonstrating conversion to paper of mixed hardwood species from tropical Africa, Asia, and South America. Also, for the same agency, the Service completed a world survey of forestry activities and deforestation problems in developing countries and established a support network to serve foreign-aid field missions in forest-related development projects; and another project involved a study of "Energy Expansion Through Biomass Production, Conversion and Use."[2]

Recent years have brought new directions. Man and the Biosphere (MAB), a program approved at the United Nations Educational, Scientific, and Cultural Organization (UNESCO), brought the Forest Service into a world investigation of the structure and function of the biosphere, changes brought by man, and effects of those changes on human populations.[3] The 1970 Stockholm conference on the environment, plus efforts of organizations like the World Wildlife Fund and International Union for the Conservation of Nature, stimulated attention to the relationship between global forestry and global future. The United Nations Environmental Program (UNEP), established as an outgrowth of the Stockholm conference with headquarters at Nairobi, Kenya, brought a continuing focus on issues of encroachment of desert on productive lands, management of tropical forests, and the relationship among forest, man, and land ecosystems.[4]

In June 1978, the Department of State and Agency for International Development convoked in Washington a United States Strategy Conference on Tropical Deforestation. Among those present, Dr. Frank Wadsworth, Director of the Institute of Tropical Forestry, related the issue to U.S. national interest. The Panama Canal, he said (which this country was then preparing to turn over to the Republic of Panama), would soon be unable to handle large ships because of siltation from continuing deforestation.

The Canal Zone has been said to encompass "the most extensive and accessible lowland tropical forest in middle America." Although only 10 miles wide and 50 miles long, the zone embraces mangrove swamps, freshwater lakes, rain forests, and open savannas, providing sanctuary to many species of flora and fauna. Each time a ship makes a transit through the canal, it draws 52 million gallons out of man-made Gatun Lake while rising from sea level. The amount of water consumed in a single day has been estimated to equal a two-week supply for the city of Boston.

Though a stable watershed is essential to the canal, farmers in recent years have invaded the basin, cutting patches of forest, often on steep slopes, burning trees and planting corn. As Dr. Wadsworth noted, 25 years earlier 83 percent of the watershed was covered with forest, but by 1978 only 32 percent was covered, with erosion, siltation, and loss of water resources as the result.

The Institute of Tropical Forestry, headquartered in Puerto Rico, had already done pioneering work in its field: protecting the largest mountain tracts on Puerto Rico through the Caribbean National Forest; operating three experimental forests; furnishing technical counsel that led to reserving 30,000 acres as the beginning of the Puerto Rico state forest system; reversing the decline of the Puerto Rican parrot, and serving as a training ground for foresters from Latin America and other tropical areas.

That 1978 Washington conference concluded that the interests of the United States, as well as of other countries, would best be served through conservation of the tropical moist forests. An interagency task force, representing Departments of State, Agriculture, and Interior, AID, and others, subsequently produced, in May 1980, a report titled "The World's Tropical Forests: A Policy, Strategy and Program for the United States." This report called for "tropical forest management" to expand both short-term and long-term economic and social benefits—including sustained-yield timber management, recreation, and scientific and educational benefits.[5] Then followed "Global Future: Time to Act," a January 1981 report to the President prepared by the Council on Environmental Quality and Department of State, showing the potential for impoverishment of world resources and degradation of the world environment based on the continuance of current policies and practices.[6]

These reports show the principal direct causes of deforestation to be conversion and use for agriculture, fuelwood gathering (since most people in undeveloped countries depend on wood for heating and cooking), and poorly managed commercial logging. The consequences of deforestation are rising prices and wood shortage, and floods due to loss of vegetation on the upland watersheds. Destruction of the forests threatens not only future wood supplies, but food production and habitability as well.[7]

Malaysia, in the heart of Southeast Asia, provides a case history of rapid development during the mid to late 1970s. Forests were being cleared and cultivated in monoculture farm crops, such as soy beans, oil palm, rubber, cocoa, coconut, and sugar cane. As elsewhere in the tropics, rare species of wildlife were deprived of their ancient habitat—in Malaysia, the Sumatra rhino, Malayan tiger, orangutan, and gau (a species of wild cattle).[8] Cutting in the most critical area for the rhino, Endau Rompin, straddling the border between the states of Johore and Pahang, became a major issue in 1977. Public pressure forced a halt to logging and classification of Endau Rompin as part of an already planned national park.[9]

"If the present rate of exploitation is not drastically reduced and controlled," editorialized the Kuala Lumpur (Malaysia) edition of the *New Straits Times* on November 22, 1979,

> Malaysia will run short of prime quality logs by 1990. Further, at the felling stage alone, damage can be as high as 45 percent; at the processing stage, the waste is about 30 percent. The timber utilization is thus under 30 percent.
>
> Concerted action to arrest this wasteful slide is imperative. It is encouraging, therefore, to hear that the Government intends to reduce the felling of timber from the present 922,000 acres to 330,000 acres—and that the state governments, who have jurisdiction over land matters, have already agreed to the plan.

Curiously, while the Forestry Department warned of the consequences of unrestrained felling and pleaded for a reduction, another branch of government, involved with commerce, was engaged in promoting the sale of Malaysian timber on international markets. It happens with the best of governments. Foreign aid programs underwritten by the United States have spurred conversion and accelerated development. Low tariffs on wood products and beef also have contributed to decline of the world's tropical forests.

For such reasons "Global Future," the CEQ report, urged that the interagency task force on tropical forestry be strengthened into a coordinated federal program and that it communicate with international development agencies. It also recommended that the Institute of Tropical Forestry in Puerto Rico and Institute of Pacific Islands Forestry in Hawaii be designated as "national centers" for tropical forestry research, education, and training, considering that few undeveloped nations have professional forestry schools or background for sound timber practices.

Other recommendations call on the National Science Foundation, the Forest Service, and the Interior Department to increase their research on the design of ecosystem reserves.

It may take some time for these proposals to mature, but they mark the trend of the future in a shrinking world. These two landmark reports note that federal resource agencies have lacked a reservoir of personnel trained for work abroad and that a corps of park, wildlife, and forestry professionals to assist other nations would serve the best interest of the United States as well as the world.

Forests in the tropics and subtropics are quite different from those in the temperate zone. Despite their lush appearance, rain forests generally grow on poor soil, not on the deep, rich soil of a temperate deciduous forest. When leaves drop from rain forest trees and decay, the released nutrients are immediately absorbed by the network of shallow roots and redeposited in the trees. It is the ability of the trees to retain essential nutrients that forms the basis of food webs—and that also makes them more vulnerable than temperate-zone forests. Once the tropical forest is cleared, the thin soil cover gives way. Regeneration is difficult.

National planners deal in terms of narrow economic payoffs—of timber, pulp, fuelwood, or a converted patch of cropland, something direct and quantifiable, preferably immediate. They speak of "developing" the tropical forests, the richest eco-zone on earth, rather than of conserving forests as the foundation of wealth and development. Standing as they are, these forests promote regular supplies of irrigation water to agriculture, protect hydro-power facilities (and such waterways as the Panama Canal) against siltation, ensure dependable supplies of high-quality water for domestic use, and support public health.

American experts can't quite impose, or even advise, management practices of temperate-zone forests for the tropics; but they can show the social and economic values implicit in healthy forests as the alternative to impoverished arid lands, deteriorated and treeless from the mistakes of centuries past.

17

WORKING WITH OTHER AGENCIES

> I have had one consistent ambition since I became Secretary of the Interior, and that has been to be the head of a Department of Conservation, of which, necessarily, Forestry would be the keystone.
>
> —Harold L. Ickes, 1940.

Interagency relationships are part of the federal way of life. The Forest Service relates to many bureaus in the Department of Agriculture, to other federal departments, to state infrastructures, counties, and cities, and to international agencies as well. Collaboration, communication, and, at times, conflicts arise in dealing with matters relating to energy, the environment, wildlife, education, labor, fire, insects, disease.

Questions need answering at all levels of the Forest Service: Who has jurisdiction? What does the law provide? Who has the knowledge? Is that knowledge sound? Is that agency still handling it?

Whatever the structure may be at one point, it is subject to change with a new administration. During the Carter administration, Secretary of Agriculture Bob Bergland established an Office of Environmental Quality, headed by a former Forest Service official, Barry Flamm. That office reached deep into the Department of Agriculture to stir an environmental awakening and conducted a series of workshops aimed at rekindling family farming in the United States. With the advent of the Reagan administration, the Office of Environmental Quality was eliminated, and its functions were spread among various agencies to increase efficiency, so it was said.

Each administration has its own ideas on how to make government work best and how to make bureaus responsive to its policies. That is not easy to accomplish. When Dr. Rupert Cutler was Assistant Secretary of Agriculture, he tried to evoke change in the Forest Service and its sister agency, the Soil Conservation Service, but he approached the

269

leaders of both bureaus as a fellow professional seeking common ground
for new directions. His successor, John B. Crowell, Jr., expressed esteem
and respect for Forest Service professionalism. In sharp contrast, Interior
Secretary James G. Watt cut a wide swath, firing virtually all bureau
chiefs and reaching into the ranks to press for his desired changes in
policy and practice.

Whether he succeeded is another question. Directives from above
sometimes achieve results, but not always. The cast of characters at the
top may change, but bureaus and the bureaucracy roll on. There is an
implicit attitude, like an unwritten rule: We will outlast them. The
administration that promises to fire thousands and streamline the system
has not yet been known to succeed. As the Czar of Russia once confessed,
"I don't run this vast country—10,000 civil servants do."

Each agency has built-in loyalties based on tradition and mission.
Give several agencies a single tract of land and ask for a prescription
on the best use. The National Park Service is likely to recommend
preservation; the Forest Service, multiple-use management; the Federal
Highway Administration, a network of roads, but the Geological Survey
and energy bureaus will urge looking for the minerals first. If there's
a free-flowing stream, the Army Corps of Engineers will opt for a
concrete dam and reservoir—though it may have to bid against the
Bureau of Reclamation and Soil Conservation Service.

If one program is completed, another must emerge. A case in point
is the Rural Electrification Administration (REA), a sister agency of the
Forest Service in the Department of Agriculture. When REA was created
in 1936, only 10.9 percent of all farms in the United States had electricity.
By 1969, more than 98 percent of the 3.1 million farms were electrified,
more than half of them through REA-backed facilities. But by no means
was the agency nearing the end of its string. From a New Deal innovator
that had given struggling agricultural cooperatives the capital to bring
electric lines to dark, Depression-ravaged farmsteads, REA had become
the money supplier to a system of affluent power cooperatives that
today produce and transmit electricity to suburbs, cities, and industry.

Lines Across the Land, a book published by the Environmental Policy
Institute in 1979, analyzed the role of the National Rural Electric
Cooperative Association. It had begun long ago as a citizen group
working for an alternative source of energy to the large investor-owned
utilities. During the 1970s, however, NRECA had consistently opposed
environmental legislation and laws to protect agricultural resources. It
had drifted into partnership with its old foes, the private utilities, with
a variety of joint ventures, including nuclear power plants. The rural
cooperative association was a formidable lobby with Congress and the
White House, but not for conservation or energy alternatives.[1]

When David A. Hamil, appointed as REA administrator in the Nixon
administration, was asked whether the job of electrifying the dwindling
number of farms might be discontinued, he said categorically, "Nothing

could be further from the truth." He had not accepted his position, said the administrator, "to see our program wither on the vine or to preside over its liquidation."

Career employees welcome such expressions when political appointees assume command of their agencies. They stimulate support and loyalty. Thus Ervin L. Peterson, Assistant Secretary of Agriculture during the Eisenhower administration, announced that he had not accepted his appointment in order "to preside over liquidation of the national forests." He spoke at a time when the National Park Service had "thrown down the gauntlet" (to quote one of its own publications) to the Forest Service, demanding that large acreages administered by the latter agency be transferred to the former. Assistant Secretary Cutler likewise supported the Forest Service position on Mineral King, though environmental groups strongly supported, and achieved, legislation transferring that valley in California from the Sequoia National Forest to Sequoia–Kings Canyon National Park.

The rules of the road are simple. The agency that does not actively and energetically pursue its goal, developing programs and political support for programs while protecting its flanks from other agencies, retrogresses in a competitive arena. The appointed administrator at cabinet or subcabinet level who wants to get his points across abets the bureau in its competitiveness and, in return, gains acceptance in the ranks.

The Rural Electrification Administration is not the only agency in Agriculture that evolved from the social conscience crusade of the 1930s and 1940s into a self-propelling bureaucracy in the postwar decades. The Soil Conservation Service (SCS), when it came into being during the Depression years, stood for salvation of the landscape. Under the late Hugh H. Bennett, it built its reputation fighting dust storms, stopping erosion of gullies, designing contour strips to hold topsoil in place, and generally looking after all land-related resources, including fish and wildlife. Restoration of the national grasslands, now administered by the Forest Service, is a tribute to the work done by Bennett and his associates.

Then times changed. Following a program ostensibly designed to develop agricultural stability in the Great Plains area, SCS paid farmers a share of the cost to drain wetlands to expand crop production and furnished free engineering aid in doing so, while other agencies in the department helped to find markets for corn and grain. Eventually, the food supply exceeded demand, but the process continued, paradoxically supplemented by generous cropland retirement programs. Worst of all, the draining of wetland upset the native life community and destroyed vast areas of waterfowl habitat. The Fish and Wildlife Service at the same time was striving to acquire and restore natural waterfowl areas, but clearly was not nearly as aggressive as the SCS in applying natural resource concepts. Finally, the so-called Reuss Amendment (named for its sponsor, Representative Henry S. Reuss of Wisconsin) prohibited SCS from furnishing further technical or financial assistance in draining key wetlands.

Public Law 566, the Watershed Protection and Flood Prevention Act of 1954, despite its constructive sounding title, gave SCS impetus for extensive and controversial stream channelization, scarcely in keeping with its earlier activities. This activity, which reached a peak during the late 1960s and early 1970s, has declined, but only after leaving significant changes in large areas of rural landscape.[2]

SCS in 1981 had a staff of 13,000 working at federal, state, and local levels to assist landowners with practices in land use. More than $15 billion had been spent from the 1930s through the 1970s in federal assistance and cost-sharing soil protection efforts, yet the problems clearly remained unsolved and barely defined before the public.

The General Accounting Office in 1977 criticized SCS for emphasizing complicated engineering projects. During 1981, a total of 638 small watershed projects were under construction, at a total estimated cost of $2.3 billion, many with little effect on serious soil erosion problems.

The Forest Service has overall leadership in the Department in forest and forest range conservation, development, and utilization, but other agencies are involved as well. SCS provides technical forestry advice and assistance to owners of farms and other lands, primarily in connection with soil conservation programs.[3] The Federal-State Cooperative Extension Service provides general education and assistance to landowners and other groups, including materials on forestry, through a grass-roots network. The Agricultural Stabilization and Conservation Service (ASCS) administers cost-sharing programs of financial assistance for reforestation, timber stand improvement, and related practices on nonindustrial private forestlands, while Farmers Home Administration conducts a loan program that includes credits for forestry and range operations.

Relations with the Interior Department

Most federal lands are administered by the Department of the Interior, which, according to some of its critics, is where the Forest Service belongs. As one of the arguments, they like to cite the fact that about twice as many bills bearing on the Forest Service are referred to House and Senate committees dealing with Interior than to committees dealing with Agriculture, and the Forest Service budget is considered and processed by appropriation subcommittees on Interior and Related Agencies rather than the subcommittees on Agriculture. The Public Land Law Review Commission in its report of 1970 proposed (as have others before and since) establishment of a Department of Natural Resources, basically an expansion of Interior to encompass the Forest Service. The Commission expressed the belief that the diffusion of policy direction as between the Forest Service and other public land agencies (National

Park Service, Fish and Wildlife Service, and Bureau of Land Management) had led to unnecessary differences, embarrassing conflicts over the use of national forestlands for national parks, public confusion, and expensive duplication of activities and programs.

Relationships between the Forest Service and Interior agencies have been characterized by pride in purpose and rivalry in jurisdiction, but there have been many examples of cooperation as well. A view of the federal role in recreation reveals an interesting story.

The work of the Outdoor Recreation Resources Review Commission, created by Congress in 1958, made it possible to study the magnitude of the entire scene for the first time. As a result of the commission's recommendations, the Bureau of Outdoor Recreation (BOR) was established in 1962 to promote federal, interstate, and regional cooperation in this field. Its functions and activities included nationwide planning and research and the preparation of a continuing inventory of recreation resources and needs.

BOR was established as an agency of the Interior Department. The first director, Dr. Edward C. Crafts (who served from 1963 to 1968), insisted that the Bureau be allowed to function as a semi-independent agency free of Interior control—"an unusual arrangement clearly recognized and supported by Secretaries Udall and Freeman." Possibly so, but Dr. Crafts was viewed at Interior with a degree of skepticism, being himself a transplant from the Forest Service.

The National Park Service had originally opposed establishment of BOR, which preempted key Park Service functions extending beyond administration of the national parks. "It is true that at first I felt the Park Service should continue to handle recreation planning, as we have done since 1936," Conrad L. Wirth, director of that agency, declared in an interview published in *American Forests* in January 1964, at the time of his retirement. "I was wrong. On further study, I reached the conclusion that an operating bureau should not handle over-all planning. To have an agency of government specifically for planning purposes is eminently sound. It has the quality of objectivity that we might lack."

BOR activities, while they lasted, had significant bearing on the national forests in a variety of ways. For one, the Land and Water Conservation Fund, which Congress established in 1964 (and which BOR administered), made funds available for (1) matching grants to the states on a 50-50 basis for planning, acquisition, and development of outdoor recreation areas and facilities, and (2) federal land acquisition for outdoor recreation purposes by the National Park Service, Forest Service, and Fish and Wildlife Service.

The Land and Water Conservation Fund opened a new era in protection of key recreation values for the public at a critical period of soaring land prices. Under the established procedures, BOR set standards; the Forest Service and other federal agencies were required to submit plans for each proposed purchase, based on "outstanding values" that might otherwise be lost, and to receive BOR approval in advance of Congressional appropriation.

The fund enabled the Forest Service to purchase significant holdings within national forest boundaries, particularly along the shores of streams and lakes. One of the earliest acquisitions was the 18,000-acre pristine Sylvania tract in northern Michigan, previously an exclusive hunting preserve.[4] In other cases, opportunities were lost, or ignored. Chief Edward P. Cliff stressed the need of large-scale acquisitions in the southern Appalachians, wherever they became possible, because of encroaching land-use conflicts. Yet in 1975, when a prime 15,000 acres of high mountain land in western North Carolina, called the "Bemis tract," became available, the Forest Service passed it by completely. Patrick Noonan, president of the National Conservancy, noted that the area embraced choice trout streams and a connecting link on the migration route of the last and largest black bear country in North Carolina outside the Great Smoky Mountains. On this basis, Noonan offered the services of the Conservancy as last minute intervenor to save the tract before it was placed on auction for subdivision. Forest Supervisor Del. W. Thorsen explained in a letter to the author the Service's disinterest:

> As to foreseeable watershed impacts from impending developments, we can only surmise. We do know that the soils in Little Buffalo and Snowbird drainages are relatively stable and not highly erodible.
> Since most tracts were sold to individuals who presumably intend to build vacation cottages in a forest setting, we believe they will want to protect the environment and not damage the watershed. Construction of access roads possibly will expose more soil to the elements than any other anticipated use. However, this damage need not be long lasting if the roads are properly designed and maintained.

A later supervisor, George A. Olson, pursued a diametrically opposite course on the same national forest, the Nantahala. Early in 1981, to a large extent through Olson's efforts, the Forest Service acquired two key tracts of 38,986 acres on the headwaters of the Tuckaseegee River near Sylva and Cullowhee at a total cost (including land, survey, title, and overhead) of $13.4 million. The larger tract, of 33,600 acres, formerly had belonged to the Mead Corporation, which had closed its mill at Sylva. The smaller one, the Bonas Defeat Tract, includes the spectacular Tuckaseegee River Gorge. The two areas embrace unique features: high elevation natural grass balds, Canadian-like spruce-fir forests, waterfalls, 90 miles of trout stream, and critical habitat for numerous wild species, from black bear to golden eagle, raven, red crossbill, and brown creeper, with unusual opportunities for backpacking, rock climbing, and hunting.[5]

These purchases were difficult to consummate because of opposition from North Carolina logging interests and a smoldering resentment in the region against the extensive federal land holdings in national forests and national park units. Despite uncertainties following the 1980 election, Congress made the funds available for these two major purchases.

During the 1980s many key tracts still need to be acquired within the boundaries of national forests. The emphasis has been on areas in the East, though needs have been identified elsewhere as well. These include four inholdings in the greater Yellowstone ecosystem of Montana and Wyoming. One of these, a portion of the Buffalo Valley, lies on the Bridger-Teton National Forest near the east entrance to Grand Teton National Park. Because of its important visual quality to both national park and national forest, the Forest Service prepared a proposal for Land and Water funding either to purchase the land outright or to acquire scenic easements. In 1980, however, funds were unavailable. Under terms of the 1964 Land and Water Conservation Fund Act, "except for areas specifically authorized by Act of Congress, not more than 15% of the acreage added to the National Forest System pursuant to this section shall be west of the 100th meridian." But increasing interest in national forest purchases in the West has made it difficult to stay within this percentage restriction.[6]

The Land and Water Conservation Fund is due to expire in 1989, but in 1981 the Reagan administration sought to weaken it badly. Environmental and recreation groups and friends of The Land and Water Conservation Fund in Congress rallied to maintain it at a significant level, at least for the moment.

In addition to administering this source of funding, BOR also played a role in settling agency differences over jurisdiction of recreation lands. These issues have included: Flaming Gorge National Recreation Area, Wyoming-Utah, with national park land transferred to the Forest Service; Picture Rocks National Lakeshore, Michigan, national forest land transferred to the Park Service; Whiskeytown-Shasta-Trinity National Recreation Area, California (authority distributed between the two agencies); and North Cascades, Washington, which saw establishment by Congress of a new national park and two new national recreation areas out of national forest lands, as recommended by BOR.

BOR figured prominently in the establishment of the National Trails System, which Congress authorized in 1968. Here again, jurisdiction over the new system became a process of cutting and fitting between agencies. Administration of the Appalachian Trail, extending 2,000 miles through the eastern mountains, was awarded to the National Park Service, though it includes more than 500 miles in eight national forests. On the other hand, administration of the Pacific Crest Trail, running 2,313 miles from Canada to Mexico, was given to the Forest Service. Likewise, agreement was made, as required by the Wild and Scenic Rivers System Act of 1968, for studying 27 rivers for possible inclusion in the National Wild and Scenic Rivers System, with responsibility for 18 rivers assigned to Interior, and for 9 rivers to Agriculture (the Forest Service).

The Carter administration brought in new ideas on how BOR should be run. It was rechristened Heritage Conservation and Recreation Service (HCRS, or "Hookers"), with certain historic preservation activities of the National Park Service transferred to it. The Reagan administration opted otherwise. Interior Secretary Watt, who had been BOR director under the Nixon administration, abolished the agency altogether and folded its functions and personnel into the Park Service.

Where Does the Forest Service Belong?

There were jokes that he didn't know one end of a cow from another, but Harold L. Ickes was distinguished as a durable public servant who made notable contributions to the field of natural resources. He was the self-styled "old curmudgeon," stubborn, unpredictable, suspicious, and temperamental. He browbeat and bullied outstanding career employees in his own department. His pet peeve, however, was the Forest Service and its unceasing effort to stay clear of his clutches.

In one move after another, Ickes sought to outmaneuver the shades of Gifford Pinchot and undo the Transfer Act of 1905. In 1935, hearings were held in both houses of Congress on legislation to change Interior's name to Department of Conservation and to empower the President to transfer appropriate bureaus to it. In 1936, Roosevelt appointed a Presidential Committee on Administrative Management, which proposed much the same. Roosevelt showed that he felt Interior should deal with public lands, Agriculture with private lands, but generally he allowed jurisdiction in the New Deal agencies to overlap, keeping his lieutenants on their toes and off balance.

Even after the transfer of forest reserves to Agriculture, Pinchot fueled the fight anew when he supported construction of a dam in the Hetch Hetchy Valley of Yosemite National Park in order to serve the water needs of San Francisco. He was a member of Theodore Roosevelt's "Tennis Cabinet." When John Muir and other preservationists pleaded with the President to protect Yosemite, Pinchot influenced the President's failure to respond. It marked the parting of the ways between Pinchot and Muir, who remarked bitterly, "We may lose this particular fight, but truth and right must prevail at last."

That concept of "truth and right" spurred the demand for new national parks and a separate agency to administer them. The Forest Service wanted to run the parks, with logging under controlled conditions, and resisted passage of the 1916 Organic Act that brought the National Park Service into being. (Pinchot, now outside the government, actively opposed formation of the new bureau.) From then on, national forests became the hunting ground of national park advocates. In 1920 alone, Congress received proposals for 30 new parks, all to be formed from lands under Forest Service jurisdiction. Chief Forester Henry Solon

Graves asserted that the new parks would result in "practical partition of the national forests" and would be counter to public interest.

The Forest Service gave up some areas willingly, others under protest. The Service opposed formation of Rocky Mountain National Park, resulting in a smaller park than proponents wanted and lacking the most spectacular features of the Colorado Rockies. It objected to inclusion of Jackson Hole in Grand Teton National Park and decried the "lock-up" of timber resources that would result with establishment of Olympic National Park. William B. Greeley, who retired as Chief in 1928 to become Secretary-Manager of the West Coast Lumbermen's Association, was a leader in this hot fight, but the national park was established, nevertheless, in 1938.

In 1934, when the Taylor Grazing Act established grazing districts on 142 million acres of public domain administered by Interior, the foresters began sustained efforts to bring about their transfer; but Interior countered by demanding the 25 million acres of untimbered grazing lands on the national forests. Ickes claimed, "Wilderness areas are fundamentally inconsistent with the operation of the Forest Service. There is only one way to ensure wilderness areas and that is by setting them up as national parks or national monuments." But this was in the very period when Robert Marshall was enlarging national forest wilderness and strengthening regulations to protect it. National forests, it could be argued, were best for dispersed, or wilderness-type, recreation, while national parks were meant for more formal, organized tourism, complete with entertainment and profit-making concessioners.

Franklin D. Roosevelt's Presidential Committee on Administrative Management issued its findings (the Brownlow Report) in 1937, proposing a Department of Conservation, as Ickes had hoped. The Forest Service may not have been able to oppose it openly, but there were no restraints on the opposition expressed by the master press agent and leader of the foresters' lobby, Gifford Pinchot himself. He talked about the built-in services to special economic interests in the Department of the Interior, which had been exposed in the Teapot Dome scandal of the 1920s. Although then over 70, Pinchot attacked the plan in scores of newspapers and magazines and enlisted the aid of farm and forestry organizations. By 1938 the Forest Service lobby had created so much turmoil, overtly and covertly, that Roosevelt felt the Department of Conservation idea might jeopardize his broad reorganization bill, and he withdrew that portion of it.

The struggle continued. In 1939, Congress adopted a watered-down reorganization bill, authorizing the President to recommend transfer of agencies subject to veto by both houses within 60 days. The following year the President actually had the Bureau of the Budget prepare an executive order relocating the national forests. The transfer was virtually complete, but before the package could be sealed and delivered, widespread Congressional opposition, mustered by the Forest Service under

Earle Clapp, appeared to tip the scale against the transfer. Ickes threatened to resign and the President tried anew. But two influential senators, Alben Barkley of Kentucky and George Norris of Nebraska, joined the opposition and Roosevelt capitulated—while Ickes fumed about the Forest Service "running wild even to the extent of defying the express orders of the President and opposing a bill which was introduced with his knowledge and consent."

The issue, put to bed during World War II, was revived during the postwar studies of the first and second Hoover Commissions on streamlining the federal government. The first commission, after considerable maneuvering, split on a plan to set up a Department of Natural Resources while accepting a proposal to transfer the Bureau of Land Management to Agriculture. Nothing came of it. The second commission in 1955 made no recommendations for changes of resource agencies, confining itself to more reachable goals.

In recurrent proposals varied arrangements have been proposed: shifting the civil works functions of the Army Corps of Engineers, Forest Service, Soil Conservation Service, Tennessee Valley Authority, and Federal Power Commission to join Interior in a new Department of Natural Resources is the heart of the all-encompassing plan, but supporters of each of these agencies propound counter arguments, and one or another is subsequently dropped from the succeeding proposal. Moving national forests to Interior, however, remains a continuing objective.

In 1963 Secretary Udall and his counterpart at Agriculture, Orville L. Freeman, sought to bring the old warfare to an end, signing a joint letter to the President: "We have reached agreement on a broad range of issues which should enable our Departments to enter 'a new era of cooperation' in the management of federal lands for outdoor recreation. This agreement settles issues which have long been involved in public controversy, we have closed the book on these disputes and are now ready to harmoniously implement the agreed-upon solutions."

At the same time the "treaty" was signed, the National Park Service was hopeful of winning the transfer of as many as 15 million acres of national forest land. A joint study into future administration of the North Cascades in Washington State was provided in the agreement of the Secretaries; but other areas the Park Service eyed with desire included the Sawtooth Mountains in Idaho, Oregon Dunes, Great Basin in Nevada, Flaming Gorge Reservoir astride the Utah-Wyoming border, and the Minarets and Bristlecone Pine areas in California. Both sides dug in, continuing the old battles rooted in the philosophic controversy between John Muir and Gifford Pinchot. Secretary Udall avoided the perpetuation of unpleasantries—he preferred to show how he had broadened the base of his department from a purely regional operation serving the West to one of national involvement and responsibility. The inclusion of new agencies, he said, such as the Bureau of Outdoor Recreation, Federal Water Resources Council, Federal Water Pollution Control Agency, and

Office of Saline Water, earned for Interior the right to be called "Department of Natural Resources."

A subsequent proposal for transfer was made by the Public Land Law Review Commission in its report of 1970. While early location of national forests in Agriculture may have been valid for their day, reasoned the commission, increasing emphasis on outdoor recreation and environmental quality justify the shift from a farm-oriented department to the public lands functions of Interior. To support its position, the commission cited varied aspects of interlocking public land administration. For example, the Bureau of Land Management holds jurisdiction over mineral and surveying activities on the national forests, while Interior controls the program of withdrawals on those forests established out of the public domain.

The PLLRC report tried to make the transfer appetizing to the Forest Service: "Along with its outstanding skills in effective administrative management of a large institution involved in public land management, the Forest Service would bring a long history of research and cooperative programs with states and private landowners." So it recommended that research on environmental quality of public lands management be intensified through existing Forest Service research programs and that cooperative forestry activities be expanded to embrace other forms of cooperation with the states.

Nixon Succeeds in Making Changes

Through all the controversy and costly studies, it would be difficult to say that significant reorganization had been accomplished to meet the changing times. In mid-1970, however, the Nixon administration announced a reorganization and regrouping of certain federal activities dealing with pollution and other environmental problems and policies, partially as the outgrowth of recommendations by the President's Council on Executive Organization, headed by Roy Ash of Litton Industries.[7]

The President's order created two new federal agencies:

1. The Environmental Protection Agency (EPA) was established as an independent superagency to deal with massive problems of air and water pollution, principally the setting of environmental quality standards and their enforcement. It brought together under one authority water quality and pollution control functions from Interior; air quality and pollution control functions and solid waste management from Health, Eduction and Welfare; pesticide control from Agriculture; and radiation regulation from the then Atomic Energy Commission.

2. The National Oceanic and Atmospheric Administration (NOAA) was established in the Commerce Department to include forecasting of pollution problems and research that must be done to identify and combat pollution in coming decades. NOAA incorporated a number of separate federal agencies, including Environmental Science Services

Administration (already in Commerce), Bureau of Commercial Fisheries from Interior, Great Lakes Survey of the Army Corps of Engineers, and branches of the Navy, Coast Guard, and National Science Foundation. These were intended only as first steps. On March 25, 1971, President Nixon sent Congress a message proposing a large-scale reorganization aimed at consolidating seven Cabinet departments and assorted independent agencies into four new ones. Among these would be a Department of Natural Resources, which would work "to conserve, manage and utilize our resources in a way that would protect the quality of the environment and achieve a true harmony between man and nature." The Department of Agriculture would have been eliminated altogether. This part of the reorganization failed to take hold.

The Environmental Protection Agency, born with promise, has suffered pains of unwieldiness and bigness, complicated legislation difficult to enforce, and continuing pressures of economy versus the environment. At its best, it represents a challenge in public administration.

The EPA experience offers some relevancy to the Department of Natural Resources proposal. For all the arguments supporting a superagency administering the federal lands, there are arguments against it as well. A single agency might have greater political power, but as things now stand the public has a choice of systems and a means of measuring values and efficiency: as between the Park Service and Forest Service in managing wilderness; BLM and Forest Service in managing grazing lands and selling timber; and between these and other agencies in providing outdoor recreation.

"As to the matter of jurisdiction," Secretary of Agriculture Henry Wallace (whose son filled the same job twenty years after him) once said in the course of a dispute with the Department of the Interior over grazing, "I am not so much concerned about who does the work as I am that it be done, done promptly and done well." Yet there is an ingrown tendency among loyal personnel of all agencies to respond to any proposal by asking first, "Is this in the best interest of the bureau?" and only then asking, "Is it in the best public interest?"

The Forest Service interacts cooperatively with Interior agencies on various common interests. Together with BLM it developed a system of increased fees charged to livestock operators grazing cattle and sheep on public lands. Jointly, they maintain the Interagency Fire Center, at Boise, Idaho. Cooperative projects are conducted with the Fish and Wildlife Service to protect rare and endangered species. One unusual project involved reintroduction of the masked bobwhite quail into southern Arizona (where it had disappeared around the turn of the century after the tall grass–mesquite plains had been devastated by cattle drives). About 1950, Tucson wildlife enthusiasts located a small quail population on a ranch in Sonora, Mexico. Several of the quail were taken by the Fish and Wildlife Service to be propagated at its research center in Patuxtent, Mexico—from there masked bobwhite were released in Coronado National Forest in 1969, with the Forest Service providing protection to nesting sites. National park and national forest personnel cooperate in joint field planning, as for protection of the grizzly bear in the greater Yellowstone ecosystem, though coordinated follow-through leaves much to be desired.

The New Eras—First Environment, Then Energy

The National Environmental Policy Act of 1969 was adopted during a rising wave of popular concern for the environment. The Democrats had called the Kennedy-Johnson administrations the "Third Wave" of conservation, the first and second waves having been the administrations of Theodore Roosevelt and Franklin D. Roosevelt. As for Nixon, his observers noted that, while he showed no personal sensitivity, he recognized the validity of environmental protection and public support for it. NEPA became the first law he signed on taking office.

Concern for environmental values became part of the charter of the Forest Service, as well as all other federal agencies. The public would have to be given the chance to express its views. Resource managers would be required to prepare and publish environmental impact statements. They would have to consider the effects of roads, logging, pesticides, and deal with new kinds of personnel and policies in other agencies. At the Fish and Wildlife Service, for instance, the emphasis would shift from predator control to protection of endangered species (including some of the same predators that had been targeted for extinction). Pesticide registration had been shifted from Agriculture—which promoted the use of poisons—to the new EPA. And there was the new Council of Environmental Quality (CEQ), a presidential agency, to assess environmental impact statements in pursuit of the new national policy.

Under its first chairman, Russell E. Train, CEQ became interested in national forest issues. It sought to influence the President to issue an executive order fixing restraints on clearcutting. Although this effort failed, the CEQ provided a forum for public interest groups to vent their feelings, for NEPA gave the public an unprecedented right to information on the environmental impact of proposed federal actions and opportunity for involvement *before* the fact.

When Train left CEQ he became Administrator of the Environmental Protection Agency and again encountered the Forest Service, this time over the pesticide issue. In February 1974, on yielding to intense political pressure brought to force his hand in lifting the ban on DDT in the Pacific Northwest Douglas-fir forests, Train sharply criticized the Forest Service and Department of Agriculture for failure to develop effective substitutes; their efforts, he said, were "almost totally inadequate—to the point of dereliction."

With the Middle East oil embargo, another powerful influence arose in government. The idea of making this country "energy self-sufficient" brought the public lands, including the national forests, under scrutiny as potential energy sources. During the tenure of James Schlesinger as Secretary of Energy under President Carter, there was speculation that control over the public lands might be shifted to the Department of Energy. Although the shift was not made, the National Uranium Resource Evaluation (NURE) project involves a federally subsidized survey of nuclear materials on public lands everywhere. In a speech delivered at Jackson, Wyoming, in 1978, William Menard, director of the U.S. Geological Survey, forecast that within 20 years all national forests would be opened to energy production, and in 40 years the national parks as well. The Geological Survey was then conducting the U.S. Mineral Resource Assessment Program (CUSMAP) "to delineate rock units and structures that control known and potential mineral and energy resources." Another project undertaken by the Nonfuel Minerals Policy Coordinating Committee, during the Carter administration, was intended, according to its chairman, Secretary of the Interior Cecil D. Andrus, to review "the mineral resource potential of federal land to help the mineral industry continue providing the supplies needed to assure the economic security of our nation."

The Forest Service administers the national forests, but mining permits are approved or denied by the Bureau of Land Management, and oil and gas development is administered by the Geological Survey. These are aspects of interagency relationships that are not widely understood, and therefore not often questioned, as perhaps they should be in light of growing pressures and inevitable conflict with other forest values.

FORESTRY BEFORE CONGRESS AND THE COURTS

> The need is for institutions that know how to expose and resolve, unsentimentally, the elemental issues in a dispute. For this task the courts are uniquely qualified. The availability of a judicial forum is a reality for the ordinary citizen—that he can be heard and that, in a setting of equality, he can *require* bureaucrats and even the biggest industries to respond to his questions and to justify themselves before a disinterested auditor who has the responsibility and professional tradition of having to decide controversies upon the merits.
>
> —Joseph L. Sax, in *Defending the Environment*

The Short-term Emphasis

The power center of American politics has focused on economic factors, with noneconomic factors such as land health and human welfare taking a back seat. Much of the record has been written by economic and political forces promoting land exploitation for the sake of securing raw materials and short-term profits.

This is nothing new. For years, through Democratic and Republican administrations alike, departments charged with maintaining and protecting the storehouse of natural resources have been the weak sisters of the federal family. Although Ronald Reagan called for reductions in spending for domestic social and resource programs, even before his time the agencies and departments involved received only a minuscule amount of the total federal budget.

During the 1960s and 1970s, environmental awareness became good politics, but there was scant rearrangement of budgetary priorities. In 1965, the portion of the budget allocated for natural resources (to be divided among the Forest Service, National Park Service, Fish and Wildlife Service, Bureau of Land Management, Bureau of Outdoor Recreation, Bureau of Mines, Geological Survey, and Federal Water

Quality Administration) was equal to 1.5 percent of the total. By 1970 it was only 1.3 percent and by 1971 even less.

The overemphasis on short-term exploitation has been confirmed repeatedly in the budgets allocated by Congress for the Forest Service.[1] In 1959, the Service formalized and consolidated short-term plans into a twelve-year schedule for the national forests, priced out by years to meet projected needs. It was approved for publication as a working plan by the Secretary of Agriculture and Bureau of the Budget (later the Office of Management and Budget); then it was revised to cover the ten-year period 1963–1972 and was submitted to Congress by President Kennedy as the National Forest Development Program.

If, however, one reviews the Development Program during the years 1963–1970, he finds the proposed level of spending for "timber sales administration and management" fulfilled to 95 percent of the total, while other phases fared poorly. "Reforestation and stand improvement" was financed at 40 percent of the proposed level, "recreation–public use" at 45 percent, "wildlife habitat management" at 62 percent, and "soil and water management" at 52 percent. When these imbalances came to light, Congress was moved to adopt the Resources Planning Act of 1974 and National Forest Management Act of 1976, presumably to insure sound planning and equity in funding of management functions.

In actual fact, by meeting planned levels in the area of timber sales only, Congress has provided *de facto* sanction for the emphasis on short-term consumption at the expense of long-term protection. Some critics have placed the burden of responsibility on the Forest Service itself. Dr. Charles A. Reich, professor of law at Yale (best known as author of *The Greening of America*), in 1962 prepared a monograph for the Center for the Study of Democratic Institutions on "Bureaucracy and the Forests," based largely on a seminar course he taught on public resources. Reich asserted that Congress had transferred its powers of legislation and policy making. Congress told the Forest Service (Reich wrote) to "best meet the needs of the American people," but left it entirely up to the Service to determine the needs (except, of course, to the extent that Congress controlled things through the budget).

"Because Congress has offered no standards or policies of its own, almost any choice by the Forest Service would be within its delegated authority," he continued. "The great danger is that an entrenched bureaucracy will become shortsighted in its perception of the public good. It may see only the needs of the next decade when planning for a century is essential. It may see only local demands when national needs cry for consideration. It may see where immediate economic gain lies but fail to see the values of 'non-economic' uses. It may care so much about today's balance sheet that it forgets tomorrow's heritage."[2]

Reich's paper was written before the passage of the Wilderness Act and other laws that have placed restraints on resource agencies. His viewpoint in the body of public opinion contributed to the introduction and enactment of these laws. Nonetheless, prior to the ground swell of popular concern, barely a handful of Congressmen spoke with knowledge or boldness on the issues. Well into the 1970s the power in Congress was controlled mostly by an alliance of southern conservatives and elderly wheelhorses of all sections, who derived their power from committee chairmanships assigned on a basis of seniority and the ability to survive.

Rowing Upstream in Congress—the Saylor Saga

Representative John P. Saylor was an exception to the rule—an exception to almost any rule. Before he died in office in 1973 at the age of 65, his constituents in mountainous western Pennsylvania had sent him to Congress for a total of twelve terms, and he was a Republican running in a predominantly Democratic area. He was regarded, and regarded himself, as a *national* Congressman, uniquely free of local pressures in dealing with natural resource issues, and for fully two decades he was a powerful, effective force. He left his mark on Congress and on natural-resource policy, taking his place in history alongside Gifford Pinchot, another Pennsylvanian, and Theodore Roosevelt as Republican conservation trailblazers.

Saylor dealt with the Forest Service and forestry in many ways. He was a member of the National Forest Reservation Commission, appointed by the President to review and approve areas for purchase under the Weeks Act, though that was the simplest part of things. His base of action in the House of Representatives was the Committee on Interior and Insular Affairs, where, through the process of seniority, he became the ranking Republican member. He commented on various occasions that only six of about forty members of the Committee were from the East. During his period, there were no Easterners at all on the Senate Committee on Interior and Insular Affairs (later the Committee on Energy and Natural Resources). The western power bloc in Congress, generally attuned to the voices of mining, oil, timber, water development, and grazing interests, controlled the Interior Committees in both houses.

"Western politics have long been imbued with the concept that natural resources are to be used, and used profitably," Saylor noted during an interview in his office in Washington in 1972. "One of the nice things about being an eastern member of the Committee is that the Interior Department doesn't have anything in my district. The Forest Service doesn't have anything in my district either. I have no Indians in my district. I can look at propositions as they are presented and call them as I see them."

Saylor was known to take strong, completely independent positions, despite his loyal Republican allegiance. He was a towering personality, standing six-feet-four, robust and yet wiry, blessed with a sharp wit and willingness for combat in uphill political battles against tough adversaries. For years he was the counterbalancing influence in the House Interior Committee to Wayne Aspinall, the powerful Democratic chairman from Colorado. Aspinall makes a case history in his own right. He was an expert in both natural resource issues and power politics, a dominant chairman of a type later scarcely to be found, who controlled the flow of legislation. Aspinall's critics considered him a spokesman for commercial interests, but generally avoided head-on conflict in view of the power he wielded. When finally defeated in the 1972 primary election, Aspinall returned to Washington representing the mining industry.

Saylor, like Aspinall, understood how to work within the system. Though hundreds of bills are introduced into each Congressional session, few are passed. Nevertheless, the landmark legislation of two decades—including the Wilderness Act, Wild and Scenic Rivers Act, Land and Water Conservation Fund Act, establishment of North Cascades National Park, and the law prohibiting shooting wildlife from airplanes—bears the stamp of Saylor's influence and determination.

On July 12, 1956, Representative Saylor declared on the House floor: "We are a great people because we have been successful in developing our marvelous natural resources; but, also, we Americans are the people we are because we have had the influence of the wilderness on our lives." Those words were part of his statement introducing the first Wilderness Bill in Congress.

It was uphill all the way. At first it was impossible to get the executive branch of government to take the bill seriously. Then, after the Interior and Agriculture Departments and the Bureau of the Budget submitted favorable reports, the House and Senate refused to act. But he never gave up. "I cannot believe that the American people have become so crass, so dollar-minded, so exploitation-conscious that they must develop every little bit of wilderness that still exists." So he declared in 1961, the year President John F. Kennedy urged enactment of a wilderness preservation bill. And finally the bill was passed in 1964, eight years after it had first been introduced.

Saylor served on the Public Land Law Review Commission, a minority influence in a body created by Congress at the behest of his chronic adversary, Representative Aspinall. The PLLRC consumed seven million dollars of public funds and five years of study and research, involving nineteen Commission members, forty-eight staff members, twenty-five advisory council members, representatives of fifty governors, sixty paid consultants, and 900 public witnesses.

After the report of the Commission had been submitted, Saylor delivered a speech in 1971 in which he gave his own evaluation. The membership, he said, had been stacked, weighted, and loaded. "There was hardly an all-out conservationist in the whole lot. On important environmental issues, I often felt I was standing entirely alone. Outclassed, outnumbered, and outgunned, the national conservation and environmental studies inherent in the study were neatly smothered, ignored, or shunted below the dominant-use philosophy which characterized the approach of the Commission's membership."

He would often express himself in such forthright, colorful language. Following are revealing samples on issues relating to national forests, as excerpted from statements on the House floor and speeches before various groups:

On the mining act of 1872: "The Constitution of the United States is nearly two hundred years old and it is still a sound, workable document. The Mining Law is nearly one hundred years old and it is an abomination and insult to the people of the United States."

On timber: "The Congress, in refusing to debate the infamous Timber Supply Act, maintained our national policy of protecting the public forests from the ravages of the timber-cutting industry. The effect of President Nixon's 'directions' to the Agriculture, Housing, and Interior Secretaries was to do by executive fiat what could not be done legislatively. Conservation, environment, ecology—that is, the public's concerns—are to be subservient to the pressures of the logging and lumber industry."

On grazing: "These bills (introduced by Representative Aspinall and Senator Gordon Allott, both of Colorado) seek to give a twenty-year renewable right to graze the public lands with little or no limitation placed on how the grazing would be performed. Under the Aspinall-Allott proposals, ranchers would be compensated for any reduction in grazing privileges. This is the antithesis of the free-roaming, cowboy, western, frontier spirit. It is the guaranteed-income plan for stockmen."

On predator control: "The war on predators has been waged with little scientific knowledge of their beneficial roles, or with little moral or ethical consideration for man's responsibility for conserving natural life as an integral part of the environment. The operations of one division of the Bureau of Sport Fisheries and Wildlife, sanctioned and sheltered by one administration after another, are sinister and contemptible. Yet it continues unleashed and virtually unchallenged."

Representative Saylor was not the only voice in the wilderness or for the wilderness. Yet he gave a new dimension to the politics of conservation. Although he received the John Muir Medal from the Sierra Club, he also was given the Distinguished Service Award from the ultraconservative Americans for Constitutional Action, showing that conservation belongs to no party and to no single point of view. He rowed upstream against the Watergate tide of his time to prove that politics could be used for good purpose.

Dealing with Congressional Committees

The Forest Service over the years has engendered considerable acceptance and respect on Capitol Hill. The Service deals with a variety of committees, which process about 600 to 800 items of forestry-related legislation a year. With the exception of appropriations, legislative activity is centered in the agency's Legislative Affairs Staff. It reviews congressional developments daily, obtains and distributes bills to other offices concerned, prepares legislative reports, and assists in developing testimony.

Of the principal committees involved, the Senate Committee on Agriculture, Nutrition and Forestry and the House Committee on Agriculture are concerned with forestry in general and with national forests other than those created from the public domain. The Senate Committee on Energy and Natural Resources and the House Committee on Interior and Insular Affairs have jurisdiction over national forests created from the public domain, grazing, mineral resources, petroleum development, mineral land laws, and water projects of the Bureau of Reclamation. As indicated earlier, about twice as many bills bearing on the Forest Service are referred to the Interior committees as to the Agriculture committees, and the budget of the Forest Service is considered and processed by House and Senate Interior appropriation subcommittees rather than the subcommittees on Agriculture. Still other committees have jurisdiction over civil works projects of the Corps of Engineers and over roads and highways, including forest highways, roads, and trails.

Even at best, the committee system in Congress has obstructed development of a broad natural resource concept and action program. Both houses are so fragmented and compartmentalized that senators or members of Congress rarely get involved in anything except their own committee business, unless it bears directly on their districts. Despite efforts at congressional reform, the rule of the road known as "congressional courtesy" keeps them in their own bailiwicks. Their lives are largely a matter of perpetual motion—attending committee hearings, caucuses, floor sessions, lunches, dinners, and receptions with constituents and lobbyists, relying on staff personnel, in their own office or in the committee, for data and guidance on legislative issues of growing complexity.

Sometimes committees operate against each other. The House Merchant Marine and Fisheries Committee may be trying to save wetlands for waterfowl habitat, while the Senate Agriculture, Nutrition and Forestry Committee authorizes the expenditure of millions of dollars to drain wetlands for agricultural purposes. Or, even in the same committee, one subcommittee may be pressing to preserve wilderness in contrast to another subcommittee striving to open wilderness to energy development.

The Symbiotic Relationship

The legislative process, despite talk about congressional prerogatives, essentially is a joint undertaking. Seldom are laws passed without participation, if not approval, of the agencies involved. Each bureau identifies its own legislative needs and prepares proposals for inclusion in the program of its department and of the administration currently in office.

The larger the budget of the agency the louder its voice in the halls of Congress, if not directly then indirectly through various private organizations and industries, the "constituency groups" that have become, for better or worse, a part of the legislative process. The Forest Service has a substantial base of built-in legislative interest and support, due to the large areas it administers in many congressional districts, the number and distribution of personnel it employs in national forests and experiment stations, the "25 percent money" returned through the states for county roads and schools, the commerce in timber and range, the commerce and recreation in tourism, and assistance provided to state and private forest interests. These attractions are not as overpowering as those offered by construction agencies, such as the Army Corps of Engineers, Bureau of Reclamation, Tennessee Valley Authority, and Federal Highway Administration, which provide irresistible channels for "putting money into the districts" and stimulating real estate values.

Virtually every bureau prepares drafts of bills for congressmen, presumably in response to requests. The agency that succeeds, however, cultivates understanding and support among key members of both houses through a variety of contacts in Washington and at the grass roots. The "industrial-military complex" is the classic example of symbiosis to sustain a high level of appropriations and influence in Congress. Or call it the "Iron Triangle"—a combine of bureaus, interest groups, and Congressional subcommittees. The "highway lobby" is known for its liaison with the Federal Highway Administration. It would be difficult, if not impossible, to identify a single federal bureau that does not have a constituency on the outside, an organization of some sort lobbying in its behalf, and always the clarion call is for an increase in appropriations.

Insofar as the Forest Service is concerned, its legislative programs are followed keenly by a spectrum of interests, ranging from citizen conservationists concerned with environmentalism and ecology to intensely commercial industries dependent upon raw materials from public lands. They are all allies of the Forest Service at one time or another—individually, but rarely collectively. Logically, this should tend to keep the agency in the middle, but logic does not determine the point of balance. Leaders of all agencies are reluctant to cross swords with the economic-political power structure and thus jeopardize their positions.

If the executive branch shares in shaping laws, it is also true that the legislative branch exercises influence in the field of administration. Any congressman who obtains appropriations for a new area or new installation considers himself entitled to a say in how it should be run. Congressional staff personnel, particularly of committees, at times overlook bureau doings with eagle eyes and form alliances of their own. They may or may not be acting at the direction of their senators and congressmen, except in the broadest sense, but their bosses plainly are relieved to have aides cope with burdensome data and details. The alert regional forester and forest supervisor keep in contact and consult with elected public officials, without ever ignoring the least or lowest staff member—who may be more sensitive to political power than his boss and find a vengeful way to strike back.

From Law to Litigation

"Without a law, you can't litigate," wrote R. W. Behan, Dean of Forestry at Northern Arizona University, in the *Journal of Forestry* of December 1981. He was urging repeal of RPA and NFMA as "a guaranteed way to help get management decisions out of the courtroom and back into the forest."

But would this actually happen? And should it?

During the 1960s and 1970s, changes in law and legal doctrine opened the courts to lawsuits of types they had not heard before. Until then the courts had limited their own role as to who can sue, what issues a court would review, the scope of the review, and the kinds of legal remedies. The Supreme Court played a key role in charting a new course, largely through the influence of Justice William O. Douglas. It was he who raised the question, "Should trees have standing?" and answered in the affirmative.

In the Hells Canyon decision, handed down by the Supreme Court in 1967, the legal issue was whether a dam should be constructed by private or public power interests. The case involved the Grand Canyon of the Snake River, the deepest gorge formed by any river in North America, which lies in the heart of national forest country astride the Oregon-Idaho border. The Court, in a decision written by Justice Douglas, questioned the wisdom of building any dam at all.

"The test is whether the project will be in the public interest," the Court declared. Among considerations that must be weighed it listed: future power demand and supply, future alternate sources of power, and "the public interest in preserving reaches of wild rivers and wilderness areas, the preservation of anadromous fish, for commercial and recreational purposes, and the protection of wildlife." The dam was not built; the Federal Power Commission proceeded anew with the Court's directive, while citizen conservationists pressed for legislation, ultimately with success, to establish a national recreation area, a wild and scenic river wilderness complex encompassing the Snake River as it flows through Hells Canyon and the surrounding lands above it.

During that period of the 1960s, a prelude to the surge of environmental litigation of the early 1970s (and the subsequent reaction against environmentalism within the body politic), citizen groups introduced suits on which the Forest Service might look with favor but could not pursue itself. In one case, the Izaak Walton League succeeded in blocking a mining syndicate from exploration in the Boundary Waters Canoe Area of Minnesota. In another, the Highlands Conservancy of West Virginia sought, and obtained, a restraining order against a coal company from cutting roads into a *de facto* wilderness area, known as Otter Creek, in the Monongahela National Forest.

Between July 1, 1964, and June 30, 1974, a total of 74 lawsuits was filed against the Forest Service. James W. Moorman, as executive director of the Sierra Club Legal Defense Fund (he later served as Assistant Attorney General in charge of the Land and Natural Resources Division during the Carter administration), was involved in a number of key cases. These included suits to block a fifty-year logging sale in the Tongass National Forest of Alaska and the proposed Disney recreational development in the Sequoia National Forest of California. In both cases the firms involved withdrew their proposals before litigation ran its course.

"When government agencies not only select which laws to enforce, but begin to write their own, citizen groups have no recourse but to the courts," Moorman declared at the American Law Institute—American Bar Association Conference on Environmental Law held in San Francisco on February 9, 1974. "These agencies convert the substantive statutory mandates that govern it into mere discretionary guidelines. The government, even when ordered by a court to obey the law, will sometimes continue on its course of evading the law."

Moorman recalled that as a young lawyer working in the Natural Resources Division of the Justice Department, before his tenure with the Sierra Club, he was advised of the first line of defense in any law suit: Raise the doctrine of sovereign immunity by motion to dismiss. This doctrine holds that the government can only be sued with its own consent—that the courts should not stop government in its tracks—or, as Moorman alleged, derives from the notion that the king can do no wrong.

He claimed further that the government blocks legal review by setting up an obstacle course of "administrative remedies."

For example, the Forest Service takes the position that before you can sue with regard to one of its decisions you must file an appeal with a forest supervisor, work your way up to the Chief of the Forest Service and finally to the Secretary of Agriculture. Thus, the Forest Service requires four appeals before they will acknowledge that your client is entitled to sue.

Even one review by an agency can be justified *only* if there is a real opportunity for reversal. A four-step appeal cannot be justified under any circumstances. The system is simply used to exhaust plaintiffs, not remedies.

We have been discussing how a lawyer must prepare to respond to the government's drive to act in a discretionary fashion, to avoid review, and to avoid compliance with court orders. Is there any way, you may ask, to solve this problem on an institutional basis, rather than in *ad hoc* litigation?[3]

This, in a sense, is the same question raised later by Dean Behan: on how to get "forest management decisions out of the courtroom and back into the forest." For the moment, consider only Moorman's response:

> There is only one way. That is for government officials to create a climate in which political pressures are not allowed to overwhelm all other elements of federal decision-making. To base decisions on discretion, while ignoring the written law, weakens the rule of law.[4]

The forest manager's discretion was the point at issue again in the 1960s in the lawsuit that blocked a proposed logging sale in the East Meadow Creek drainage of White River National Forest, Colorado, on the grounds that a *de facto* wilderness ought to be studied for possible inclusion in the Wilderness System along with the adjacent Gore Range–Eagle Nest Primitive Area. The major purpose of the Wilderness Act, a federal judge in Denver declared, was to remove absolute discretion from the Secretary of Agriculture and the Forest Service by placing the ultimate responsibility for wilderness classification in Congress. The East Meadow Creek decision expanded that concept to include not only primitive areas, but also adjacent roadless areas.

Such environmental constraints by Congress and the courts were destined to limit the actions of the Forest Service and other resource agencies. In an affluent society, with a large body of well-educated citizens, such matters as natural beauty, pollution of air and water, vanishing wildlife, and vanishing open space, became significant political issues, challenging the traditional economic power center on a steadily rising curve—at least until the reaction set in with arguments of energy and jobs versus environment. Court decisions and laws such as the National Environmental Policy Act of 1969 evoked many changes and forced agencies to review their progress.

The Monongahela decision—*Izaak Walton League* versus *Butz*—was the major case involving timber cutting, leading ultimately to revision of the 1897 Organic Act and to the NFMA planning process. Another significant case was brought by the Texas Committee on Natural Resources in 1976. This group asserted before Judge William Wayne Justice that national forests in Texas were being mismanaged, in violation of the Organic Act, Multiple-Use–Sustained-Yield Act, and NEPA.

The citizen group had urged a delay in a commercial sale proposed for the Four Notch Area of the Sam Houston National Forest pending a wilderness review and had been denied. The suit charged that forest resources were being damaged by overemphasis on clearcutting. The judge issued a temporary restraining order, affecting twenty-one current sales, pending a courtroom trial.

During the trial, in December 1976, the government benefitted from intervention by timber companies with a considerable stake in the national forests. The Texas Committee introduced a variety of witnesses, including Dr. W. Frank Blair, University of Texas ecologist; Charles

Stoddard, a former director of the Bureau of Land Management and author of a textbook on forestry; Gordon Robinson, a forestry consultant to the Sierra Club; and various specialists in recreation, ornithology, botany, and economics. Their testimony clearly impressed Judge Justice. On March 24, 1977, he enjoined the Forest Service from any new clearcut logging sales in Texas. He found that clearcutting on the massive scale pursued in Texas violated the Multiple-Use–Sustained-Yield Act, by impairing recreation, wildlife, and soil conservation, by increasing the hazard of fire, and probably by increasing the depredation by insects and disease. He also concluded that the Forest Service had violated NEPA by failing to file an environmental impact statement comparing the effects of clearcutting with those of selective logging.

The government appealed forthwith. It asserted that Judge Justice had erred by ordering a single environmental impact statement for all four national forests in Texas. More pertinent to the issue of management prerogatives and principles, the appeal charged the court had improperly substituted its judgment concerning the environmental consequences of even-aged timber for that of the Forest Service and Congress—and especially so "where matters of scientific judgment are concerned."

The U.S. Court of Appeals in 1978 unanimously overruled the lower court. The Appeals Court declared the recently passed NFMA was intended to allow clearcutting to continue until the Forest Service could adopt new guidelines, and thus the suit was premature.[5]

This reversal was a signal victory for those who object to litigation in resource management, who feel that projects planned by the executive branch and funded by the legislative should not be subject to frustration by the judicial branch affording venue to any citizen who disagrees with justification of such projects. As Assistant Secretary of Agriculture M. Rupert Cutler declared before the Texas Forestry Association (one of the intervenors in behalf of the government) on May 19, 1978: "Winning that appeal means that the people have won the right to have their forests in Texas managed in the best way possible. Under NEPA, a court is not empowered to substitute its judgment for that of the concerned agency as to the environmental consequences of its action, especially where matters of scientific judgment are concerned."

This viewpoint, a sharp digression from Cutler's normal pattern, is at variance with that expressed by Dr. Joseph L. Sax, professor of law at the University of Michigan in his well-known book, *Defending the Environment*. Sax saw an urgent need for repudiation of traditional reliance on government professionals, for a shift in the center of gravity of decision-making, and especially for an enlargement of adversary processes through litigation. "One wonders," he wrote, "why government briefs, devoted in such cases to assuring judges that administrative discretion should be trusted implicitly, are so ready to assume that judges themselves are such fools or such puppets they cannot be trusted to make a rational distinction between substantial and frivolous claims."[6]

In his *Journal of Forestry* article calling for repeal of RPA/NFMA, Dr. Behan discussed various aspects of decision-making via management process, Congress, and the courts. In a companion piece, James W. Giltmier, a professional staff member of the Senate Committee on Agriculture, Nutrition and Forestry (later on the staff of Senator John Melcher of Montana), defended the legislation with conflicting reasoning.[7] Behan wrote,

> Without a law, you can't litigate. From the perspective of the various forest users, repealing the law removes the necessity for litigation. They can then pursue their several objectives with means and tactics that are certainly less expensive and potentially more effective.

He asserted that requirements of legal proceedings impose a distortion on the planning process, citing in particular the plan of the Lolo National Forest, the first one scheduled for completion under NFMA. It is estimated it would have cost nearly one million dollars in preparation. Behan decried the consequences:

> The legal challenge, lodged by the Inland Forest Resources Council on behalf of the forest products industry, was professional, massive, and immediate. Is this the first example of an imperfect plan—and therefore an illegal one? If so, there is just one rational response for Daniels [the forest supervisor] to make: He has to redouble his effort, hire more planners, expand the interdisciplinary team, refine his data with a more comprehensive and more sensitive inventory, consider more alternatives, and run FORPLAN until he chokes the computer. And he will have to document every step to be ready for litigation later."

For Behan the key is to effect better public administration without placing too much faith in either planning or statute:

> I believe a new pattern of public forestry is in order, in which the forest manager sees his task as actively solving public problems, not passively executing public laws. It will call for initiative, courage, imagination, and the fashioning of unique and independent decisions to fit unique and independent situations, often through mediation and interaction with user groups.

That is one approach. Giltmier, on the other hand, writing from the legislative viewpoint, defended RPA as being instrumental in getting higher appropriations for the Forest Service by warning Congress of potential loss in benefits without adequate funding. More important:

> In both laws, Congress made a series of tough and complicated political decisions, and those decisions were made in order to keep the debate over land management out of the courts, as close to the forests as possible, and to give land managers the freedom to manage within certain limits.

The alternative would have caused the suspension of most timber harvesting in the national forests because of the Monongahela lawsuit, or the suspension of clearcutting, as proposed by Senator Jennings Randolph. . . . My own study has indicated that the Forest Service plans so that it can manage better. There is nothing wrong with that, the anticomputer Luddites notwithstanding. Clearly the planning process is teaching Forest Service people to have a better understanding of the interrelationship of the resources.

But what about the question of professional judgment versus legislative action or litigation in the courts? Here is a part of Giltmier's critique:

Congress did give land managers a lot of discretion about how they do their job, and that's good. But what kind of institution functions in such a way that the junior professionals tell the bosses how to run things? . . . What do foresters know about the place of the National Forest System in the scheme of federal spending priorities? What do foresters know about personnel ceilings or travel restrictions, except that they affect their ability to manage, and there is nothing they can do about them?

The notion that management decisions should only be made by foresters is bunk. For example, consider the differences in policy direction taken by former USDA assistant secretary M. Rupert Cutler, on one hand, and the present assistant secretary, John B. Crowell. Yet both men did or will accomplish most of their management objectives while staying within the law.

Lawsuits From Diverse Corners

James Giltmier asserted in his *Journal of Forestry* article that fewer than five lawsuits related in any way to RPA had been filed through the end of 1981. But there were plenty of other cases, with the likelihood of more to come during the rest of the 1980s. Often they reflect the struggle over wilderness and pressures for energy development.

• In two separate cases, the State of Minnesota and several associations of property owners challenged the jurisdiction of the federal government in the Boundary Waters Canoe Area and the constitutionality of the law establishing the wilderness there. But in September 1981, the Eighth Circuit Court of Appeals upheld the government. Among other points, the court ruled that the property clause of the Constitution granted the federal government authority to restrict snowmobile and motorboat use and the right to first refusal in acquiring inholdings.[8]

• In January 1980, the Southern Appalachian Multiple Use Council, representing the forest products industry, filed suit in an effort to block designation of wilderness in eastern national forests—specifically, 20 areas comprising 83,000 acres withdrawn from harvest because they were recommended for wilderness classification or withdrawn administratively for "further planning." The Multiple Use Council challenged the "assumed power and authority of the Secretary of Agriculture to designate additional areas of the national forests as study areas and to manage such areas during the period of study as wilderness areas." This suit was decided in District Court, in Asheville, North Carolina, in April 1981, by dismissing all plaintiffs except one individual having access claims to the area.

• In 1981, Citizens for Alternatives to Toxic Herbicides filed suit in federal court in Boise, Idaho, challenging an environmental impact statement covering a proposed plan to spray 60,000 acres of northern Idaho with toxic herbicides. The group alleged a lack of adequate data to evaluate the efficacy of the spray programs.

• Boise Cascade Corporation filed suit against the Forest Service during 1981 in an effort to get the agency to spray large areas of the Boise and Payette National Forests for spruce budworm. Boise Cascade owns about 150,000 acres of timberland adjacent to those two national forests and claimed the Forest Service was spraying only narrow buffer strips.

• Kansas-Nebraska Natural Gas Co., Petroleum Exploration, Inc., and several other parties brought suit in 1981 against the Bureau of Land Management and the Interior Department for the right to lease oil and gas properties within the Teton Wilderness in northwest Wyoming. The suit contended that the BLM violated the Wilderness Act when it refused to grant a noncompetitive lease, since the act authorizes such exploration until the start of 1984.

• The Sierra Club, on the other hand, asked a federal court to block the government from permitting oil and gas exploration in the Palisades, "wild and ruggedly beautiful federal land" in national forests in the Snake River Mountain Range of Wyoming and Idaho. The club charged that the Forest Service violated the law by not preparing an environmental impact statement before recommending that leases be issued. This suit was dismissed, the court finding that an EIS was not required.[9]

Strong crosscurrents and conflicts among executive, legislative, and judicial branches have been expressed over the issue of mineral exploration in classified wilderness. An earlier chapter on energy resources noted the contention over the Bob Marshall Wilderness. In May 1981, the House Interior Committee invoked a clause in the Federal Land Policy and Management Act granting power to a single committee of Congress to ban leasing in wilderness by declaring a state of emergency; the Committee used that provision to bar leasing in the Bob Marshall and adjacent Scapegoat and Great Bear Wilderness Areas. One month later the Mountain States Legal Foundation filed suit challenging the Committee's constitutional authority.

Presently the issue emerged as a struggle between Interior Secretary Watt and members of Congress over whether to encourage energy development in wilderness. In November, the scene shifted when the Bureau of Land Management issued oil and gas exploration leases covering 9,000 acres of the Lincoln National Forest, including a portion of the El Capitan Wilderness. In an editorial on November 20, the *Washington Post* commented that while the Wilderness Act allows leasing until 1984, no previous Secretary of the Interior had seen the need to do so until likely nonwilderness areas were explored and developed: "In the now-familiar style that seems to ask for a fight, Mr. Watt issued the leases without the legally required environmental impact study, without public notice and opportunity for public comment and without first informing Congress."

The Interior Department insisted, however, that the leases were issued by regional BLM officials on the recommendation of the Forest Service and without consulting Washington. Watt himself noted that the leases were issued with a Forest Service stipulation that the surface of wilderness could not be disturbed without an environmental impact study.

Democratic and Republican members of the Interior Committee responded with plans for a resolution barring mineral leasing in wilderness. Their move was buttressed by a study released by the Wilderness Society, based on Geological Survey data, showing that designated and potential wilderness areas together contain just 2 percent of U.S. oil reserves and 1.6 percent of gas reserves. Yet another study, issued in December by the congressional Office of Technology Assessment, found that the "extent of increased market demand, not the availability of additional federal leases, is expected to determine the amount of coal that will be produced." The study identified billions of tons of coal reserves already leased but not yet mined.

Some kind of legislation, or legislative review, appeared called for. Under FLPMA a congressional committee may be empowered to rescue wilderness with a declaration of emergency, but the Interior Secretary, whoever he may be, has the legal right and authority to determine the scope and duration of any ban on leasing. Lawmaking, even at its best, is likely to further litigation. Providing the facts, the effects, and the options is a useful role that resource managers will be called on to perform to clarify issues and fulfill their responsibilities.

19

FORESTRY EDUCATION OF YESTERDAY AND TOMORROW

> The professional forester of the future must be a man who is well grounded in the principles of forest land management, together with the underlying arts and sciences on which such management is based. At the same time he must have the breadth of view and the understanding of the economic, social and political world in which he lives to enhance his competence not only as a professional man but as a citizen and an individual.
> —Samuel T. Dana and Evert W. Johnson, in *Forestry Education in America Today and Tomorrow.*

Early Forestry Schools

Forestry education and the Forest Service have been very closely linked ever since Pinchot's time. Pinchot himself was instrumental in founding the Yale School of Forestry, now the country's oldest forestry school in continuous operation. Yale provided one-half of the professional staff of the Forest Service during its formative period (and six of the first ten Forest Service chiefs).[1]

Cornell, however, was the first college to offer a professional forestry curriculum with a bachelor's degree. The program there started in 1898 under Bernhard E. Fernow, who had headed the Division of Forestry in the Department of Agriculture, but it was interrupted for seven years in 1903 when the state withdrew support in a dispute over Fernow's cutting practices, and the New York State College of Forestry subsequently was shifted to Syracuse. From 1911 to 1936 both Cornell and Syracuse had professional forestry schools, but Cornell's was closed in 1936. The Biltmore Forestry School also was established in 1898, directed by Dr. Carl Alwin Schenck, the German *Forstmeister* in charge of George Washington Vanderbilt's estate outside Asheville, North Carolina. Biltmore was a one-year program to teach practical forestry to the sons of lumbermen. Pinchot had been instrumental in bringing Schenck to Biltmore (where he himself had been brought earlier by Frederick Law Olmsted, the famous landscape architect), but they had a falling out. Nevertheless, Schenck, the sole instructor, was admired by his students, who assisted him as forester for Biltmore Forest and accompanied him on trips to Europe every second year until the school closed in 1913.

The idea of the graduate school at Yale was conceived by Pinchot in 1900, while he was Chief of the Division of Forestry in the Department of Agriculture, in collaboration with his closest crony, Henry Solon Graves, the Assistant Chief. "I recall saying one evening to Pinchot: 'If you and your family will give an endowment for a Forest School at Yale,' " Graves noted many years later, " 'I will go up and run it.' This was in line with my usual modesty in suggesting my own name for positions." Pinchot's father, James Pinchot, not only gave $150,000 to found the school but opened the family estate in the Poconos for use as the summer field camp. Other early schools included Michigan State and the University of Michigan, founded in 1902; Maine and Minnesota, in 1903; Iowa State, Oregon State, and Harvard, in 1904.

Pinchot played a continuing role in forestry education. In 1909, while Chief of the Forest Service and President of the Society of American Foresters, he called the first national conference on forestry education, laying the basis for standards in instruction and involvement of SAF as the recognized accrediting agency.

In his book, *Professional Forestry in the United States*, Henry Clepper describes the early period of forestry education:

Although the first two schools (Cornell and Biltmore, both of short life) were headed by German foresters, professional forestry education in the United States was characterized from the start by a fresh and independent spirit influenced by the subject matter, teaching methods, and educational customs of western Europe. With the urgent need for training men for a profession just coming into existence and lacking educational traditions of its own, the early schools were quite different from their European prototypes. Forestry education, like the whole conservation movement, was markedly influenced by the crusading fervor of Gifford Pinchot. An essential goal of that crusade was to reverse the trend of timber destruction and to bring about the management of millions of acres of publicly owned woodland acquired and being acquired by the federal government and the states. Forestry educators entered into this program with a zeal that persists even seven decades later.

Moreover, during the first three decades of professional education, there was concentration on those subjects that best served American needs. For example, a major influence on all schools was the written examination given by the U.S. Civil Service Commission to recruit junior foresters for positions in the federal government. On approaching their graduation, students were encouraged to take the examinations, whether or not they aspired to careers in federal service, and many schools kept detailed records of the examination ratings of their graduates. A controlling, if not acknowledged, force in subject matter teaching was its contribution to the ability of students to pass examinations. The prestige of some schools was based on this factor.[2]

The Forest Service down through the years has been a major source of employment for forestry school graduates, a source of faculty members with practical as well as academic experience, and a source of considerable funding, although the forest products industry has played a significant role since World War II. The Forest Service, in fact, has ties with more than sixty schools through provisions of the 1962 McIntire-Stennis Act (authorizing federal support for forestry research at land grant colleges administered by the Cooperative Research Service of USDA). To indicate the value of this support, according to Dr. Richard A. Skok, Dean of Forestry at the University of Minnesota, it is not uncommon for a school to derive 50 percent or more of its budget from research-directed funds.

Criticism and Change

Forestry education has undergone various phases of change and been subject to serious criticism from both outside and inside its ranks. Some of it has been sharp and bitter.[3]

Guy M. Brandborg (former supervisor of the Bitterroot National Forest), speaking at a seminar at the University of Montana in February 1970, charged:

There must be something lacking in the training of forestry school graduates because, as an example, even now they are defending the overcutting practices as well as soil and resource destruction that is occurring in the Bitterroot. This is taking place despite known social and economic consequences upon residents of the area. It should be appropriate, therefore, to inquire if the public can justify continuing to spend hard-earned money for the training of foresters in our publicly financed institutions if, as graduates, they are unwilling to manage land in a manner that will pass these resources to the next generation in a more productive condition.

Vernon Bostick, formerly a Forest Service employee and an educator, in a letter to the author of March 1971, wrote:

Forestry schools have changed considerably since I was an instructor in forestry and range management at Washington State College, and since I was a forest ranger on the Carson National Forest. In those days sustained yield and selective cutting were our guiding principles. We trained foresters for government service and we taught conservation. I used the Copeland Report and Senate Document 199 as texts in my classroom.

What the forestry schools want is more jobs for the ever-increasing number of forestry graduates. There are as many students taking forestry at Colorado State now as the total enrollment when I attended CSU. To broaden employment opportunities, foresters are now trained for private industry instead of as conservationists.

As for criticism from within the educational ranks, Charles H.W. Foster, Dean of the Yale School, now called the School of Forestry and Environmental Studies, in an interview in 1980 declared:

> There has indeed been considerable dissatisfaction with forestry education on all fronts. Industry feels graduates are not like the technical foresters of old, that they're not prepared for timber cruising and other field work. Public agencies, on the other hand, feel they haven't had enough exposure to other disciplines.
>
> About 20,000 students are enrolled in forestry and related schools, graduating about 5,000 each year. But there aren't 4,000 job openings, leading to an oversupply of professionally trained people.
>
> Forestry schools are in some disarray, particularly since the rise of environmental awareness, not knowing what they should do to cope with it, facing competition with other programs. Clearly, educators are going through serious reflections. Foresters need to expand in human dimensions, to be people-related in order to deal with new constituencies.

Certain changes have been made. *Forest Resource Management: Decision-Making Principles and Cases*, published in 1979, represents the work of 33 authors drawn from forestry schools, public agencies, business concerns and, to a certain extent, from outside the timber-forestry circle. It was the end product of a long-term Forest Curriculum Development Project. In the preface the four principal authors (William A. Duerr, Dennis E. Teeguarden, Neils B. Christiansen, and Sam Guttenberg) define their objective: "A major aim is to focus the work upon integrated forestry: the creation and use of all the forest values from scenery to wood. We have tried to see forestry as a system of interacting variables and also as part of a larger system that in the final analysis is the sum of human experience."

Experience at the University of Florida illustrates an adaptation of this approach. The School of Forestry began in 1937 with an under-graduate major in forestry. During the 1960s the school added a distinct separate major in wildlife, with modest options in range management. Along with other forestry schools, Florida during the 1960s strengthened requirements in biological sciences, physical sciences, mathematics and communications—but at the expense of some applied forestry courses, including a summer field camp. In March 1970 Dr. John L. Gray, Director of the School, told a meeting of the Southern Section of the Society of Range Management:

> At this stage, my opinion is that the professional forester being turned out by most of the 46 colleges and universities offering undergraduate majors in forestry is prepared to manage forested lands, primarily but not exclusively for timber, and to learn fiber farming, or factory forest management. Generally, he is not well prepared to become a river basin planner, for example, or the manager of a land company with range, timber, water, recreational and wildlife resources, or enterprise possibilities.

The following year, 1971, the School of Forestry changed its name to School of Forest Resources and Conservation and modernized its curricula with a stronger foundation in ecology, management science, policy, and communications. Nevertheless, in September 1972, in a paper before the Southeastern Section of the Society of American Foresters, Dr. Gray declared:

> Some current problems between foresters and wildlife biologists in my own state and even in our own school are evidence of a problem here. The lack of understanding and effective cooperation in the interest of forest resource owners, the public and the images of both professions are at times almost appalling. For example, we spend a lot of time talking to (but not listening to) one another about the relative merits of diversity of undisturbed natural ecotypes versus wall-to-wall slash pine. We ought to be working together to identify and objectively evaluate some realistic modifications in each system of management.

Many forestry schools entered the decade of the 1980s in transition, endeavoring to broaden definition, curricula, and faculty, interacting with other divisions and schools on campus. Some feel more positive about progress than others.

Dr. Richard W. Behan, Dean of Forestry at the University of Northern Arizona, for example, in a letter of October 1981:

> I do think we are undergoing a revolution in forestry education. The single-use obsession with biological tree production is just about dead.
>
> At our school we have three 16-credit blocks stretched over three semesters of integrated "seamless" instruction. Everything is team-taught, rather than in single courses. We try to look at the forest as a single entity, not as "timber," "range," or "wildlife" in separate boxes. And we try to teach the handling of that entity as an enterprise in management, not biology. We are still working out some of the concepts and practices, but the forestry education we offer at NAU today is light-years away from my own graduate experience.

Dr. John H. Ehrenreich, Dean of the College of Forestry, Wildlife and Range Sciences at the University of Idaho, shares much the same view. At Idaho, "integrated resource management" is the stated goal. No matter what their major, students are offered a grounding in other disciplines taught at the college. Then their training is capped with courses in modeling, with inclusion of social factors in drafting management plans.

"We strongly emphasize ecology, including human ecology, and quantitatives as a framework to prepare them for the future—for the time when every office will be equipped with a computer," Dr. Ehrenreich explained in an interview in June 1982. "Nevertheless, a high degree of our program is field-oriented." A 7,000-acre school forest is located only 15 minutes from the campus at Moscow. Other field camps include a location in the heart of the River of No Return Wilderness, the largest designated wilderness outside of Alaska.

Though once strictly a regional forestry school, strongly focused on timber production, Idaho has broadened extensively. Following visits to the People's Republic of China by two professors in Wildland Recreation Management, William McLaughlin and Gary Machlis, the Government of China sent two of its professionals to Idaho in 1980 for training programs in wildland recreation and national park development.

Dr. Benjamin A. Jayne, Dean of the School of Forestry and Environmental Studies at Duke, is less certain, even though his school is considered one of the most progressive, in a letter to the author of August 1981:

> Substantial changes have taken place during the past ten years or more in both public and private forest practices. Although marked changes in the education of foresters have also taken place, I am personally sometimes dismayed at the slow pace of change during the past 25 years. When compared with professional education in both engineering and business, I believe that we have gained little, but rather may be getting further behind.

Once concerned almost exclusively with problems of managing southern forests, Duke has shifted to addressing problems of managing natural resources, though still using the South as a resource for study. A graduate school offers degrees in both Master of Forestry and Master of Environmental Management, based on three areas of teaching and research: natural resource science, economics/policy, and systems science. The Duke Center for Resource and Environmental Policy Research concerns itself with land-use planning, energy systems policy, international environmental policies, water resource policy, and historical, cultural and ethical considerations in resource and environmental policy.

Duke features a cooperative "3-2" program, designed to produce a balance between liberal and technical education. It involves an initial summer session plus two full academic years at Duke, with 60 academic units, and the remainder (covering three years) at another school. On completion, the student earns a bachelor degree from the undergraduate institution and a Master of Forestry, or Master of Environmental Management, from Duke.

Dr. Carl Reidel, Director of the interdisciplinary Environmental Program at the University of Vermont, notes that while current forestry educators are better equipped with technical expertise, they lack experience in the field, and that curricula suffer from reduction or elimination of field training camps. But this is only part of his criticism. In a challenging 1977 speech to the Northern California chapter of the Society of American Foresters, he declared:

> How can foresters be trusted to manage a natural preserve when they can't identify the birds or wildflowers on the area, much less understand

304 FORESTRY EDUCATION OF YESTERDAY AND TOMORROW

their special environmental requirements? How can foresters even be trusted to mark timber for intermediate cutting if they are insensitive to the practical problems of felling a tree or locating a ski trail on a rocky side hill?

For those of us several decades out of college, this return to "good plumbing" will mean more intensive retraining. And for more recent graduates, it means continuing education and on-the-job training. For future forestry eduction, it will mean a return to field training camps and supervised internships. That will require some major changes in curricula, perhaps replacing some advanced analytical studies with applied training for those foresters who intend field careers upon graduation. We simply cannot train every forester in advanced computer sciences and biochemistry if we are also to produce foresters with credibility as land managers.[4]

Forestry education in this country, dating from Yale, Cornell, and Biltmore, is more than eighty years old. Accreditation of forestry schools by the Society of American Foresters began in 1935, when fourteen of the then twenty schools, or 70 percent, were accredited. By 1980 a total of forty-three of the fifty-two forestry schools, or 83 percent, were accredited—and there actually were 2.6 times more forestry schools than in 1935. Though some foresters feel that accreditation can and should limit the number of degree-granting forestry schools, Dr. Richard A. Skok, Dean of Forestry at the University of Minnesota, is not among them. "Accreditation or reaccreditation has provided a leverage for the marginal school where the institution itself was willing to respond to negative reports and strengthen its program," contends Dean Skok.

Most universities have moved aggressively into systematic reviews for all their programs. As university enrollments are anticipated to decline in the years ahead, retrenchment and allocation are terms becoming all too familiar to those of us dealing in academic administration. Thoughtful and forthright accreditation reports will have much higher currency in the years ahead both for forestry school faculties and administrators and for the decision makers in the higher reaches of the universities.

In "A Frank Appraisal of Forestry Education," presented before the New England Section of the SAF in March 1978, Dean Skok raised several key points. Forestry education programs, he said, have been under the dominance of agricultural faculties and administrators who may lack understanding, yet hold control over curriculum, course offerings, recognition of faculty, and admission standards for students. This creates still another problem in holding on to field instruction. On this point Dr. Skok declared:

I know of few, if any, other aspects of the forestry curriculum that meet with such unified acceptance in the profession. Such instruction for our students comes at a high cost. The large student contact hour time required of the faculty, the need to work with small groups of students, and the locational factors all contribute to this. These higher costs when subsumed in the average dollars to educate a junior or senior in forestry for one year are difficult to justify to those unfamiliar with forestry when comparisons are made between programs within an institution.

To what extent should forestry courses be designed to meet the needs of potential employers? Even the best of schools have difficulty in detachment from the job market in the forest products industry. As Dean Skok sees it:

> Close interaction between employers and the schools is an essential means of guidance in curricular development. Advisory committees that review curricula, invitations to potential employers to visit the campus to speak and chair seminars, and conferences exploring educational needs represent some of the ways in which we need to more actively and assuredly maintain relevance and responsiveness. But our responsibility in guiding the profession in its evolution through education must extend beyond the employers. We need to continually assess the needs of all who are impacted by our profession's performance and what this means in the preparation of our graduates. We have not devised and applied effective processes to assure this happening.

Dr. Skok noted the surge in women forestry students from a negligible level in the 1960s to nearly 20 percent by the late 1970s. Nevertheless, he said, "we have not been successful in attracting minorities into forestry programs nationally, despite sincere efforts in this area by some public agencies, schools, and industry. The agenda for the future calls for us to seek creative and imaginative means and to assure accessibility to our programs for those from social and ethnic groups that are nearly nonrepresented in the profession of forestry at this time."[5] (Aspects of minority forestry education are discussed in Chapter 4.)

A brief review of the program at Colorado State University will bring this discussion close to the Forest Service. CSU has long been known as the "ranger school" or "ranger factory," a training ground for the Forest Service, National Park Service, and other public agencies. The first forestry course was offered as early as 1904, and the Forestry Department—which has since grown into the College of Forestry and Natural Resources—was organized in 1915. In the Department of Forest and Wood Sciences the newest major, Natural Resources Management, includes required course work in social sciences and humanities along with the normal forestry and natural resource subjects. In a senior course, Integrated Resource Management, students in four disciplines—recreation, wildlife, range, and forestry—deal in practical problems on the nearby Roosevelt National Forest; in the process of preparing an environmental impact statement, they are obliged to conduct public hearings, then negotiate and advocate among themselves.

"While this is a traditional source of new employees for federal agencies and we try to keep faith with the market, I'm not sure whether the curriculum is too constrained," commented Dr. Jay M. Hughes, Dean of the College of Forestry and Natural Resources (and an alumnus of the Forest Service) in a March 1982 interview.

> Possibly this reflects a tendency to seek pat rules in determining the outcome of issues on public lands.
>
> Perhaps we should be concerned with questions such as: Is the right kind of talent moving up? Or is there a gap, a vacuum, in the kinds of professions involved? Are Forest Service people getting sound training to handle public involvement? Do we know how to assemble data to make sound professional decisons? Or are we rooted in crisis management?

It is now generally conceded that students with different kinds of backgrounds, including undergraduate studies in liberal arts, are acceptable in forestry schools. They may lack the basic jargon, but the same may be said of principal executives of major timber companies, few of whom have training in forestry.[6]

Forestry training has been criticized as narrow, designed to measure values in terms of commodity production, and as turning out "wood merchants" rather than land protectors; yet forestry schools are an important part of the system of managing the nation's resources. Thus the self-examination and response to criticism, in whatever degree, represent progress in the public interest.

"Changes occur over time and we seldom have the capability to foresee the nature of these changes," observes Dean Skok of Minnesota in urging diversity in rules that the forestry profession imposes on forestry education and in rules that schools impose on individual students, and diversity among schools in their approaches.

One never knows how things will come out. In *Whose Woods These Are*, published in 1962, this author recorded the experience of a college-trained forester in the early years of the century, when most rangers were country-bred woodsmen and cowboys who didn't know silviculture and didn't wear uniforms. In that world of westerners, young easterners had hard going. Yet this ranger came to New Mexico from the sophisticated East (it was said that he had political influence as well as a desire to see the great open spaces) and proceeded to commit a grave antisocial indiscretion—he told one cow outfit he had seen another branding its calves. This caused his boot heels to be shot off in a saloon; to vary his step, he was roped and jerked up and down a high post. Next day he was told he must kill somebody to restore his authority as a government agent in the community. But he refused: that would, he said, be just another western crudity, while the subtler eastern revenge would be to cheat his enemies in a poker game. His relationship with the community seemed to worsen with time and an investigator was sent from Forest Service headquarters to probe his behavior. But by then his intrepidity had won many friends, and even his enemies resented outside bureaucratic meddling in the public life of a local official, college bred or not.

20

THE INDIVIDUAL'S ROLE AND GOAL IN PUBLIC FORESTRY

> I was determined to broaden the horizons of Forest Service people. I thought we were too hidebound, didn't have broad viewpoints, and were so wrapped up with our own agency we couldn't see what other people were doing. I don't know whether I succeeded, but I tried.
>
> I was trying to avoid the penalties of bigness. When an organization gets big we lose the sense of service. People who never looked at the clock stopped at eight hours.
>
> —Richard A. McArdle, former Chief, in recollections of the 75th anniversary of the Forest Service, 1980.

In the original edition of this book, published at the beginning of the decade of the 1970s, the final chapter titled "Toward Leadership in Environmental Forestry," made somewhat the same point as did Chief McArdle in his 75th anniversary recollections. The need to guard against narrowness is not exactly new. That last chapter quoted a warning delivered by Chief Ferdinand Silcox in 1933 against the bureaucratic danger of foresters becoming satisfied with their own decisions and permeated with a holier-than-thou attitude. The Forest Service, Silcox declared, must keep wide open the channels by which citizens could see for themselves and judge decisions, actions, processes, and their effects.

The alternative to involvement and alertness of the public is surrender to the unrelenting political pressures of the special interests. This has been demonstrated over and over and over again from Pinchot's time to this day. It was the reason for his "going public" over the heads of his superiors during the fight over Alaska coal lands with Interior Secretary Richard A. Ballinger.

The need for openness and involvement was plain again during the "Wild West Show" staged in Congress during the 1950s by Senator Pat

McCarran and Representative (later Senator) Frank Barrett in behalf of the livestock industry. For every pressure applied by the stockmen's associations, the American Forestry Association, Izaak Walton League, Wilderness Society, and many other groups responded with a vigorous counterbalance. Major magazines and newspapers of the country rallied to the defense of the Forest Service, though some would later become its severe critics.

The options between openness and acquiescence to powerful interests were evident anew during the late 1960s, in the fight over the National Timber Supply Bill. Only an outcry of the people succeeded in saving their lands from single-use allocation. And the same holds true during the 1980s, a time of sharpening conflict over a shrinking resource base.

Technical experts do not like to be told what to do with lands under their supervision. Still, on public land this can and will happen. And so it should, for no land-use program can succeed or long survive unless it is consistent with public interest or welfare, and has public support. The difficulty is that foresters, like any group of technicians, deal largely with men (and women) experienced in their own fields. Competence is judged to a great extent within the bounds of their professions, which goes back to their training, admittedly limited in scope. Today groups of Americans representing a nonprofessional, environmental point of view have become instrumental in shaping national policy.

To cite a paragraph from the last chapter of the 1971 edition:

Professor Lawrence Merriam of the University of Minnesota has suggested that the new forester join some of these groups as a participant; they might help him overcome some of his biases and offer him an opportunity to educate others. At the same time, in an age when providing wood is only one use of the forests, the agency can help itself through internal reorganization designed to bring ecologists, biologists, recreational planners, social welfare specialists, women, and racial minority spokesmen into decision-making positions in the hierarchy.

To some extent, limited or otherwise, that has been done. The final page of "Toward Leadership in Environmental Forestry" bore an optimistic note: "The Forest Service is in transition. Its plans point toward a better balance among programs, and it has expressed an intent to devote major attention to involving the public it serves. There appears to be a groundswell of feeling throughout the organization that a new phase is emerging."

Whether that new phase has quite emerged during the past decade may be subject to question, but the feeling for it is clearly widespread in the ranks. More foresters recognize that the central issue is not simply management in some technical sense, based on formula and training, but the selection and determination of qualitative values, as each individual channels them through his or her own mind. Land use, after all, embodies both science and philosophy, but the philosophy is more essential, and fundamentally a personal experience.

The Forest Service is an institution, but not a monolith. It is composed of people and ideas, old and new and changing; of currents and cross-currents, filtering upward and downward through levels of authority, with varying degrees of impact and influence. The stereotype image of the hardline green-shirt forester with "green blood" and chain saw on his arm may or may not be valid, but this forester is not alone in the organization. The biologist or hydrologist or some other professional on a national forest or a ranger district is apt to view the same scene with an entirely different perspective—and if he is courageous and competent, his ideas may prevail. By the nature of his or her broadened training at the kind of schools cited in the previous chapter, the young, or even the old green-shirt forester, assuming him to be the line officer, may want those ideas to prevail.

Any large institution reflects forces at work within. Loyalties tend to be directed to specific interests and activities. As the Weitzman Report indicated, timber management carries a lot of weight, like the first among equals. A loyalty may be channeled to State and Private, or Recreation, or Wilderness, and perhaps that's as it should be, among people who believe in what they're doing. Then there are loyalties shared, as among those who studied at the same school. ("Old Joe must be okay. He went to the U. of XYZ.")

What it comes down to, however, is the individual and his philosophy and personal goals. Can he be satisfied with job security or must he have something more to fulfill a legacy for his time on earth? Dealing with forest resources gives one a unique opportunity of stewardship that transcends by far the reward of security.

In reviewing the statements of others cited in the pages of this book, two in particular (besides those appearing at the opening of chapters) bear relevance to the philosophy of public forestry in these times. In Chapter 1, John R. McGuire, former Chief of the Forest Service, is quoted as follows:

> Those on the side of an increase in the pace of timber harvesting [on the national forests] argue that the current rate produces too low a return on the timber investment and delays the replacement of slow growing, overmature trees by a faster growing, more juvenile inventory. They also point to benefits from stabilizing employment and dampening the rising trend of stumpage prices with larger federal timber sale offerings. They typically urge substitution of an economic model for the biological model on which the concept of sustained yield policy is based.

The second statement, appearing in Chapter 19, relates to the above but is broader still. Dr. Richard A. Skok, Dean of Forestry at the University of Minnesota, is quoted on the value of interaction between employers and forestry schools:

But our responsibility in guiding the profession in its evolution through education must extend beyond the employers. We need to continually assess the needs of all who are impacted by our profession's performance and what this means in the preparation of our graduates.

These concepts can be applied like guidelines, reminders that certain ideas are larger than others, and that some should stand above one's technical training. In public policy and public administration, pressures of all kinds are very real, but the first step in coping with them is to shape a personal philosophy based on recognition of historic values, social values, and assessment of long-term impacts of one action as opposed to another.

Gifford Pinchot pioneered the scientific approach to forestry and conservation in America. Fundamental in his system was government control to protect natural resources. Yet he also appealed to the business community regarding conservation, demonstrating to the lumberman that the practice of sound forestry principles was good business practice. His combination of careful planning, use of government power, and appeal to business interests enabled Pinchot to attain results unmatched in U.S. resource policy or practice.

Pinchot appealed to the idealist and materialist. But is such an approach possible today, when industrial democracy, based on the small independent manufacturer and merchandiser and a strong agrarian influence, have given way to huge corporate and urban structures in society? If such an approach is still desirable, how can parameters be drawn to delineate between socially acceptable and valid practice and the unacceptable and invalid? Does it make sense, for instance, to export to Japan large-scale shipments of prime Pacific Northwest logs, while domestic sawmills close for lack of supply, and then require the harvest of public forests, which might better be allocated to other uses?

Is the forester's primary role in society "to keep the wood bin full," especially when it comes to filling it from public forests? Some will ascribe a "timber shortage" to three major failures: the refusal of small landowners to open their land for timber production, the withdrawal of timberland for national parks and wilderness areas, and the failure of the government to cultivate the national forests more intensively. Turn these "failures" into successes and the wood bin would be full—until demand again outstrips supply. But can there be other reasons for the "shortage," and other solutions?

It is said that if the *New York Times* were de-inked and recycled, instead of being burned, that in itself would save nearly 36 square miles of mature Canadian forests each year. But unrestricted use of these raw materials, "throwaways," and disposal by burning are more the American way. Logging, milling, homebuilding, publishing, and other such pursuits, after all, emerged during an age when forests were plentiful, even a nuisance. Those days are plainly over, as evident by efforts to eliminate waste and increase efficiency in the woods and mills.

Should the public forester and the Forest Service advocate recycling and the elimination of waste? While considerable progress has been made, should there also be a comprehensive program, a national goal, to achieve these ends?

In 1981, Dr. E. O. Wilson, of Harvard University, responded to a question posed by the *Harvard Magazine* concerning the most serious problem of the Eighties:

> The worst thing that can happen—will happen—is not energy depletion, economic collapse, limited nuclear war, or conquest by a totalitarian government. As terrible as these catastrophes would be for us, they can be repaired within a few generations. The one process ongoing in the 1980s that will take millions of years to correct is the loss of genetic and species diversity by the destruction of natural habitats. This is the folly for which our descendants are least likely to forgive us.

Forests and grasslands are significant repositories of natural habitats. The more complex and diverse the forest or grassland community the greater its stability, and its productivity in the long run. This scientific view is wholly compatible with the philosophic love of earth that attracts many personnel to the Forest Service. Given the role national forests can play in safeguarding genetic and species diversity, as well as watershed and soil values, should these public lands continue to be regarded as another accessible part of the raw materials pool? For that matter (to revert to the point raised by Dr. Skok), in this shrinking world should any landowner, large or small, public or private, be able to determine the use of his estate solely for his own benefit without regard for what his actions do to others and for society as a whole?

The Forest Service may or may not have devoted as much energy and leadership to facing these questions as one might wish, yet it should provide the forum, especially for those within its ranks. The agency has been assigned a broad mission: "To be a leader in assuring that the Nation maintains a natural resource conservation posture that will meet the requirements of our people in perpetuity." This goes far beyond forestry, though forestry is fundamentally sound in itself. As Pinchot pursued it, forests were for use, but Pinchot's forestry implies frugality and conservatism, based on respect for natural systems, coupled with prevention of waste and loss, rather than on sheer expediency.

"What this generation needs is another Pinchot," it is sometimes said, a statement inevitably followed by the rejoinder, "But Pinchot wouldn't last very long." While both observations may be correct, there is plenty of room for people inside and outside of government who feel impelled to think through the issues, find the answers for themselves, and then strive to bring them to fruition, whatever the odds. It's the Forest Service tradition—as typified by the ranger, far from headquarters, ready to sink or swim on his own—to do things this way. Times change, as Chief McArdle observed, and bigness induces the blight of blandness.

But individuals at all levels with minds of their own still never look at the clock, and they can make the difference.

Such individuals not only never look at the clock; they never look at only one year on the calendar. To them, as trustees, the horizons of the future become their pervading goals.

AFTERWORD

Michael Frome was very considerate in offering me the chance to get in the last word in his revised and updated version of *The Forest Service*, and I appreciate the opportunity. The offer was not without risks to him as an environmental writer and critic. Over the years, we have sometimes agreed, but quite as often disagreed about the best management of the national forests and the Forest Service's other programs. The dialogue has been candid and constructive even in our disagreements. After reading Mike's manuscript, I noted two characteristics that leave little room for rebuttal: (1) He has recognized that there is a diversity of viewpoints on many issues, even though he clearly expresses his own strong environmental advocacy, and (2) the book is a provocative, fact-filled work that should stir its readers to greater consideration and personal involvement in important resource issues. The latter characteristic may be its most notable service.

The book also reveals that Mike has experienced a strong critic-friend relationship with the Forest Service, softened by an expression of hope for the future. He is critical of the Forest Service, as well as many of the policies of the various administrations under which it has served over the years—including the present. In some respects, Frome regards the Forest Service as a prodigal son—capable of making many mistakes but equally capable of righting them by returning to the path he perceives would have been followed by Gifford Pinchot, the first Chief of the Forest Service. (His concept of what Gifford Pinchot would do today is subject to considerable debate, of course. For example, the author has been a long-term, vocal opponent of clearcutting as a timber harvest technique. Yet, as he indicates, Gifford Pinchot advocated clear-cutting under certain conditions.)

The author leaves no doubt about his feeling that the Forest Service over the years has been overly responsive to the commodity interests representing timber, grazing, drilling, and mining. But he has given specific recognition to some who vigorously argue that the Forest Service has been too responsive to environmental concerns.

This more balanced treatment has added a welcome dimension to the first version of the book written in 1971. Two of the themes in the first book were the "horrors" of clearcutting and the perceived inadequacies of the agency's public involvement efforts of the time. This more comprehensive update has provided a better perception of the subject of clearcutting by recognizing the diversity of views across the nation about when, where, and how clearcutting should be used as a harvest technique. And instead of window dressing, public involvement is now recognized by Frome as an important factor in the Forest Service's decision-making process—one that has matured in the decade of the seventies. We agree.

We think the nation itself has matured in that critical decade between the first and second editions of *The Forest Service*. Debilitating confrontation on environmental issues has been significantly replaced by the far more fruitful hard work of hammering out reasonable solutions that Gifford Pinchot would have heartily endorsed. He expressed it this way: "There are many great interests in the National Forests which sometimes conflict a little. They must all be made to fit into one another so that the machine runs smoothly as a whole. It is often necessary for one man to give way a little here, another a little there. But by giving way a little at present, they both profit by it a great deal in the end."

Pinchot would certainly have nodded knowingly if told that the principle was almost unanimously endorsed by the American people but putting it into practice doesn't occur without some disagreement and conflict. Such conflicts occur both in terms of the basic role national forests should play in meeting future needs and among uses for specific sites in those forests. In the past, disagreements largely involved differences between major user groups, such as recreationists and miners. More recently, the clashes have occurred between members within user groups, such as backpackers against trailbikers, or motorboat users against float boaters, even though they share the common interest of outdoor recreation.

Because it is the agency that must make the decisions, the Forest Service is the lightning rod of such conflict. One of the reasons is the important but sometimes difficult multiple-use concept under which the Forest Service operates. It is a concept encompassing wise use—from utility to preservation. To some people, utility and preservation often start from philosophical bases of incompatibility.

The formulas for resolution become very complex when one realizes that the national forests have approximately 230 million stockholders representing many different views. And they have the right to be heard because they are not only collecting the benefits but are paying the bills for management. They represent different needs and aspirations in relation to the use of their land. Some demand the tangibles, such as timber, forage for red meat, minerals and oil, and ample water. Others are more interested in the less tangible values of recreation, wildlife, wilderness, and natural beauty. The national forests have the capability of providing substantial amounts of each. Although most would argue in principle that "balance" is needed, there is frequent disagreement as to what represents "balance."

Thus, it is not surprising that the climate is frequently right for disagreement among these millions of "stockholders." It is expressed to their "board of directors"—the Congress—and also to their executive officers, from the President on down to the district ranger. Particularly, too, since 1960, the courts have been frequently called upon to resolve disagreements.

As the population of American "stockholders" grows, interest and controversy seem to grow even faster.

However, like Frome, we are optimistic about the future. The bright side of the coin is that the bulk of Forest Service decisions throughout the country have engendered support from the populations they affect. These decisions that are made under the Forest Service system of decentralized management receive little attention because satisfactory resolutions are not contested and, thus, are not brought to public attention.

Unfortunately, the opposite is not true. Controversial decisions immediately take the limelight, particularly if there are many players representing strong interest groups on different sides of the question. The intensity of some specific disagreements leaves the impression with many that controversy is the hallmark of the agency.

That isn't the case. I believe that every Forest Service employee can hold his or her head high because, even though unheralded or quickly forgotten, the agency's day-to-day performance over the years has added up to an enviable record in resource management and public service that enjoys much public support.

A prime reason is the integrity of Forest Service people. As my predecessor, John R. McGuire, said once, "The Forest Service has never had a major scandal"—and that record is still intact after 78 years. McGuire and every Chief before him set a standard for me, and all other Forest Service employees, of honesty and integrity in seeking to offer their very best for the public good.

That is not just a Forest Service judgment. As cited by the author in this book, a Pennsylvania State University study in 1981 selected the Forest Service as one of ten public and private organizations in the U.S. with model qualities that can improve understanding of how organizations can be successful. All ten are described as well-managed and highly productive for a sustained period of time. One of the characteristics noted for all ten was a demonstrated commitment to the highest ethical and moral standards.

This discussion of integrity brings me to one section which I think could leave the reader with a misconception of the agency's internal workings. It deals with the so-called "whistleblower" program. The author's presentation leaves the reader with the impression that the Forest Service is not tolerant of divergent opinions within its ranks.

We do, in fact, welcome opposing points of view. Such differing views contribute important information, as well as indicators of values that may be embraced by different users. Opposing viewpoints can lead to improved decisions, in some cases, through creative solutions that meld the best of seemingly different values.

When the facts and opinions are in, however, the "buck" has to stop somewhere, or paralysis results. If an employee cannot accept the decisions, then problems arise that may only be resolved in a climate of turmoil—through grievances, or voluntary or involuntary reassignments.

Because of Privacy Act restrictions, I cannot rebut the circumstances of the three cases cited by the author in Chapter 3. However, I can attempt to draw a distinction between them and "whistleblowers." These were not whistleblowing cases to counter "waste, abuse and fraud." Rather, they were cases of substantive disagreement by staff members with the decisions that were made—after consideration of their objections and the other facts and opinions.

Within any organization there will occasionally be disagreements, and it is not always possible to resolve them to the satisfaction of all concerned. However, I can state unequivocally that many changes have been made by the Forest Service as the result of both formal and informal suggestions from employees and the public, whether they involved formal "whistleblowing" or not.

Despite disagreeing with some of the author's conclusions, such as this one, I'm certain a thoughtful reader will quickly recognize two continuing, positive themes on which both Mike and I agree: The Forest Service is a proud agency with a proud tradition, and it has the capacity to manage change in stride with the changing public needs and desires for commodities, amenities, and environmental quality.

This book, *The Forest Service*, should not be considered an objective history, but a retrospective from an environmental advocate's vantage point, reduced to human scale by expression of many individual points of view. This approach contributes to its readability and provocativeness. What's more, the reader can sense a new note of hope for Mike's "prodigal son," a theme largely missing from the 1971 edition of the book.

Special credit, I think, should go to former Chief McGuire, who contributed to this book and who was the Forest Service's "point man" in the Environmental Seventies. He was largely responsible for helping with Congressional action that mandates a long-term comprehensive planning process for the agency. And it is on this foundation that renewed hope for the future rests. I am honored to have the opportunity to carry on the effort that has put the Forest Service in the forefront of the kind of planning that institutionalizes reasoned anticipation and advance problem solving as the alternative to crisis management.

NOTES

Chapter 1

1. John H. McGuire, "The Outlook for the National Forests," Horace M. Albright Conservation Lecture, University of California, College of Natural Resources, Berkeley, 5 February 1981, p. 7.
2. William F. Hyde, *Timber Supply, Land Allocation, and Economic Efficiency* (Baltimore: Johns Hopkins Press, 1980), p. 56. Gen. Ref.
3. Lee Smith, "The Neglected Promise of Our Forests," *Fortune*, 5 November 1979, p. 110.
4. Ibid., p. 124.
5. USDA, Forest Service, "1980 Report to Congress on the Nation's Renewable Resources," F.S. Publication Fs-347, 1980, p. 21.

Chapter 2

1. Theodore Roosevelt, *Autobiography* (New York: Scribner's, 1913), p. 290.
2. Carl Schurz, "The Need of a Rational Forest Policy," Sec. of Interior Annual Report Public Lands, 46th Congress, 2nd Sess., 1879 Ex Doc 1-5, Serial Set 1910, pp. 26–29.
3. Harold K. Steen, *The U.S. Forest Service—A History*, 2d ed. (Seattle: University of Washington Press, 1977), pp. 5–8, 23–46, 15, 16, 19, 71, 78–80, 95.
4. John Wesley Powell, *Report on the Lands of the Arid Region of the United States*, (Washington, D.C.: Government Printing Office, 1879), pp. 10–15. This report resulted in establishment of the Irrigation Division of the U.S. Geological Survey in 1888 and passage in 1903 of the Reclamation Act, which provided funds for irrigation purposes. However, Powell's idea of restricting settlement in the arid lands stirred resentment and since then has been generously overlooked.
5. Powell urged preserving the country's natural wealth against exploitation and monopoly control. He opposed the cattle trusts and disposition of the public domain. Similar forces were at work in the East. In 1885 the legislature of New York established the Adirondack Forest Preserve in portions of the Adirondacks and Catskills and reinforced its sanctity with a constitutional amendment of 1894. The amendment, adopted in reaction to scandals of fraudulent land deals and bribery in Albany and to illegal logging on state forests, reads as follows: "The lands of the state, now owned or hereafter acquired, constituting the forest preserve as now fixed by law, shall be forever kept as wild forest lands. They shall not be leased, sold, or exchanged, or be taken by any corporation, public or private, nor shall the timber thereon be sold, removed, or destroyed." Though

318

tested in court, the constitutional amendment has been consistently sustained and remains in force.

6. Gifford Pinchot, *Breaking New Ground* (New York: Harcourt, Brace & Co., 1947), p. 116.

7. Carl Alwin Schenck, *The Biltmore Story* (Saint Paul: American Forest History Foundation, 1955), pp. iii–v, 138.

8. Theodore Roosevelt considered Pinchot the foremost leader in the struggle to coordinate governmental forces in the conservation of all natural resources. Pinchot, he said, was "practically breaking new ground."

9. "Gifford Pinchot and the Conservation Ideal," *Journal of Forestry* 48, no. 2 (February 1950), pp. 83–86, contains the text of an address by Mrs. Gifford Pinchot at the dedication of the Gifford Pinchot National Forest, October 15, 1949. It includes these excerpts:

> Conservation to Gifford Pinchot was never a vague, fuzzy aspiration. It was concrete, exact, dynamic. The application of science and technology for the purpose of enhancing the life of the individual. The very stuff of which democracy is made.
>
> It dealt with research, with programs to improve country life, with electrification of farms, with rural education. It dealt with equality of opportunity, with control of monopoly. The list is a long one.
>
> Most important, it dealt with the conservation of natural resources as an international problem affecting issues of permanent peace. . . .
>
> With his deep insight into the well-springs of democratic action, with his abiding concern in the ethical and spiritual bases of American life, Gifford Pinchot provided the initial leadership in applied conservation.

10. Robert Marshall, *The People's Forests* (New York: Smith and Haas, 1933), pp. 77–78.

11. *Management Practices on the Bitterroot National Forest—A Task Force Appraisal*, published by Region 1 at Missoula, p. 15. William A. Worf, Chief, Division of Recreation and Lands, chairman of the task force, was quoted in the *Daily Missoulian* of May 12, 1970: "The critics were right. There was a 13 percent oversell of Ponderosa pine in 1966–69. The hard fact we face is the Forest Service simply can't serve the public well without a better understanding of what the public wants. We need comments, advice, and criticism from the American people."

The Bitterroot has indeed been the subject of considerable criticism in modern times. This national forest, established by presidential proclamation in 1897, covers 1,575,959 acres—460,812 in Idaho and 1,115,095 in Montana. Public expressions of concern in the mid-sixties focused on management plans for logging and road building in the Magruder Corridor, covering 173,000 acres between the Selway-Bitterroot Wilderness and Salmon River Breaks Primitive Area. Taking note of the extensive public criticism, and of the floods of 1964 and 1965, Secretary of Agriculture Orville Freeman in September 1966 appointed a citizens committee (headed by Dr. George A. Selke, former Chancellor of the University of Montana and former Commissioner of Conservation of Minnesota) to review management plans and their possible effects on soils and fisheries. The committee reported in April 1967 and Secretary Freeman essentially adopted its recommendations, providing that: management of the Corridor promote water quality and flow conditions to maintain and enhance the salmon and steelhead fisheries; timber harvesting be deferred pending comprehensive evaluation of commercial values; roading be deferred; a long-range recreation plan be prepared; research on erosion and sedimentation be strengthened, and a new and more integrated management plan be prepared.

A University View of the Forest Service (Government Printing Office, 1970) details yet another controversy on the Bitterroot. This report was prepared for the U.S. Senate Committee on Interior and Insular Affairs by a select committee of the University of Montana, at the request of Senator Lee Metcalf. On December 2, 1969, Senator Metcalf had written Dr. Arnold Bolle, Dean of Forestry: "I am especially concerned, as are my constituents, over the long-range effects of clearcutting, and the dominant role of timber production in Forest Service policy, to the detriment of other uses of these national resources." Dean Bolle appointed a committee with himself as chairman. The committee found grievous shortcomings on the Bitterroot: most notably that multiple-use management did not prevail as the governing principle, a condition deriving from a bureaucratic line structure archaic and undesirable.

12. U.S., Congress, Senate, Committee on Interior and Insular Affairs, "*Clearcutting Practices on National Timberlands,*" (Parts 1, 2, 3), hearings before the Subcommittee on Public Lands (the Church committee), 92d Cong., 1st sess., 5–7 April, 7 May, 29 June 1971. These include testimony of more than 90 witnesses, plus additional statements submitted for the record. Materials incorporated in these documents include the complete text of *Forest Management in Wyoming,* pp. 1116–1199 in Part 1, the 1971 report by the Wyoming Forest Study Team. The 13-page summary by the subcommittee stated:

> The Subcommittee does not question that under appropriate conditions clearcutting is a necessary, scientific, and professional forestry tool, nor does it believe Congress should legislate professional forestry practices in public land management any more than it does engineering practices for the Bureau of Reclamation or medical practices for the Veterans Administration. However, if these practices lead to basic questions of acceptable environmental impacts, national policy objectives, and conformance with existing statutes, Congress should take a look.

The "Church Guidelines" suggested that clearcutting should not be used on federal lands where:

a. Soil, slope or other watershed conditions are fragile and subject to major injury.
b. There is no assurance that the area can be adequately restocked within five years after harvest.
c. Aesthetic values outweigh other considerations.
d. The method is preferred only because it will give the greatest dollar return or the greatest unit output.

Clearcutting, according to the guidelines, should be used only where:

a. It is determined to be silviculturally essential to accomplish the relevant forest management objectives.
b. The size of the clearcut blocks, patches or strips be kept at the minimum necessary to accomplish silvicultural and other multiple-use forest management objectives.
c. A multidisciplinary review has first been made of the potential environmental, biological, aesthetic, engineering and economic impacts on each sale area.
d. Clearcut blocks, patches or strips are, in all cases, shaped and blended as much as possible with the natural terrain.

13. *Plumbline,* newsletter of the Western Wood Products Association, carried the following item on page 1 of its issue of 5 January 1972:

> Letters are urgently needed as soon as possible to the Secretary of Agriculture and the Secretary of the Interior documenting the economic effect on individual mills if an order were issued banning or sharply reducing clearcutting on federal land, according to Executive Vice President Wendell B. Barnes. James R. Turnbull, NFPA executive vice president, reports that under pressure from the Council on Environmental Quality, the Administration is giving serious consideration to an executive order to be issued by the President which would move clearcutting as a silvicultural practice down to a last resort, bottom-of-the-totem pole position in forest management.

Industrial Forestry Association Facts in its issue of January 11 documented efforts to kill the proposed executive order. On January 8, an industry group met at the office of Secretary of Agriculture Butz with the Secretary and officials of Agriculture and Interior, including Chief Edward Cliff and Associate Chief John McGuire. "We emphasized its dangers to the nation's supply of homebuilding materials by subjecting agencies to harassment by environmentalists who would appeal and sue to prevent projected clearcut sales, employment and the precedent for extending controls on private lands. We *all* agreed on undesirability of such an order and that no one could operate successfully under that proposed." Secretary Butz agreed to arrange a meeting at the White House with John Whitaker, of the Domestic Council. Meantime the industry leaders were given a chance to read the draft executive order. "It was everything we feared—sacrifice of forestry on the altar of vote-catching." The meeting at the White House was held January 10. It included Russell E. Train, Chairman of the Council on Environmental Quality, along with Chief Cliff and Associate Chief McGuire. It marked the end of the proposed executive order.

Secretary Butz was later quoted as calling concern over clearcutting the result of "irrational faddism." H.R. Glascock, of the Society of American Foresters, in a letter published in the *Washington Star* on February 21, 1942, asserted: "The recent emotional treatment of clearcutting in the press amounts to journalistic overkill, which may thwart the public interest in forest management according to biological fact."

Clare Conley, Editor of *Field & Stream*, saw things differently. In an editorial in the April issue of his magazine (76, no. 12, p. 4), he spelled out some of the provisos of the draft and reviewed what happened to it:

> January 7, 1972—The final draft of an Executive Order to be announced by President Nixon was typed up and sent in secret to the U.S. Forest Service, the Council on Environmental Quality, and to the Bureau of Land Management. It put strict limits on the use of clearcutting—"Except for limited experimental and research purposes not associated with commercial timber sale," the order said. Recreation and scenic values were to be protected. Harm to wildlife was to be kept to a minimum and the beauty of the areas was to be protected. Waters associated with a clearcut were not to be laden with eroded soil.
>
> Furthermore, the order required future timber sales to have specifications for the operations of the timber companies. The maximum use of merchantable timber was to be made and the amount destroyed as slash was to be reduced. Cutting, skidding, or yarding was to be controlled near lakes and streams. . . .
>
> January 13, 1972—An official announcement released this day to the press stated that no Executive Order pertaining to clearcutting would be issued by President Nixon. In referring to the affair, Sen. Fred Harris of Oklahoma said that it once again proved an old political axiom that public interests win in public; and private interests win in private.

Merrill G. Hastings, Jr., publisher of *Colorado Magazine*, cited another proviso in an editorial in the July-August 1972 issue of his magazine: "The Presidential Order stated, 'Timber sale planning procedures will make adequate provision for obtaining and taking into account the views of citizens.' At last, the people to whom the national forests belong were going to have a chance to be heard. The proposed order no longer gave top priority to just those who cut trees for profit" (p. 1).

14. U.S., Congress, House, Committee on Agriculture, *National Timber Supply Act of 1969*, hearings before the Subcommittee on Forests, 91st Cong., 1st Sess., 21–23 May 1969, pp. 1–3.

15. The Advisory Panel on Timber and the Environment was appointed September 2, 1971. A summary of its 117-page report was made public by the White House on September 24, 1973. Dr. Stephen H. Spurr, then president of the University of Texas (formerly a longtime forestry professor and later president of the SAF), presented his views as a member of the Advisory Panel in "Timber and the Environment," *Journal of Forestry* 72, no. 1 (January 1974), p. 9.

> De-emphasis on timber growing in favor of recreation and protection-forest uses on as much as a quarter or a third of the national forests could be accomplished with only a minimum impact on total timber production. I do not imply that all low-site lands should be taken out of timber production. Neither, for that matter, should land withdrawals be restricted to low-growth capacity lands. Quite clearly, representative samples of our best timber lands should be in natural area reserves. Both withdrawals of forest lands from timber cutting and the zoning of other forest lands to limit timber production must be done block by block, acre by acre, by professionals applying national and regional policies in the field.
>
> What I have argued for is a substantial increase in the withdrawal of national forest lands from timber production and restriction of timber harvesting on still further acreages of national forest lands which should continue to be managed under the multiple-use concept but should be managed primarily for outdoor recreation and related uses rather than for timber production. At the same time, I and other members of the panel see greater potential in more intensive management of the better commercial forest land, in both private and public ownership.

Chapter 3

1. Kenneth A. Gold, "A Comparative Analysis of Successful Organizations" (Study prepared for the Office of Personnel Management, July 1981), pp. 5–6. The study was conceived in the fall of 1980 to deal with declining productivity in both private and public sectors. Gold wrote: "The adoption of pre-packaged productivity improvement plans by basically unhealthy organizations may succeed in some cases but may, in many other instances, simply deflect attention from the basic management principles that have enabled some organizations to remain healthy and highly successful throughout the vacillations in American productivity." Thirty organizations were selected as candidates for case study. Then ten were selected, studied, and visited. These were: Arena Stage, regional production theatre, Washington, D.C.; City of Charlotte, North Carolina; Dana Corporation, manufacturer of truck parts and industrial products, Toledo, Ohio; L.L. Bean, manufacturer and mail order distributor of outdoor clothing and equipment, Freeport, Maine; City of Sunnyvale, California; Time Inc., communications company, New York; Customs Service, Passport Office, Forest Service, all federal bureaus of Washington, D.C., and field offices; and the Hewlett-Packard Corporation, Palo Alto, California.

2. Herbert Kaufman, *The Forest Ranger: A Study in Administrative Behavior* (Baltimore: Johns Hopkins Press, 1960), p. 236.

See also "The Forest Ranger Revisited," by Christopher K. Leman, a paper prepared in 1981 for delivery before the American Political Science Association. The author undertook his research while at the office of Resources for the Future, with field review of the five ranger districts Kaufman had studied in the 1950s. Leman observed that the district ranger's position had become more like the job of the forest supervisor of the 1950s. "They delegate—they are expected, and even required, to delegate—to subordinates as many of the routine tasks as they can. But they still have to be in the woods much of the time, and their knowledge of their areas is consequently great even when they are relatively new to their districts."

3. The number of temporary employees rose appreciably during the 1970s, with limitations imposed by the Office of Management and Budget on permanent full-time employees. With personnel ceilings and limitations, contracting and consulting also became more significant. Several types of workers, including Youth Conservation Corps, Job Corps, and Senior Community Service Employment, have been available to the Forest Service, though funded elsewhere.

4. U.S., Congress, Senate, Committee on Agriculture and Forestry, *Forest Service Reorganization and Forestry Programs,* hearings before the Subcommittee on Environment, Soil, Conservation and Forestry, 93d Cong., 1st sess., 26 and 27 June 1973, pp. 358–359.

5. Missoula (Montana) *Missoulian,* 13 August 1979, p. 12.

6. Christopher K. Leman in "The Forest Ranger Revisited," comments with reference to performance standards:

> It is still not uncommon for a forest supervisor or (as in the northeastern region in the mid 1970s) a regional forester to force district rangers to publicly commit themselves to meeting their timber sales target. Field behaviour still depends on the reporting, monitoring, and review of performance in meeting quantitative goals.
>
> One reason for the continued dominance of the timber goal is that many other goals are not stated in quantitative terms. One solution would be to make more effort to state other values quantitatively (p. 34).

7. Sidney Weitzman, "Lessons from the Monongahela—An In-Service Analysis Based on Interviews with Forest Service Personnel" (1977), p. 30.

8. The Socially Responsive Management plan for the Rocky Mountain Region was initiated by the Foundation for Urban and Neighborhood Development (FUND) to apply in daily operations, long-range planning, and special project assessment activities, together with courses of study for Forest Service personnel emphasizing practical exercises and presentations.

9. Margaret A. Shannon, in "Sociology and Public Land Management" (*Western Wildlands*, Winter 1981, pp. 3–8), provides insight into the gap between sociologists and resource professionals. She served as social scientist on the interdisciplinary planning team of the Lolo National Forest during 1979 and 1980 and writes:

> Until 1976, the sociologist's land management role was limited to predicting social impacts of proposed resource-development projects (or other management decisions). My own discussions with both resource managers and agency sociologists indicate that social scientists have been marginalized and isolated within the agencies to a significant degree. Partially, they are perceived as threatening the conventional wisdom of the agency. Spawned from justified paranoia, but more directly from insecurity in the resources field, this feeling has prevented sociologists from effectively participating in decision processes. In addition, they have wrapped themselves in an illusory cloak of scientific objectivity and disdained all contact with the decision process. Thus, the structure of the EIS process as practiced simply legitimated conventional decisions. The sociologists were isolated from critical decisions, in fact chose isolation to maintain what they perceived as their scientific credibility.
>
> In some ways the social scientist has a special charge in public land-management agencies. The purposes of land management are found in social institutions. Social claims to a resource include the meaning of that resource to a person, or a social group. To understand the forest and the land we must understand society. To resolve competing and conflicting resource claims on public lands we must see the forest as others do, and then seek areas of agreement in their diverse perspectives. We must also try to minimize or mitigate undesirable social impacts, following a political social impact analysis.

10. The Appalachian Mountain Club maintains about 300 miles of the Appalachian Trail, twenty backcountry shelters, and ski touring trails from Pennsylvania to Maine. Various programs supplement paid trail crew and shelter caretakers. In the Adopt-A-Trail program, started in 1981, volunteers do basic maintenance on their own time and at their own pace, with AMC providing paint, some tools, and training. A lot of work is done through AMC chapters and from AMC camps on work trips.

11. The Forest Planning Model, FORPLAN, as explained in "FORPLAN Overview" (USDA, Forest Service, Land Management Planning, April 1981), does the following:

> Calculates the ranges of goods, services, budgets, and their flows through the RPA horizon.
> Helps to identify opportunities for resolving issues/concerns by varying output levels.
> Calculates the long-term sustained yield and base harvest schedule.
> Selects/displays cost efficient combination of management prescriptions by land area by time; to accomplish a given set of objectives.
> Allows the analysis of management opportunities at the forest, for describing production possibilities to National Forest System portion of regional guide for RPA assessment.
> Helps to determine how to implement the RPA program in conjunction with local and regional considerations, by providing overall long-range direction for incremental implementation through the yearly budgeting processing (p. 27).

12. Robert A. Entzminger, Director of Information, Eastern Region, provided the following comment, dated January 8, 1982, on the Ferguson case: "There are very few people now working in the Regional Office that were here during the time Mr. Ferguson's case was active. In the hasty review that I made of the case file, one of the major concerns of the Forest Service was that Mr. Ferguson did not use internal channels to call attention to his professional opinion. There have been all kinds of reviews and investigations of this case."

13. Zane G. Smith, Jr., Regional Forester of the Pacific Southwest Region, provided the following comment, dated January 19, 1982, on the Montagne case:

> In retrospect it does not appear to have been so much a case of whistleblowing as it was a case of persistent disagreement on many subjects over several years, pursued in a very public arena. The Ballinger Canyon ORV and blunt-nosed leopard lizard issue was well aired and publicly debated. There was no attempt to hide anything, just lots of honest differences of opinion. NEPA works and served us well through that issue. Ultimately, we had to make a decision and proceed. At that time it became necessary in any organization for everyone to pull

together and carry out that decision. This was not happening. Monty disagreed with the decision, could not support it, and put much of his energy into discrediting the forest.

The fact that Monty disagreed in this manner was not the reason for his reassignment but the effects were. It appeared from the record that Monty had lost his effectiveness in a field capacity on the Los Padres. The manner of his dogged disagreement with and pursuit of the issue had alienated peer wildlife biologists and fellow workers. He was perceived by many of these as becoming detrimental to overall wildlife management needs on the Los Padres and was unable to be effective in interdisciplinary team situations. At the same time he had strong skills that could have been utilized effectively in a regional staff capacity.

We have continued to upgrade the management of Ballinger Canyon. We have moved from 'open' ORV use under the original ORV plan to fencing of potential critical habitat for the hybrid blunt-nosed leopard lizard and to prohibition of hill climbing and restriction of use to designated trails only. On-site administration of use has been intensified. Now ORV use continues as a viable but restricted recreation opportunity in a manner which is allowing rapid and obvious recovery of vegetation in the area.

I found two references in your manuscript which I believe are inaccurate. It states that Monty Montagne and four other biologists discovered the presence of the rare blunt-nosed leopard lizard in Ballinger Canyon. No blunt-nosed leopard lizards have ever been found in the canyon. A hybrid has been found, but hybrids are specifically exempt from the Endangered Species Act. The Blunt-nosed Leopard Lizard Recovery Team did, for a time, have interest in designating the canyon floor as critical habitat and in having the hybrid designated along with the species. However, after considered review and evaluation no national forest land on the Los Padres was proposed for designation as critical habitat by the team. The species was verified as not occurring there through detailed surveys. U.S. Fish and Wildlife Service ruled that the hybrid was not to be considered for designation.

It states further that Montagne was dismissed despite strong support from the scientific community. He was not dismissed, he was given a directed reassignment and chose to retire.

14. USDA, Forest Service, Audit Report 899-33-SF, Forest Road Construction, Lolo National Forest (Missoula, Montana, August 30, 1979), p. 2.

15. Following his ordered reinstatement, with two years back pay, Arthur Anderson told the *Daily Missoulian:* "It's not going to be easy to work for the agency again. If they put me in a corner and say, 'Just look out the window,' I'm not going to be very happy about it."

Everett L. Towle, Deputy Regional Forester of the Northern Region, provided the following examples of internal critics who were heeded:

• Robert Peterson strongly and vocally supported most of Anderson's allegations of mismanagement and appeared as a witness in his behalf at the Merit Systems Protection Board hearing. Following Anderson's departure from the Superior Ranger District, Peterson requested and was granted a transfer to Missoula for personal reasons. There he was promoted competitively to a GS-9 position.

• Ed Prinkki raised many questions during 1976 and 1977 about the engineering organization on the Lolo National Forest. A classification appeal and a major grievance went to the national level. The result was a complete reorganization of engineering on the forest. Since that time Prinkki continued to work as a civil engineering technician and received a $300 cash award for performance in 1980.

• Chuck Tribe, while working on the Lolo National Forest during the 1970s, wrote his Congressman that planned airconditioning of the forest supervisor's office headquarters would be a waste of money. The airconditioning plan was cancelled as a result. Tribe later was placed in charge of the Lolo Planning Team, a key role in developing the first national forest management plan under NFMA.

Chapter 4

1. Paula J. Williams, "The Role of Women in Forestry in the United States" (Paper submitted to the Student Essay Competition on "The Role of Women in Forestry" for the 8th World Forestry Congress, Jakarta, Indonesia, October 16-18, 1978), pp. 1–7.

2. Patricia Seubert, "Women in the U.S. Forest Service: A Study in Attitude" (Report for Advanced Natural Resource Policy, University of Vermont, 1982), pp. 7, 12.

3. "Blacks and Careers in Forest Resources" (Proceedings of the Tuskegee Forest Resources Council, Tuskegee Institute, Alabama, January 17–19, 1979).

4. E. Delmar Jaquish, "Forest Service Outreach to Nontraditional Audiences" (Report prepared for USDA Forest Service, 1981), p. 1.

5. In 1978 the Society of American Foresters formed a Committee on Minority Action to encourage recruitment of minorities and support minority programs in educational institutions. This committee gathers and disseminates data and aids minorities locate financial assistance.

Chapter 5

1. Lawrence W. Rakestraw, *A History of the United States Forest Service in Alaska* (Anchorage: A cooperative publication of the Alaska Historical Commission and Alaska Region, Forest Service, with assistance of the Alaska Historical Society, 1981), pp. 3, 5.

2. Chief Cliff's letter to the *Times* was dated December 1, 1971, and appeared in the newspaper's edition of January 10, 1982. It was a lengthy letter not published in its entirety. The following was omitted from the printed version:

> Increases in timber harvest that occurred over the years were justified by better utilization, improved technology, and increased growth. . . . Forests, like air and water, are a national heritage that must be used, improved, and conserved in balance. All three are subject to damage by excessive use resulting from unlimited population growth. The Forest Service is dedicated to managing the nation's forests wisely, enhancing their amenity values so that future generations can continue to benefit from their use and enjoyment.

Officials besides the Chief were involved in answering media critics. In a letter dated April 27, 1972, Paul E. Neff, Director of Timber Management, protested an editorial in *Field & Stream*:

> To say clearcut logging is more harmful than a forest fire is an overstatement indeed. An uncontrolled fire destroys most of the forest resources including timber; logging, properly conducted, utilizes the resources and causes minimal, temporary damage. Moreover, it provides Americans with the lumber and forest products without which the nation could not exist.
>
> The preference of lumber companies has little to do with the actual determination of the cutting system to be used on national forest timber sales. Rather, these determinations are made by a professional review of the many physical and biological factors relating to a forest.

Whether sales on the Monongahela, in particular, were "properly conducted" with adequate "professional review" has been subject to considerable question. In his paper, "The Forest Ranger Revisited," Christopher K. Leman made the point that agency regulations and directives are porous with loopholes:

> National and regional directives laid down various guidelines, but these were often ignored. For example, in the Monongahela National Forest clearcuts over 20 acres were to include special safeguards, and none over 100 acres were to be allowed. But the directives allowed so many silvicultural and administrative exceptions that clearcuts of 200 or 300 acres became common. Even though they clashed with the spirit of Forest Service policy, these controversial practices were technically within the guidelines (pp. 22, 23).

3. The 1970 Forest Service team was reconstituted for another look in 1982, following objections by citizens of Wyoming and the state Fish and Game Department to logging plans of the Bridger-Teton National Forest near Pinedale. This later conflict is dealt with in Chapter 7.

4. "Forest Management in Wyoming—Timber Harvest and the Environment on the Teton, Bridger, Shoshone and Bighorn National Forests" (Report prepared by Wyoming Forest Study Team for USDA Forest Service, 1971), p. 39. In "Landslide Hazards Related to Land Use Planning in Teton National Forest, Northwest Wyoming" (Forest Service

Intermountain Region, Ogden, Utah, 1971), Robert G. Bailey, a Forest Service hydrologist, wrote:

> In this area, landslides have been reactivated where road construction has further aggravated the slide conditions by oversteepening the uphill slope by the road-cut and placing a load of unstable fill on the downhill slope.
> In steep topography, the soil and rock mantle are in precarious balance with the physical factors of the environment, and any material change in the environment results in instability. Two of the principle methods by which man usually increases instability in mountainous terrain are by road construction and logging activities.
> . . . only a few research studies have been made to determine the effects of these practices on deep-seated slope failure. A review of the literature on this subject, however, reveals the overwhelming consensus which maintains that *forests do in fact play an important soil protective role and that clearcutting can promote not only erosion, but also landsliding* [emphasis added]. Literally hundreds of commercial logging operations indicate that this conclusion is well founded. Too often, however, failure to recognize the limits of disturbance allowable beforehand in areas of inherent instability has led to seriously deteriorated water quality (p. 3).

5. "Forest Management in Wyoming," p. 3.

6. U.S., Congress, House, Committee on Agriculture, *National Timber Supply Act of 1969,* hearings before the Subcommittee on Forests, 91st Cong., 1st sess., 21–23 May 1969, p. 169.

7. A. Claude Ferguson, "In the Matter of the National Forest Enabling Act and National Forest Land in Indiana" (Statement presented to the National Forest Land Acquisition Joint Study Committee of the Indiana Legislature, Tell City, Indiana, September 5, 1979).

8. Funds channeled to counties for schools and roads can be significant. For fiscal 1982 leading recipients were: Oregon, $73 million; California, $35 million; Washington, $27 million; North Dakota, $10 million, Idaho, $9 million, and Montana, $8.1 million.

9. It is now recognized that natural fire, generally started by lightning, often plays a positive role in ecosystem development, in the same way as do rain, sun, and wind. Some plants, such as lodgepole pine and aspen, thrive in areas that have been burned. Lodgepole, in fact, becomes vulnerable to mountain pine beetles after 80 years or so; thus fire helps remove infested trees. Fire encourages aspen to send up new sprouts from the root systems of mature trees. It helps to remove and control competing vegetation that otherwise would ultimately crowd out mature aspen. And ash serves as a fertilizer, returning nutrients to the soil.

Contrary to widespread belief, study has shown that few animals are killed by fire. Many species remain in the vicinity of a burn because of concentrated food sources, such as insects and subsequent new plant growth.

Chapter 6

1. Gifford Pinchot, *Breaking New Ground*, p. 118.

2. Subscribing to Senator Randolph's prediction, Richard W. Behan wrote in *Forest Planning* (June 1981): "The people opposed to clearcutting in the Bitterroot and Monongahela National Forests didn't want planning. They wanted to see greater environmental protection" (p. 3).

3. U.S., Cong., Senate, "National Forest Environmental Management Act," hearing before the Subcommittee on Environment, Soil, Conservation and Forestry, November 20, 1976, p. 2. Important reading matter to insure understanding of RPA/NFMA include the following: "A Citizen's Guide to the Forest and Rangeland Renewable Resources Planning Act." Issued by USDA, Forest Service in cooperation with the Conservation Foundation, Washington, 1981. R. Max Peterson, "Updating RPA: Master Plan for Managing Our Resources," *American Forests*, April 1981; "Agenda for Forest Productivity in the 1980s," Prepared by National Forest Products Association, Washington, 1981; "Choices—RPA: Planning for the Nation's Renewable Resources." A discussion guide to the film of the same name. Conservation Foundation, Washington, 1981; "A Conservationist's Guide to

National Forest Planning." Issued jointly by National Audubon Society, Sierra Club, and Wilderness Society, Washington, 1981; *Forest Planning*, Journal of the Nationwide Forest Planning Clearinghouse, Eugene, Oregon.

4. John V. Krutilla and John A. Haigh, "An Integrated Approach to National Forest Management," *Environmental Law*, Lewis and Clark Law School, Portland, Oregon, 1978. Resources for the Future reprint No. 156, p. 375.

5. USDA, Forest Service, "An Assessment of the Forest and Rangeland Situation in the United States," 1980, pp. iii–xvii; USDA, Forest Service, "A Recommended Renewable Resources Program—1980 Update," 1980 (Gen. Ref.); USDA, Forest Service, "The 1980 Report to Congress on the Nation's Renewable Resources. Final Environmental Impact Statement Under the Forest and Rangeland Renewable Resources Planning Act," 1980.

6. "National Forest System Land and Resource Management Planning. Final Rule," *Federal Register*, September 17, 1979, pp. 53, 928.

7. "National Forest System Land and Resource Mgt. Planning," *Federal Register*, February 22, 1982, pp. 76–78.

8. "Alpine Lakes Management Unit Direction and Wilderness Proposal," issued by Snoqualmie and Wenatchee National Forests, Seattle, Washington, June 1973.

9. USDA, Forest Service, "Alternative Goals, 1985 Resources Planning Act Program," December 1981, pp. 1–3.

10. The Congressional mandate to market 4.5 billion board feet per decade from the Tongass National Forest in Alaska makes interdisciplinary planning difficult. This is evident from a letter from Darrel Tracy, a landscape architect, to Thomas Barlow, specialist in national forest issues for the Natural Resources Defense Council. The letter, published in *Amicus Journal* (Winter 1982, pp. 35–38), indicates the conflict between the roles of production-oriented professionals and other specialists within the agency:

> I arrived in Sitka assigned to the Chatham forest management area as an architectural technician on April 1, 1975, just in time to witness the last stages of planning for the 1976–1981 timber sale. It was immediately apparent to me that in order to manage the visual resource honestly, the Forest Service would have to adopt more costly logging methods and that landscape architects would have to be given greater professional recognition. . . . Theoretically, planning for the 1981–1986 period in which I was fully involved should have seen a marked improvement. In response to public criticism the Forest Service had already come to accept the notion, at least in writing, that other values mattered.
>
> We had two soil scientists, one hydrologist, one wildlife biologist, one fisheries biologist, two landscape architects, and no archaeologists. Foresters and engineers, on the other hand, could be counted by the dozens.
>
> This supposedly was the beginning of the era of "interdisciplinary teamwork," but in reality the recommendations of resource specialists were considered and used where it was convenient—except for the rock-and-ice wilderness proposals everything would eventually be cut.
>
> Jim Knode was the only person on Chatham's planning team with a formal education in planning. To everyone else it was a game in which you read the manual, followed directions, played politics, went to meetings and tried not to be discovered. Management decisions were made before they came to your shop and you merely covered them with the paperwork.
>
> What it amounts to is that the 1981–86 plan reflects minimal concern for aesthetics. Although the nation requires "visual quality objectives" as part of every plan, I had to fight like hell to get the barest mention of what these might be in the final environmental impact statement. Layout crews are now putting the plan on the ground and making changes at will. There is no visual resource monitoring, because the single Chatham architect is busy with other plans.

Chapter 7

1. Speculative bidding becomes understandable by recognizing that 60 percent of the price of finished lumber derives from relatively fixed costs of milling, logging, and hauling. Timber purchased at low price and left standing while inflation spirals upward is like money invested in a mutual fund. The market went sour, however, in the late 1970s, reflecting high interest rates.

The forest products industry was divided on how to cope with the issue of unmet contracts. Some advocated legislation to allow purchasers to turn back sales with minor penalties. Others believed further relaxation would only encourage future speculation, drive prices still higher, and punish purchasers who had continued to operate and complete sales at a loss during the period of depressed markets.

In its announcement of April 1982, the Forest Service said it would develop a policy to permit contract prices to be adjusted automatically in the event of future severely depressed markets.

2. Pinchot made a tour of the Forest Reserves as a special agent of the General Land Office. His assignment was to report on conditions and needs, forests, relationships to lumbering, agriculture, mining, grazing, commerce, and settlement. In his "Report on Examination of the Forest Reserves" (National Agricultural Library Document no. 99.61) he wrote:

> Bighorn Reserve: Forest management in this Reserve requires first of all protection against fire. The system of clearcutting in strips will probably best accomplish the desired results in this Reserve where reproduction is vigorous and the soil and climate are adapted to forest growth (p. 53).
>
> Cascade Reserve: Actual forest management will (occur) . . . where the character of the reproduction indicates either clearcutting in strips or a system of selecting felling extremely localized (p. 74).
>
> Lewis and Clark Reserve: Clearcutting in strips is indicated as a system of management most likely to assure the safety of the forests until more careful studies can determine the best methods of increasing the proportion of valuable trees. Wider strips in lodgepole pine, or even clearcutting, over blocks of some size in the old forests of yellow pine may be safe and profitable (p. 86).

The first reference to clearcutting in a Forest Service *Manual* appeared in the edition of 1911 (p. 114).

3. John L. Gray, Director of the Pinchot Institute for Conservation Studies, in a memorandum of March 1982 to the author:

> Foresters prior to the 1940s preached and attempted to practice the individual tree selection system of silviculture and selective cutting based on European thinking which involved mainly management of spruce, fir, beech and other shade-tolerant species. U.S. forestry research and experience were slow to recognize that the southern pines, Douglas-fir, yellow poplar and many other principal U.S. commercial species were shade-intolerant, light-seeded and could not be successfully reproduced by natural means or otherwise except in full sunlight and often with exposure of mineral soil through prescribed burning.
>
> I spent 32 years as a professional in the southern pine region and can verify the above from observation and from research publications with reference to the southern pines, yellow poplar, etc. My experience with eastern oaks is less extensive but some of these also cannot be successfully reproduced under individual tree selection silviculture.
>
> Ecological and silvicultural research evidence overwhelmingly supports the above. Controversy occurs in deciding on optimum size, shape and distribution of clearcuts.

4. Sidney Weitzman, "Lessons from the Monongahela" p. iii, p. 5, p. 11. Weitzman shares Gray's view (see note 3) to a large degree. He interviewed West Virginians and assessed their objections: "They were concerned not so much with clearcutting as a silvicultural tool, but how these clearcuts were applied, i.e., the large size and concentration of clearcut areas. They also were concerned about the adverse effects of clearcutting on selected game species, on visual impact, and watershed values (p. 1)."

"Lessons from the Monongahela Experience," or the Weitzman Report, of 1977, has never been widely circulated, though it reflects favorably on an agency able to examine its own mistakes with objectivity. "The purpose of this study was not to fix blame or to determine who was right or wrong," as Weitzman wrote. "It concentrates on Forest Service actions and behavior in response to many conflicting pressures. Only by focusing on what it might have done differently can the Forest Service analyze its attitudes and behaviour patterns in response to those conflicting interests (p. iii)."

Weitzman dealt with evolution of policy and process:

As Forest Service operations became more complex, the directives system grew in size and complexity and attempted to provide adequate coverage for the new technical age—as it did in the earlier custodial era. This modus operandi may no longer suffice.

Ecosystems have different silvicultural, economic, and environmental constraints. Their management requires highly qualified people rather than reliance on directives designed for technicians (p. 32).

5. *Management Practices on the Bitterroot National Forest* (May 1970), pp. 1, 2, 8–9. The task force that Regional Forester Rahm appointed to review management on the Bitterroot National Forest (referred to in Chapter 2) found the same narrow outlook that Weitzman had cited on the Monongahela. "There is an implicit attitude on the part of the Bitterroot National Forest that resource production goals come first and that land management considerations take second place," the task force warned in its report of April, 1970.

6. Thomas W. Power, "Alternative W and the Rare II Process in Montana: A Brief Economic Analysis" (Study prepared for the Montana Wilderness Coalition, September 1978).

7. Natural Resources Defense Council, *Giving Away the National Forests*, Washington, June 1980, pp. 29–31.

"Cost Benefit Analysis: A Tricky Game" (*Conservation Foundation Letter*, December 1980) helps to explain difficulties in getting to the core of costs and benefits:

> During the very process of quantification and analysis, many assumptions and judgements must be made which can skew the results. These judgements may be based on intentional bias or on the subjective subconscious. Either way, they undercut the very attributes that recommend cost-benefit analysis. . . .
> Results may depend heavily on the choice of a time frame or discount rate, and on whether average or incremental figures are used. Thomas E. Hamilton, of the U.S. Forest Service, notes a few of the techniques for justifying large program expenditures: Reforest an area and take credit for all the trees, including the ones that grow normally. Compare the cost of a gypsy-moth eradication program with the whole value of the forest protected, then come back in five years and do it again. The possibilities in any program are limited only by the imagination (p. 2).

Subsidizing the Timber Industry: The economics of national forest mismanagement (Cascade Holistic Economic Consultants, General Forestry Report No. 4, Eugene, Oregon, March 1980) provides an even more critical view than NRDC. Referring to unattributed costs of producing timber:

> These include the costs of developing and maintaining tree seedling nurseries for reforestation; protection of the forest from fire, insects, and disease; and the costs of hiring wildlife biologists, soil scientists, and other specialists to review timber sales to insure that other resources are not permanently damaged by timber harvesting. Although many of these costs would not be incurred were there no timber sale program, the Forest Service budget is too vague to easily determine their exact level.
> All of these costs are appropriated by Congress, and are paid for out of the U.S. Treasury and ultimately by the taxpayers. Without paying these costs, the timber industry receives the benefits of the public forest programs by purchasing the timber and selling wood products for profit. These benefits are significant, since a corporation which can depend on the U.S. Treasury to pay the costs of timber management, even where those costs far exceed the benefits, has a major advantage over a company which must pay those costs out of its own income. For example, Louisiana-Pacific, which cuts most of its timber from federal land, had net profits in 1979 equal to 20% of its gross sales—a figure almost unheard of in a highly competitive industry (p. 4).

8. Marion Clawson, "An Economic Classification of U.S. 'Commercial' Forests," *Journal of Forestry* 79, no. 11 (November 1981). Clawson provides valuable background data and analysis:

> At one time the Forest Service established five productivity sites: Class I, capable of producing 165 or more cubic feet per acre annually: Class II, 120 to 165; Class III, 85 to 120; Class IV,

50 to 85; and Class V, 20 to 50 cubic feet. Recent reports have dropped these designations of site classes, using instead the number of cubic feet per acre per year as the designation. More seriously, from my viewpoint, the agency has consolidated the two most productive site classes into a single one, i.e., over 120 cubic feet per acre annually. This change deprives analysis of valuable information and has the effect of downgrading the importance of the most productive forestlands, where I think the greatest potential lies for increased wood growth.

See also Clawson's *The Economics of National Forest Management* (Washington, D.C.: Resources for the Future, 1976).

9. Carl R. Puuri and Raymond G. Weinmann, "Continuing Education and Certification of Silviculturists in the USDA Forest Service," *Journal of Forestry* 79, no. 4 (April 1981), p. 209.

Certification requires preparation of an actual prescription for timber harvest acceptable to an examining panel, which usually incudes a forest supervisor, a certified silviculturist, the regional silviculturist, a personnel officer, and professors at a university conducting the continuing education program.

A directive on silvicultural practices, issued in the Forest Service Manual on January 28, 1981, includes the following:

> The preparation of prescriptions requires a high degree of professional skill by a competent silviculturist. The silviculturist must be familiar with local climatic and edaphic conditions, the silvics of the tree species concerned, the land management and resource direction pertaining to the area, the available management alternatives, and pertinent Forest Service Manual instructions and policies. It is imperative that these requirements be met in every area in which a silvicultural practice is performed. Experience through tenure may often be necessary to deal with complex resource situations.

10. The Siuslaw policy statement can be found in the August 1974 Siuslaw National Forest Supplement No. 1, Chapter 8210 of the Forest Service *Manual*, available from the Siuslaw National Forest, Corvallis, Oregon.

Chapter 8

1. USDA, Forest Service, "The Nation's Range Resources, A Forest-Range Environmental Study," Washington, D.C., 1972, pp. 19, 37.

2. "Synopsis of Range Conditions on the Tonto National Forest," Tonto National Forest, Phoenix, 1979.

3. Bruce Hronek, "The Need for Thoughtful Vegetative Management on the Tonto National Forest," Tonto National Forest, Phoenix, 1978, pp. 5–7.

4. USDA, Forest Service, *The National Grasslands Story*, 1964, p. 1.

5. "Management Plan, Rolling Prairie Planning Unit," Custer National Forest, Billings, Montana, 1975, pp. 5, 25.

6. "Management Situations, Thunder Basin National Grassland," Medicine Bow National Forest, Laramie, Wyoming, 1981, pp. 1–2.

In May, 1982, the National Wildlife Federation, Montana Wildlife Federation, and Northern Plains Resource Council filed suit to force the federal government to examine what they called "potentially devastating" effects of what would be the largest coal lease sale ever undertaken. Involved were 13 tracts of land covering approximately 32,000 acres in the Powder River Basin of Montana and Wyoming. The suit centered on federal laws requiring development of comprehensive land-use plans in advance of coal lease sales. It charged the Department of the Interior with failure to evaluate competing resources, including agriculture, grazing, and wildlife. "This sale could provoke radical social and environmental change over the next 50 to 100 years in a region already experiencing ravages of profound change," declared Dr. Jay D. Hair, executive vice-president of the National Wildlife Federation. "I agree with those western leaders who are asking federal land stewards to help the region plan for orderly change before rushing pell-mell into wasteful and harmful development. We are not convinced the coal is needed for domestic markets within the next decade, or longer."

7. Frances Moore Lappé, *Diet for a Small Planet* (New York: Ballantine, 1975), pp. xvii, 3, 10–13.

8. Jean Hocker and Story Clark, "Jackson Hole: Protecting Public Values on Private Lands," (Report for Jackson Hole Project of the Izaak Walton League of America, Jackson, Wyoming, 1981), pp. 1–4, 9, 48–50, 80–81.

Chapter 9

1. U.S., Cong., Senate, Committee on Energy and Natural Resources, *River of No Return Wilderness Proposals*, Part 1, hearings before the Subcommittee on Parks, Recreation, and Renewable Resources, May, 1979, pp. 2–22.

2. USDA, Forest Service, "The Forest Service Role in Outdoor Recreation," Washington, D.C., 1978, pp. 1, 10–13.

3. Max Peterson, "National Forest Trails—New Frontiers," *National Parks Magazine* 54, no. 10 (October 1980), pp. 5–7.

4. R. Max Peterson, "Looking at Recreation Through Forest Service Eyes," *Parks and Recreation* 16, no. 3 (March 1981), pp. 42–47.

5. In a statement of March 25, 1982, on 1983 budget proposals for federal recreation programs, the American Recreation Coalition declared:

> The era of tight-budget restraints at federal, state and local levels is impacting recreation. A mainstay of recreation land acquisition and federal development at the state and local levels, federal grants from the Land and Water Conservation Fund, has been at least temporarily ended. From California to Idaho to Massachusetts, property tax limitations have forced dramatic recreation program cutbacks at the state level. Toledo, Ohio, has eliminated its parks department. Campgrounds at national parks and forests are opening later in the spring and closing earlier in the fall. Interpretive programs are being cut back and brochures describing facilities and features in national parks cannot be reprinted once current stocks are exhausted.
>
> Even though public recreation program budgets constitute a minor part of the total spending on recreation—currently less than five percent—the programs they fund are of disproportionate importance. Without public softball fields, the market for bats, gloves and softballs would be much smaller. Without public campsites, sales of tents and RVs, along with a wide array of ancillary equipment, would undoubtedly decline. In a very real sense, public recreation sites and programs are catalysts for recreation in America.

6. In May 1982 Frederick deHoll, Supervisor of the Los Padres National Forest, reported that law enforcement officials during 1981 made a total of 560 marijuana seizures on national forest lands in California. These resulted in the removal of 84 tons of marijuana plants valued at about $250 million, though representing only about 35 percent of the marijuana grown on national forest lands during the year.

Supervisor deHoll said growers go to elaborate lengths to produce their crops in remote locations. Innocent forest visitors, Forest Service personnel, and their families have been threatened when approaching too closely. Besides taking precautions against human visitors, growers saturate the areas around their plantations with arsenic rodent poisons since rodents are strongly attracted to marijuana.

7. David Sheridan, "Off-Road Vehicles on Public Land" (Report prepared for Council on Environmental Quality, Washington, D.C., 1979), pp. 7–12, 36–43.

8. H. G. Wilshire, "The ORV Phenomenon-Management-Impact—The Ballinger Canyon Designated Motorcycle Trail System," in Richard N.L. Andrews and Paul F. Nowak, eds., *Off-Road Vehicle Use: A Management Challenge* (Ann Arbor, Mich., 1980), pp. 78–79.

9. Legislation passed by Congress in October 1978 enlarged the Boundary Waters Canoe Area from 1,045,000 acres to 1,084,000 acres and placed the entire BWCA in wilderness—in contrast to "portal" and "interior" zone classifications which had previously been in effect. The law immediately reduced motorized use from 64 percent of the water surface to 34 percent, with provision for further phase-out reductions to 25 percent. Snowmobile use was reduced to five specified corridors, with phase-out of three corridors by the end of five years, leaving just one each at the east and west ends of the BWCA.

10. R. Max Peterson, "Looking at Recreation through Forest Service Eyes," *Parks and Recreation* 16, no. 3 (March 1981), p. 43.

Chapter 10

1. In April 1982, Dr. Timothy Clark, a wildlife biologist of Idaho State University, raised hope that the black-footed ferret may be making a comeback. Contrary to belief that the species was reduced to only one or two living specimens, Dr. Clark reported finding at least 22 alive and breeding on the plains near Cody, Wyoming. It was the highlight in a decade of monitoring the ferret by Dr. Clark. He and a team of assistants visually identified 10 animals, then followed tracks, marks and droppings to obtain evidence of the existence of 12 more. As usual in such cases, he would not name the area as a safeguard against disturbance.

The black-footed ferret, *Mustela nigripes*, is slim, wiry and weasel-like, with black mask, tail, and feet, growing to about two feet in length and two pounds in weight. Once common in a range extending from North Dakota to Texas, it declined with eradication of the prairie dog, its principal food source. Ferrets generally pursue a solitary course, but congregate in the breeding season.

2. The red-cockaded woodpecker, once common in the South, has been thoroughly reduced in numbers through loss of mature pine forest habitat and was declared an endangered species in 1970. See Robert G. Hooper, Andrew F. Robinson, Jr., and Jerome A. Jackson, "The Red-Cockaded Woodpecker: Notes on Life History and Management" (published by Forest Service Southeastern Area, Atlanta, in 1979), pp. 1–6, where it is reported that older pines are needed to provide roosting cavities and extensive pine and pine-hardwood forests to meet foraging requirements. The publication outlines management steps to aid the bird: (1) retain existing cavity trees; (2) provide trees for new cavities; (3) provide adequate foraging habitat; (4) control hardwoods in the colony site, and (5) provide future colony sites.

3. USDA, Forest Service, "Forest Service Policy, Goals and Objectives for Wildlife and Fish Habitat Management in the 1980s," Washington, D.C., 1980; USDA, Forest Service, "Managing for Wildlife and Fish Habitat: A Framework for the '80s" (Gen. Ref.), Washington, D.C., 1980; National Wildlife Federation, "Wildlife and Our National Forests—A Citizen's Guide to Commenting on National Forest Plans," Washington, D.C., 1981; "Evaluation of Forest Service Wildlife and Fish Programs in the Intermountain, Pacific Northwest and Southern Regions." A report to the Forest Service by the Wildlife Management Institute, Washington, 1979 (Gen. Ref.).

4. David Phillips and Hugh Nash, *The Condor Question, Captive or Forever Free?* (San Francisco: Friends of the Earth, 1981), pp. 37–45.

5. George Edgar Lowman, "Endangered, Threatened and Unique Mammals of the Southern National Forests" (Report prepared for Southern Region, Forest Service, Atlanta, 1975), p. 104.

6. E. Burnham Chamberlain, "Rare and Endangered Birds of the Southern National Forests" (Report prepared for Southern Region, Forest Service, Atlanta, 1974), pp. i, 95–98.

7. Bill Schneider, *Where the Grizzly Walks* (Missoula, Mont.: Mountain Press, 1978), pp. 147–150.

8. Charles J. Jonkel and Christopher Servheen, "Bears and People—A Wilderness Management Challenge," *Western Wildlands* 4, no. 2 (Fall 1977), pp. 23–24. The authors write: "The cumulative effects of increased backcountry use, combined with the effects of expanded development in non-wilderness grizzly range (i.e., oil and gas leasing, logging, road building, subdivision development, livestock grazing, water impoundments, etc.), inevitably increase the number of man-grizzly contacts, and hence the possibility of behavioral changes. These interactions usually end with the bear losing. Essentially, cumulative disturbances carried too far tend either to push grizzlies to the limits of their ability to avoid man and his developments, or cause the gradual deterioration of the bears' natural habits."

9. "Guidelines for Management Involving Grizzly Bears in the Greater Yellowstone Area," Intermountain Region, Forest Service, Ogden, 1979, pp. 1–2, 13, 69.

10. Douglas and Elizabeth Chadwick, "Living with Mountain Goats," *Defenders of Wildlife* 49, no. 4 (August 1974), pp. 267–271.

11. George Lowman, "Endangered, Threatened and Unique Mammals," p. 7.

12. Carl R. Sullivan, "Forests and Fishes in Southeastern Alaska," *Fisheries* 5, no. 5 (September-October 1980), pp. 2–6.

13. *Federal Register,* September 19, 1979, pp. 53928–53999.

14. The policy statement appears in the Forest Service Manual, Section 2600. See also Colorado Division of Wildlife, "Today's Strategy . . . Tomorrow's Wildlife," Denver, 1977.

15. Anne and Paul Ehrlich, *Extinction* (New York: Random House, 1981), pp. 86, 70.

16. "California Wildlife and Their Habitats: Western Sierra Nevada," Pacific Southwest Forest and Range Experiment Station, Berkeley, 1980, pp. iii–1.

17. The spotted owl, *Strix occidentalis,* inhabits mixed forests dominated by large conifers, scattered in widely dispersed enclaves from British Columbia to Southern California and from central Colorado to the Sierra Madre of Mexico. The northern race, *Strix occidentalis caurina,* is associated with the old growth forest, dense and diverse with big trees—Douglas fir, Western red cedar, hemlock, Sitka spruce, plus rhododendron and herbaceous plants covering the ground.

Eric Forsman, reportedly the first person to examine a nest of spotted owls in the Northwest since 1926, became interested while working as a Forest Service guard in the Cascades in 1969. He heard a pair of owls calling at night from woods near his guard station and learned to lure them close by imitating the call. Later, as a graduate student at Oregon State University, he conducted a spotted owl study, supervised by Howard Wight, leader of the Cooperative Wildlife Research Unit, and funded by the Forest Service. It began in 1972 to answer three questions: What are the owl's range and habitat requirements? What is the effect of forest management? How can the species be perpetuated? In the course of a laborious hunt in mountainous areas of wet forest conducted over a two-year span, Forsman learned that spotted owls maintain and utilize traditional roost trees year after year, requiring terrain with mature forest dominated by Douglas firs 200 to 600 years old—a dense, multilayered forest with canopy of old growth foliage at the top and layers of shade-tolerant trees and other plants underneath. He recommends a staggered rotation of long cutting cycles to assure diversity of plants and perpetuation of old growth timber. "We really need more low-elevation forest areas set aside as wilderness on biological and ecological grounds."

Chapter 11

1. For general reading on wilderness see, Michael Frome, *Battle for the Wilderness* (New York: Praeger, 1974); John C. Hendee, R. C. Lucas, and G. H. Stanley, *Wilderness Management* (Washington: Forest Service-USDA, 1981); Aldo Leopold, *Sand County Almanac* (New York: Oxford University Press, 1949); Robert Marshall, *Alaska Wilderness* (Berkeley, Calif.: University of California Press, 1970); Roderick Nash, *Wilderness and the American Mind* (New Haven, Conn.: Yale University Press, 1973); Sigurd Olson, *The Singing Wilderness* (New York: Knopf, 1957) and *Wilderness Days* (New York: Knopf, 1972).

2. In determination of historic roles, both Carhart and Leopold have their champions as to which did what first. There was no conflict between them, however, and each made his own pioneering contribution.

Leopold expounded broad scientific concepts not yet fully recognized or applied. In "Wilderness as a Land Laboratory," in *Living Wilderness* 6, no. 6 (July 1941), p. 3, he wrote:

When prairie dogs, ground squirrels, or mice increase to pest levels we poison them, but we do not look beyond the animal to find the cause of the irruption. We assume that animal trouble must have animal causes. The latest scientific evidence points to derangements of the *plant* community as the real seat of rodent irruptions, but few or no explorations of this clue are being made.

Many forest plantations are producing one-log or two-log trees on soil which originally grew three-log and four-log trees. Why? Advanced foresters know that the cause probably lies not in the tree, but in the micro-flora of the soil, and that it may take more years to restore the soil flora than it took to destroy it. . . .

All wilderness areas, no matter how small or imperfect, have a large value to land-science. The important thing is to realize that recreation is not their only or even their principal utility. In fact, the boundary between recreation and science, like the boundaries between park and forest, animal and plant, tame and wild, exists only in the imperfections of the human mind.

3. "Wilderness: Part of Oregon is Fading Away," *Earthwatch Oregon* 11, no. 5 (August/ September 1979), pp. 14, 15. This editorial includes the following:

> By 1964, increasing pressure to preserve some of the rapidly disappearing wildlands gave rise to the Wilderness Act, but in Oregon the original thirty million acres of forested wildlands had already shrunk to twelve million. And by 1971 another five million acres were opened to road building and logging. Estimates since 1971 indicate that another 1.3 million acres have been lost to development. Wilderness designation has protected just 1.2 million acres—much of it marginal land unsuited to commercial forestry.
>
> Oregon's thirty million acre forest has been liquidated at a pace that would surprise even the most ardent developers. Just 4.5 million acres—small parcels mostly, between the logging roads that push farther into the high country every season—remain as they were: wild and undeveloped.

4. Samuel T. Dana and Evert W. Johnson, *Forestry Education Today and Tomorrow* (Washington, D.C., Society of American Foresters, 1963), p. 11.

5. Evans, in his letter to Worf, and other critics assert that Forest Service activity in behalf of wilderness was a bureaucratic defense of "turf" against the threats of Harold L. Ickes, Secretary of the Interior, and the National Park Service. George Marshall, brother of Robert Marshall, sought to clarify this point in a letter to the author of January 25, 1972 after reading the first edition of *The Forest Service:*

> You speak of Ickes' desire to take over national forests and wilderness areas into his proposed Department of Conservation being a spur to the enlargement of wilderness areas in national forests. I am familiar with this theory and there may be something to it, as well as the one that the Forest Service established wilderness areas (primitive areas at that time) to protect its empire against the National Park Service and Department of Interior. However, it should be kept in mind that there were foresters, such as Aldo Leopold and Robert Marshall, and probably L. F. Kneipp and Carhart, who saw wilderness protection and classification as important public policy and an important forest policy, regardless of any rivalry there might be between government departments and bureaus. In any case, it was Robert Marshall, rather than Ferdinand Silcox, who proposed not only the enlargement of primitive areas (they were not called wilderness until 1939) and especially recommended a large number of new primitive areas. Silcox, in most instances, at least, followed Bob's recommendations just as John Collier and Ickes followed his recommendations for wild and roadless areas on Indian lands. Bob had a very strong influence on wilderness and certain other policies of the Forest Service during the 1930s, including the years prior to his rejoining the Service as Chief of the Division of Recreation and Lands. This is somewhat beyond the scope of your book, but I thought I might as well mention it while on the general subject.

6. For more information about wilderness management in the Forest Service, see "Technology Transfer Planning for Strengthening Wilderness Management in the East" by Joseph W. Roggenbuck and Alan E. Watson. This report, prepared in 1981 by two faculty members of the Department of Forestry at Virginia Polytechnic Institute, for the Pinchot Institute for Conservation Studies and Southeastern Forest Experiment Station, covers a study of 29 Forest Service managers of wilderness, their perception of problems and information needs.

In general, the study found Eastern national forests lagging behind the West in innovative management. Most managers were formerly timber specialists with mixed views of management and multiple use. Some considered it a conflicting single use—"Preservation not conservation." Most had no wildlife objectives, believing that wilderness legislation prevented them from taking wildlife management actions. Although Forest Experiment Stations conduct research designed to improve wilderness management, few were aware of it. Few had moved to establish carrying capacity and only three (Boundary Waters, Minnesota; Linville Gorge, North Carolina; and Great Gulf, New Hampshire) had instituted use limitations, despite Forest Service regulations requiring that wilderness plans provide for limiting and distributing visitor use to safeguard "values for which wilderness areas were created."

Crowding was reported as a significant problem, the use unevenly distributed because of steep slopes and dense vegetation. In Shining Rock Wilderness, in North Carolina, for example, over 50 percent of all overnight users camp in one small meadow. The study noted that land uses—including timber harvesting, highways, mining, second home development, motorization—impinge on the natural character of wilderness, making management necessary to strike a balance between some use and unacceptable environmental change.

The VPI researchers found 31 designated wilderness areas in the East, all but four added to the system since 1974. The East contained only 17 percent of total acreage in the system, with two areas—Boundary Waters and Everglades National Park—accounting for three-fourths. RARE II identified 2.3 million roadless acres in 23 Eastern states. Proposals of the Carter administration would have doubled the amount of Eastern wilderness to 567,000 acres.

7. William A. Worf, "Two Faces of Wilderness—A Time for Choice," *Idaho Law Review* 16 (1980), pp. 423–437.

8. Richard G. Walsh, Richard A. Gillman, and John B. Loomis, *Wilderness Resource Economics: Recreation Use and Preservation Values* (Denver: Department of Economics, Colorado State University, and American Wilderness Alliance, 1982), p. 7.

The authors show 10 million acres of potential wilderness in Colorado being derived from 36 percent of the state in public ownership, including 6.5 million acres administered by the Forest Service, 1.3 million acres by the Bureau of Land Management, and approximately .5 million acres by the National Park Service.

Respondents in the survey were asked to indicate the maximum they would be willing to pay for protection of wilderness areas in Colorado, contingent on the amount of wilderness protected. They were asked to state percentages attributable to their own current recreation use and a series of "preservation values." Surprisingly, they listed protection of water quality, air quality and wildlife habitat ahead of recreation.

Another approach to the same question is taken in the report, "Evaluating the Potential Impacts of Oil and Gas Leasing on the Wilderness Resource—A Philosophy and a Proposed Process," prepared by George D. Davis, of the Panhandle National Forests, Coeur d'Alene, Idaho, in 1982. He identifies 22 individual wilderness values in five distinct categories, naturalness, ethical, psychological, recreational, and other, and proposes applying these values as part of a rating system in the environmental analysis of oil and gas leasing proposals within designated and potential wilderness. He also provides historical background of both wilderness protection and oil and gas leasing, with a cross section of legal and judicial decisions.

9. David Phillips and Hugh Nash, *The Condor Question—Captive or Forever Free?* (San Francisco: Friends of the Earth, 1981), p. 177.

10. Dick Smith and Frank Van Schaik, *Mountains and Trails of Santa Barbara County* (Santa Barbara, Calif.: McNally and Loftin, 1962), p. 9.

Chapter 12

1. Besides areas in the Overthrust Belt and Inter-Mountain West, wilderness and proposed wilderness elsewhere were also targeted for exploration. In July 1981, the Forest Service issued a Draft Environmental Assessment recommending issuance of oil and gas leases in the Ventana and Santa Lucia wilderness areas near the Big Sur Coast of California. In September, BLM, on recommendation of the Forest Service, actually issued two oil

leases for the Capitan Mountains Wilderness in New Mexico. In October, the supervisor of the Ouachita National Forest recommended that a portion of the Caney Creek Wilderness in Arkansas and two areas under wilderness review be made available for oil and gas leasing.

2. George D. Davis, "Evaluating the Potential Impacts of Oil and Gas Leasing on the Wilderness Resource—A Philosophy and a Proposed Process" (Report prepared for Panhandle National Forests, Coeur d'Alene, Idaho, 1982), pp. 3, 57–58.

3. USDA, Forest Service, "Mining in the National Forest System," Washington, D.C., 1979, pp. 1–7.

4. David Sheridan, Hardrock Mining on the Public Lands (Washington, D.C.: Council on Environmental Quality, 1977), pp. 1–4, 6–7, 10–11, 27.

5. "Managing the Mineral Resources of Eastern Region National Forests," Eastern Region, Forest Service, Milwaukee, Sept. 1979, pp. 5, 8, 9.

6. Secretary Freeman, even before passage of the Wilderness Act, responded to a request from Congress with this comment: "Authority now exists under which mineral leases can be issued for leaseable minerals in the wilderness, wild and primitive areas either under the Mineral Leasing Act of 1920 or the Mineral Leasing Act for Acquired Lands. It is the policy of this Department to recommend against, and the policy of the Department of the Interior to withhold, the issuance of mineral leases in these areas unless directional drilling or other methods can be used which will avoid any invasion of the surface of the wilderness, wild or primitive area."

7. Parallel to the objection of the Wyoming congressional delegation to proposed exploration in wilderness areas of that state, the entire Florida delegation, both Senators and all 15 members of the House, opposed attempts to strip-mine phosphates in Osceola National Forest. The phosphates are largely in swampy wetlands. Scientists warned of difficulties in restoring the groundwater aquifer, a source of north Florida's water supply and of support to stands of cypress trees. Osceola's phosphate deposits were said to represent 3 percent of Florida's potential.

8. "Bridger-Teton Oil and Gas Drilling Procedures." "Bridger-Teton Prospecting Permit Procedure." Both issued by the Bridger-Teton National Forest, Jackson, Wyoming, 1980, pp. H-6, H-7, H-8, H-12, H-13.

9. Kevin C. Gottlieb, "Rigging the Wilderness," Living Wilderness 45, no. 156 (Spring 1982), pp. 13–16. The author reported that the Reagan administration had taken the following actions:

> • Opened more federal lands for exploration and production while reducing the time period and agency budgets for environmental review;
> • Accelerated the schedule for offshore leasing of gas and oil;
> • Lifted a moratorium on oil and gas leasing on 6 million acres of land acquired by the government for military purposes;
> • Directed the National Park Service to change regulations in order to allow more mining of hard rock minerals, as well as oil and gas, in five designated national recreation areas in the West;
> • Revoked the withdrawal of 20 million acres of federal land and is reviewing all other withdrawals. In all, the Administration intends to review more than 50 million acres of withdrawals.

10. USDA, Forest Service, "A National Energy Program for Forestry," Washington, D.C., 1980, pp. 1, 8.

11. "Forest Biomass as an Energy Source" (Report of a Task Force, Richard L. Doub, Chairman. Society of American Foresters, Washington, 1978), pp. 4–5, 14, 19–20.

12. Ronald J. Slinn, "A Fact Sheet on Wood as a Raw Material and for Energy" and "Wood Energy in the U.S. Pulp and Paper Industry" (Both issued by American Paper Institute, New York, 1981).

13. USDA, Forest Service, "Tree Biomass—A State-of-the-Art Compilation," 1981, pp. 1–2. This review includes the following comment: "Until recently, forest products were made principally from the main stem, called the merchantable bole, of trees. Decreasing tree sizes, increasing extraction and processing costs, new product development, increasing competition for wood, and the energy crisis have all focused attention on extending timber

utilization to include the total aboveground portion of the tree. Tree crowns can be utilized as fuel for domestic and industrial purposes or for fiber for pulp and paper and composite board products, as a soil amendment or mulch in agriculture, and a bulking agent in municipal sludge composting." In 1980 a National Tree Biomass Compilation Committee was formed to improve research and assessments of biomass throughout the nation.

14. An enthusiastic expression on the potential of wood is found in "Energy for the Eighties—A Problem Solved," by Morton S. Fry, issued in 1981 by Miles W. Fry & Son, Inc., of Ephrata, Pennsylvania, a nursery which has done development with hybrid popular clones:

> Wood, the principal source of energy for the human race for thousands of years, is becoming the front runner in the race to find an alternate source of energy. There are now clones of poplar hybrids, a product of the work done by geneticists of the United States Forest Service, which have astounding growth. These trees can be placed on close spacings with six to seven thousand trees per acre. They are self regenerative and can be reharvested every two years after the initial cutting. We have, through the use of these trees, the potential of replacing all our liquid fuel requirements, replacing all our oil and nuclear power plants, stopping the flow of petro-dollars into foreign hands while doing an about-face on our balance of trade. We can bring about the recovery of the dollar, the strengthening of the economy, a broader tax base and make a balanced budget a possibility. The American farmer or landowner is not only capable of feeding this nation but of fueling it as well. (p. 1)

However, H. V. Decker, of the New York Department of Environmental Conservation, in "Wood Energy—Just a Word of Caution," *Northern Logger* 27, no. 9 (March 1979), pp. 3, 40, 42, counters:

> Perhaps all those millions of tons of raw energy are indeed waiting for us, just for the plucking, so to speak. But are we all really convinced that wood is the answer to OPEC; the ultimate energy source, our key to energy independence forevermore? . . .
> We all welcome "new" uses for our renewable resources that provide additional outlets for wood and wood products. Such enlightened utilization does offer incentives to apply management to these valuable forest resource lands, encourage employment and enhance the economy of the state and nation. Nevertheless, we owe to ourselves, and those who will follow us, that we make the wisest and best use of these resources placed in our trust, that we do not leap blindly and without restraint into each new technology that happens along and that, in the final analysis, our long-term forestry goals and objectives must not be jeopardized by shortsighted management practices designed to meet and exploit every new, but untested, market.

15. Hiram Hallock, "Opportunities in Sawing Mechanics Research" (Paper presented at Forest Products Utilization Research Conference, 1979), pp. 1, 4.

Hiram Hallock, a Forest Service official involved in the Research and Sawmill Improvement Program, estimated in 1979 that the country had used about 2 billion, 667 million cubic feet of logs the preceding year to produce 20 billion board feet of softwood construction-dimension lumber. He said this was about 375 million cubic feet more logs than really necessary. The country thus had lost the equivalent of 2 billion, 800 million board feet of lumber, enough to build 280,000 homes (based on 10,000 board feet per single home). Hallock's figures were based on more than 700 SIP studies of softwood lumber mills, involving determination of the volume of logs sawn, board foot lumber output, sawing variation, planing allowance and kerf width, with shrinkage estimates based on species and moisture content at the time of planing. Hallock attributed resource waste to old equipment and seat-of-the-pants judgement comparable to "the way grandma cooked."

16. "Increased Use of Felled Wood Would Help Meet Timber Demand and Reduce Environmental Damage in Federal Forests" (Report of the General Accounting Office, Washington, D.C., 1973), pp. 2, 9, 10, 29.

Another aspect is covered in a study by the Forest Resources Economics Research Group of the Forest Service covering the years from 1950 to 1980. In 1979, a total of 1,835 million cubic feet of softwood timber and 300 million cubic feet of hardwood timber was exported from the U.S. These volumes convert to 11.5 billion board feet of softwood lumber and 1.8 billion board feet of hardwood lumber.

17. "Age of Wood/Extending the Timber Resource" (Forest Products Laboratory, Madison, 1979–80), pp. 5, 7; "Age of Wood: Effective Timber Utilization" (Forest Products Laboratory 1981), pp. 4, 6; "Energy Efficiency in Light-Frame Wood Construction" (Forest Products Laboratory, 1979); USDA, Forest Service, "Extending Wood Supply Through Research," Washington, D.C., 1980.

Chapter 13

1. "Program of Research for Forests and Associated Rangelands" (Proceedings of national and regional planning conferences, USDA, Forest Service, Washington, 1978), pp. i, 1–2, 16–18, 20, 26.

2. "Forest Products Research in the Forest Service," *Forest Products Journal* 29, no. 10 (October 1979), pp. 56–60.

3. "A Directory of Research Natural Areas on Federal Lands," Federal Committee on Ecological Reserves, USDA, Forest Service, Washington, 1977. Gen. Ref. The Federal Committee was established in 1966 and published the first Directory in 1968. It was largely informal until new interest was generated in 1974 with assistance from the National Science Foundation and Council on Environmental Quality. Besides participating federal agencies, national organizations have provided significant support. They include the American Institute of Biological Sciences, Conservation Foundation, Ecological Society of America, National Parks and Conservation Association, Society of American Foresters, Society of Range Management, Institute of Ecology, Nature Conservancy, Wilderness Society, and Wildlife Society.

4. Charles Cushwa, *Naturalist* 20, no. 1 (Spring 1969), pp. 1, 2–3.

5. D. Kimball Smith, "The Knowledge Flows from Hubbard Brook," *Yale Alumni Magazine* (April 1982), p. 18.

6. F. Herbert Bormann and Gene E. Likens, "The Nutrient Cycles of an Ecosystem," *Scientific American* 223, no. 4 (October 1970), pp. 92–101. See also "Publications of the Hubbard Brook Ecosystem Study," compiled by Phyllis Toyryla, Hubbard Brook Experimental Forest, West Thornton, New Hampshire, 1981.

Principal investigators of the Hubbard Brook Study are G. E. Likens, Cornell University; E. H. Bormann, Yale University; R. S. Pierce, Forest Service; and R. T. Homes, Dartmouth College. Bormann and Likens both were at Dartmouth when they started in 1963. The Research Work Unit Title is "Impact of Forest Management on Nutrients in Soil and Water." The mission is "to evaluate forest management practices with respect to conserving soil nutrients and protecting stream water quality; and to determine how nutrient cycling affects the forest-stream ecosystem."

A report of October 1977 includes the following:

> Our recent studies have documented the changes in stream chemistry that can be expected as a result of contemporary clearcutting, including stripcutting. However, the impacts of whole-tree harvesting, shelterwood systems, and other forest management practices could be different because of changes in operating procedures, equipment, and amount of timber removed. Changes in chemical parameters of streams such as major cations, anions, pH, and conductivity for selected examples of various harvesting operations will be determined. . . . As the Hubbard Brook Experimental Forest has recently been designated a Biosphere Reserve, we will continue baseline monitoring of the forest environment, concentrating on precipitation quantity and quality (including acidic components) streamflow, stream water quality, physical and chemical soil characteristics, and vegetation.

7. D. Kimball Smith, "The Knowledge Flows from Hubbard Brook," p. 18–19.

8. Jerry F. Franklin and Dean S. DeBell, "Effects of Various Harvesting Methods on Forest Regeneration" (Reprint from Even-Age Management Symposium, August, 1972, School of Forestry, Oregon State University, Corvallis), pp. 35, 37.

9. J. H. Wikstrom and S. Blair Hutchison, "Stratification of Forest Land for Timber Management Planning on the Western National Forests" (Report prepared for Intermountain Forest and Range Experiment Station, Ogden, 1971), pp. 2–4, 22.

10. "The Forest Service Needs to Ensure that the Best Possible Use is Made of its Research Program Findings," General Accounting Office, Washington, January 1972, pp. 23, 24.

11. Leon S. Dochinger and Thomas A. Seliga, "Acid Precipitation and the Forest Ecosystem," *Bioscience* 26, no. 9 (October 1976), pp. 564–565. See also "Workshop Report on Acid Precipitation and the Forest Ecosystem (First International Symposium)," Northeastern Forest Experiment Station, Broomall, Pennsylvania, 1976, pp. 1–2.

Chapter 14

1. "America Grows on Trees—the Promise of Private, Nonindustrial Woodlands." National Forest Products Association, Washington, 1980. Gen. Ref.

2. Leon S. Minckler, *Woodland Ecology: Environmental Forestry for the Small Landowner* (Syracuse: University of Syracuse Press, 1975), pp. 7, 35, 51, 119–137.

3. "Improving Outputs from Nonindustrial Private Forests" (Report prepared for Society of American Foresters, Bethesda, Md., 1979), pp. 4, 6.

4. David A. Gansner and Owen W. Herrick, "Cooperative Forestry Assistance in the Northwest," Northeastern Forest Experiment Station, Broomall, Penn., 1980, pp. 1–6.

5. The 1945 California Forest Practice Act was often criticized as being administered by the industry it was intended to regulate. The 1973 Act requires that a majority of the State Forestry Board represent the public, with background and qualifications in some phase of natural resource and land use broader than timber harvest and production.

6. Indiana Department of Natural Resources, "Indiana Forest Resource Plan." Indianapolis, 1981, pp. i, 9–10, 32–33.

7. Jerry J. Presley, "Opportunities for Wildlife Management through State and Private Cooperative Forestry Programs" (Paper presented at Wildlife Management on Private Lands Symposium, reprinted by Missouri Department of Conservation, Jefferson City, 1981), pp. 2–3.

8. Pennsylvania Bureau of Forestry, "Annual Report," 1980; Bureau of Forestry. Department of Environmental Resources, "Forestry Programs and Services," Harrisburg, 1980.

9. James B. Harrell, "Florida's Urban Forestry Program," *Journal of Arboriculture* 4, no. 9 (September 1978), pp. 202–206.

10. City Forestry Division, "Baltimore's City Forest," Baltimore, 1981, pp. 1–5.

11. "Urban Forestry White Paper," by a committee of the Society of American Foresters, Gene W. Gray, Chairman, in *Journal of Forestry* 78, no. 5 (May 1980), pp. 298–300. Some cities are willing to pay. In Seattle, a 1968 bond issue financed planting of about 15,000 trees. Since 1970, Seattle has planted about 20,000 street trees and has landscaped more than 50 new green spaces under "Operation Green Triangle."

Chapter 15

1. Dale R. Bottrell, "Integrating Pest Management" (Report prepared for Council on Environmental Quality, Washington, 1979), pp. 62–63.

2. IPM Coordinating Committee, "Progress Made by Federal Agencies in the Advancement of Integrated Pest Management" (Washington, D.C.: Council on Environmental Quality, 1980), pp. 1, 2; "Integrated Pest Management for Forest Insects: Where Do We Stand Today?" (Reprint of papers presented at a technical session of the Forest Entomology Working Group, Society of American Foresters, Washington, 1979). Gen. Ref.

3. William E. Waters, "The Case for Forest Entomology," *Journal of Forestry* 68, no. 2 (February 1970), pp. 73–77.

4. Whiteford L. Baker, *Eastern Forest Insects* (Washington: USDA-Forest Service, 1972), pp. 1, 3, 13.

5. USDA, Forest Service, "A Recommended Renewable Resources Program—1980 Update," F. S. Publication 346, p. 92.

6. USDA, Forest Service, "Pesticide Use on National Forest System Lands," Washington, D.C., 1981, pp. 2–4.

7. Robert van den Bosch, "The Cost of Poisons," *Environment* 14, no. 7 (September 1972), pp. 18–31. Robert van den Bosch, "Pesticides: Prescribing for the Ecosystem," *Environment* 12, no. 3 (April 1970), pp. 20–25.

8. Robert W. Campbell, "The Gypsy Moth and its Natural Enemies," USDA, Forest Service, Washington, 1975, p. 1.

9. Daniel Smiley, *Gypsy Moth and Man: A Story of Mutual Accommodation*, revised in 1980, with supplement in 1981 (New York: Mohonk Trust, Mohonk Lake, 1975), pp. 1, 7–8, supplement pp. 1, 4. Smiley's conclusion after the 1981 season: "Healthy forests are unlikely to be seriously damaged by the gypsy moth. Health is in part synonymous with diversity of species. A large area planted to a single species is an open invitation to an equally large buildup of pests—and not just gypsy moths. Spraying is worse than useless, as it delays the ultimate accommodation. Diversity is especially important on poor sites, where trees already are under stress and susceptible to disease or infestation."

Nevertheless, the 1982 season saw no let-up in the campaign to control the gypsy moth. "MAKING A KILLING" read a headline in the business section of the *Philadelphia Inquirer* of May 16, 1982. As the subhead explained, "Gypsy-moth Scare Profitable for Some." While Union Carbide would not disclose to the *Inquirer* its sales figures for Sevin, Chevron Chemical Co. said it expected sales of Ortho to double to more than $12 million, primarily because of the gypsy moth.

10. *Maine Environment* (April 1981), p. 2.

11. USDA, Forest Service, "Final EIS Proposed Cooperative 5-year (1981–1985) Spruce Budworm Management Program for Maine," Northeastern Area, Broomall, Penn., 1981, pp. i–iii.

12. James O. Nichols, *The Gypsy Moth* (Harrisburg, Penn.: Pennsylvania Bureau of Forestry, 1980), p. 26. This booklet discusses issues relative to use of *Bacillus thuringiensis*. In one sense it is costly, since it may require two applications per year. It kills 50 to 75 percent of caterpillars, which may result in the need of retreatment the following year. BT is adversely affected by sunlight, high temperatures, and high humidity. These problems still need solving. On the other hand, BT affects only leaf-eating caterpillars. Unlike broad-spectrum chemical materials, BT does not kill enemies of pest species, thus reduces need of subsequent treatment.

The microbe BT is formulated into a number of "biological" insecticides (under such trade names as Turicide, Biotrol, and Dipel), all generally referred to as BT. *B. thuringiensis* is a spore-forming bacterium in the genus *Bacillus*, producing a diamond-shaped crystal toxic to about 150 insects: they are susceptible because their innards have the proper alkalinity and/or enzyme necessary to dissolve the crystal. It does not kill bees, nor does it affect humans, rats, mice, swine, rabbits, dogs, chickens, pheasants, and partridges.

13. Jan M. Newton, *An Economic Analysis of Herbicide Use for Intensive Forest Management* (Eugene, Ore.: Northwest Coalition for Alternatives to Pesticides, 1979), Gen. Ref.

14. Tom Waters, "Forestry Workers: Inform Yourselves," *NCAP News* 2, no. 1 (Summer 1980), pp. 12–14. According to this article, during August 1979, eighty temporary crew members and crew bosses working on the Detroit Ranger District of the Willamette National Forest refused to expose themselves to herbicides by cutting roadside brush sprayed with 2,4-D and other chemicals. When management threatened to fire them, they threatened to resign. Management dropped the issue and none of the crew was denied rehire rights. Nevertheless, the author wrote: "The Forest Service and other agencies have a history of exposing their employees to harmful substances. Some of these substances, after further research, have been banned. These substances often cause some strange symptoms while they are used, but government managers belittle their reactions."

Chapter 16

1. Anne and Paul Ehrlich, *Extinction* (New York: Random House, 1981), pp. 159–160.
2. USDA, Forest Service, "The Forest Service and International Forestry Programs," Washington, 1980. Gen. Ref.
3. Francesco di Castri and Jane Robertson, "The Biosphere Reserve Concept: 10 Years After," *Parks* 6, no. 4 (January/February/March 1982), pp. 1–6.
This article notes that biosphere reserves have served as media of cooperation among global powers and between the advanced and developing nations. It cites three key future targets:

> First, for the decision-makers, who must be convinced of the long-term necessity and global responsibility of establishing and maintaining conservation areas;
> Second, for the "conservationists," that conservation must be backed by scientific research and sound management;
> Third, for the authorities responsible for conservation areas, that conservation is not possible without the understanding and support of the local people, and the importance of environmental education in this respect.

The MAB program was launched in 1971, emphasizing the need to conserve entire, representative ecosystems. Subsequently a global classification was developed to evaluate the biosphere reserve network and identify "gaps." By December 1981, a total of 209 reserves had been designated in 55 countries. They were national parks and nature reserves already established to protect rare or unique landscapes or species. The coverage admittedly was patchy, with large gaps needing to be filled in the areas of boreal woodlands, tropical humid forests, warm arid lands, and coastal zones.
4. The need of international cooperation was stressed at a "session of a special character" conducted by the United Nations Environment Programme at Nairobi in May 1982, commemorating the tenth anniversary of the celebrated first UN Conference on the Human Environment at Stockholm. In his report on "The State of the World Environment 1972–1982," Mostafa K. Tolba, Executive Director of UNEP, declared:

> During the 1980s, international cooperation is likely to be needed for global and regional environmental monitoring, and its extension by the adoption of authoritative, critical assessments. Cooperation will also be important for action to combat soil degradation and provide guidance on the ground to people endeavoring to develop land in areas vulnerable to erosion and diminishing productivity, for continued efforts against desertification, for the improvement of the human environment through better planning of settlements and services, for action against the diseases that degrade the lives of so many people, for continued development of less polluting technology that is cost-effective and suited to the developing world, and for action to conserve world genetic resources.

5. U.S., Department of State, "The World's Tropical Forests: A Policy, Strategy and Program for the United States" (Report to the President by a U.S. Interagency Task Force on Tropical Forests, Washington, D.C., 1980, pp. 15–19, 26–47.
6. Council on Environmental Quality, "Global Future: Time to Act. Report to the President on Global Resources, Environment and Population," Washington, D.C., 1981, pp. xxxviii–xxxix.
7. Gerald O. Barney, Ed., *The Global 2000 Report to the President: Entering the Twenty-First Century* (New York: Pergamon Press, 1981), vol. 2, pp. 117–120, 125–133. "The world faces an urgent problem of less of plant and animal genetic resource," warned Global 2000, initially prepared as a government study commissioned by CEQ. "An estimate suggests that between half-a-million and two million species on earth could be extinguished by 2000, mainly because of loss of wild habitat, but also in part because of pollution. Extinction of species on this scale is without precedent in human history."
8. Norman Myers, "Tropical Rain Forests: Whose hand is on the axe?" *National Parks Magazine* 53, no. 11 (November 1979), pp. 9–12.

9. An editorial in the *Malayan Naturalist* of January 1978, published by the Malayan Nature Society at Kuala Lumpur, includes the following:

> Endau-Rompin is the most important conservation issue to have arisen in Malaysia. For the first time the general public has become actively involved in a matter concerning the continued existence of this country's rich and invaluable natural history. The maintenance of blocks of virgin forest has deep implications with regard to climatic stability, water supplies and erosion. Through a long, complex chain of cause and effect, uncontrolled land clearance can lead eventually to a decrease in the standard of living of rural and urban peoples, both rich and poor.
>
>   Such effects are acknowledged. However, the temptations of quick profit are always present and the conflict of short and long-term interests came to a head in the Endau-Rompin issue....
>
>   Endau-Rompin caught the public attention and became a newsworthy item. Banner headlines on the story appeared on the front page of the papers, and numerous letters, most of them in support of its effect when the Pahang State Government agreed to stop issuing any further logging licenses in the Park.
>
>   This campaign should be a lesson to industrialists, civil servants, politicians and all those who are involved in making decisions determining the future of Malaysia. Malaysians in all walks of life are beginning to be concerned about the well-being of their environment. Such public feeling cannot be ignored. Neither can anyone be indifferent to the consequences of wanton destruction of our natural assets.

Chapter 17

1. Jack Doyle, *Lines Across the Land* (Washington: Environmental Policy Institute, 1979). Gen. Ref.

2. Channelization deepens and widens a stream through its watershed in order to facilitate the drainage of upstream areas. Critics have complained about disruption of riparian, or riverside, zones. In a normal streamside, woodland vegetation benefits birds, insects, and fish. It tends to slow the flow of water and hold layers of silt. In a channelized stream, silt is likely to clog the channel and negate the purpose of the project.

3. The Soil and Water Resource Conservation Act of 1977, taking into account the extensive wind erosion on the Great Plains and conversion of nearly a million acres yearly to nonfarm uses, called for a detailed survey of U.S. soils. An environmental impact statement and draft program completed by SCS in 1981 to comply with the Act proposed federal block grants to the states; development of local and state soil and water conservation programs; tax incentives for conservation measures, and targeting assistance at critical areas.

4. The Land and Water Conservation Fund was established in 1964 with 90 percent of its revenue derived from fees paid to the federal government for offshore oil and gas leases. Congress authorized an expenditure of $900 million annually for federal and state land acquisition. Since the first project in 1965, more than 27,000 projects in 14,000 communities have benefitted from the Fund. These include areas in national, state, county, and community parks, monuments, lakeshores, wild and scenic rivers, trails, historic sites, parkways, and wilderness. Congress has never appropriated funds to the full authorized level; during the last five years of the 1970s the actual appropriated funds averaged about $500 million.

On his appointment as Secretary of the Interior, James G. Watt (who had served as Director of the BOR during a portion of the Nixon administration) asked Congress to rescind $250 million already appropriated under the Fund for fiscal 1981 and to appropriate only $45 million for 1982. Congress responded by rescinding $90 million for 1981 and appropriating $149.2 million for 1982. For 1983 the Administration proposed only $69.4 million to be available to federal agencies only (and then to be used exclusively to pay for pending or anticipated court awards to landowners), with no funding for the state matching grant program.

5. The Forest Service conducted ceremonies in New Hampshire on December 6, 1980, to mark the acquisition of three significant parcels of land as additions to the White Mountain National Forest. These three tracts, totaling 15,350 acres, were former resort holdings at Waumbek, Bretton Woods, and Sandwich Notch, all identified during the

mid-1970s as worthy of protection. The Society for Protection of New Hampshire Forests, Appalachian Mountain Club, and Nature Conservancy were instrumental in seeking Congressional recognition of these projects.

6. Richard D. Hull, Director of Lands of the Forest Service, in a letter of September 19, 1980.

Among other areas identified as "acquisition needs" are two parcels totaling 570 acres of the Silvertip Ranch within the Absaroka-Beartooth Wilderness near Yellowstone National Park.

7. For detailed treatment of reorganization during the Nixon administration, see John C. Whitaker, *Striking a Balance* (Washington, D.C.: American Enterprise Institute for Public Policy Research, 1976, Chapter 3, "New Federal Institutions for Energy and Environment"), pp. 43–71, follows the Ash Council recommendations, including the proposal to establish a Department of Natural Resources.

Chapter 18

1. Former Chief John McGuire comments, after a review of the text of this chapter:

> The "overemphasis" has been even more apparent in the budget proposed by successive administrations (see, for example, the various House hearing records showing comparisons of Forest Service budget requests with Department and OMB allowances).
>
> In the Eisenhower years, the Forest Service was permitted to publish the Timber Resource Review in 1958, but without long-range program recommendations. The Kennedy people finally submitted the National Forest program, but refused to submit long-range proposals for State and Private Forestry and Forestry Research.
>
> Forest Service planning efforts to achieve greater balance among management functions in the 1960s and early 70s led to passage of RPA in 1974. Nevertheless, each administration (through OMB control) has resisted the idea of proposing expenditures in future years and seeking appropriations with less than immediate payoff.
>
> The chapter fails to give enough emphasis to the importance of NEPA. For example, it was Section 103 of NEPA that gave the Forest Service enough leverage with the Department and OMB to proceed with RARE I and with mining regulations. And it was primarily NEPA which gave citizens additional entry to the courts.

2. Charles A. Reich, *Bureaucracy and the Forests* (Santa Barbara, Calif.: Center for the Study of Democratic Institutions, 1962); reprinted in *Center Magazine*, January/February 1975, pp. 3–4, 6.

3. James W. Moorman, "Bureaucracy v. the Law," *Sierra Club Bulletin* 59, no. 9 (October 1974), pp. 7–10.

4. The following comment is provided by J. Lamar Beasley, Deputy Chief for Programs and Legislation, and Mark A. Reimers, Director of Legislative Affairs of the Forest Service, and by James Perry, of the Office of General Counsel of the Agriculture Department:

> Jim Moorman is quoted as saying the Forest Service administrative appeals procedure is an obstacle to judicial review. In fact, it is primarily the environmental groups who utilize the Forest Service administrative appeals procedure to contest timber sales and other development projects.
>
> The quote regarding the Forest Service appeals procedure at 36 CFR 211.9 and 19 is not correct insofar as four appeal levels being required. In fact, no more than two levels are required, depending upon the starting point, with the possibility of discretionary review for a third. Public records show the Forest Service has, on a number of occasions, reversed its decisions in the course of administrative appeals procedures.
>
> Regarding Forest Service discretion, a very limited concept is advanced. In fact, the Multiple-Use Sustained-Yield Act of 1960 provided the Forest Service with a very broad grant of discretion to determine the best use of land to meet the needs of the American public for the purposes described therein. This wide range of discretion has been gradually restricted by environmental laws and other related land-affecting statutes, culminating in the National Forest Management Act of 1976.

5. Regarding the suit brought by the Texas Committee on Natural Resources, Beasley, Reimers, and Perry write: "The Court of Appeals, in overruling Judge Justice, ruled on more than the text suggests, including upholding the environmental impact statement filed by the Forest Service. It found Judge Justice's requirement of a programmatic impact statement to be in error and found the District Court in error for overruling the Forest Service decision to proceed with clearcutting. Finally, the plaintiff's petition certiorari to the Supreme Court was denied."

6. Joseph L. Sax, *Defending the Environment* (New York: Knopf, 1972), pp. 107, 112, 197.

7. Richard W. Behan, "RPA/NFMA—Time to Punt," and James W. Giltmier, "In Response to RPA/NFMA—Time to Punt," *Journal of Forestry* 79, no. 12 (December 1981), pp. 802–805.

8. Minnesota's challenge of Forest Service action in administering the Boundary Waters Canoe Wilderness Act was based on the federal government's assertion of jurisdiction over water surfaces. The State and others filed a petition for certiorari with the Supreme Court which was subsequently denied.

9. Other cases cited by the Forest Service in which environmentalists unsuccessfully challenged agency action on NEPA and related issues include *Southeast Alaska Conservation Council, Inc. v. Bergland et al.*, and *Skokomish Indian Tribe et al., v. Beaubien et al.* The Alaska case involved timber sales activities in a drainage reportedly having outstanding multiple use values. The Indian Tribe case challenged Forest Service timber cutting activities within the Shelton Sustained Yield Unit, the site of a long-term logging sale on the Olympic National Forest.

Chapter 19

1. Luke Popovich, "Professors and the Profession, Fostering Change at Yale," *Journal of Forestry* 75, no. 1 (January 1977), pp. 30–33.

2. Henry Clepper, *Professional Forestry in the United States* (Baltimore: Johns Hopkins Press, 1971), pp. 123–125.

3. Brock Evans, Vice-President of National Audubon Society (and formerly Northwest representative of the Sierra Club) in a letter written to William A. Worf on January 6, 1982, expressed the following view of forestry education as it relates to the Forest Service:

> . . . I, too, like so many, had always assumed that the Forest Service really believed in wilderness and wanted to protect as much of it as possible. But in my journeys throughout the Northwest states and Alaska, it always seemed that the agency was opposed to citizen proposals for wilderness no matter where they surfaced. Why was this so, I asked myself. And finally I came to the conclusion that it has a great deal to do with the teaching that the decision makers in the Forest Service receive in forestry school.
>
> As you know, the emphasis—certainly in the years when the present and past decision makers were in school—was heavily on the timber production aspects of forest management: mensuration, logging engineering, various aspects of silviculture, and so on. The idea that trees were anything but a crop, that species other than commercially valuable trees were anything but "weeds," the idea that recreation, much less wilderness, had any place at all in forest management if it interfered in any way with getting the logs out, was played down to say the least. This has been the overwhelming emphasis of all forestry curricula that I know of or have heard of, certainly in all the Western forestry schools, for a very long time. That must be where it all began, and that certainly explains the attitude of past and present Forest Service decision makers.

4. Carl H. Reidel, "Creating the Future" (Paper presented before the Northern California chapter, SAF, December 3, 1977).

5. Richard A. Skok, "A Frank Appraisal of Forestry Education" (Paper presented before the New England chapter, SAF, March 9, 1978).

6. Another type of forestry training is offered by the Sterling Institute, at Craftsbury Common, Vermont. The Institute's "Grassroots Project" is a one-year college program of intensive study in conservation occupations of agriculture, forestry, and wildlife management. Other programs include the Academic Short Course in Outdoor Leadership, Rural

Affairs, and Rural Resource Management. "Sterling believes that its educational approach has particular importance in areas relating to people and the natural environment," according to the Institute's statement of principles. "It believes that solutions to many problems can be reached only by melding technical knowledge with a broad understanding of the human experience and that neither the narrow technician nor the uninformed idealist can find lasting solutions." Forestry activities include: "Management of lands for continuous production of firewood, Christmas trees, sawlogs, pulpwood, wildlife habitat, and maple syrup. Skill development with chainsaw, axe, and crosscut saw. Tree identification and forest measurements. Log cabin construction."

GLOSSARY AND ABBREVIATIONS

AFA	American Forestry Association: a public organization founded in 1875 for the advancement of intelligent management and use of forests and other natural resources.
AUM	Animal Unit Month: a measure of rangeland forage use, by one horse, one cow, or five sheep per month.
BLM	Bureau of Land Management: an Interior Department agency. As successor to the old General Land Office, it administers what remains of the public domain (almost entirely in the West). BLM activities relate to mineral claims on all public lands and the Outer Continental Shelf.
CEQ	Council on Environmental Quality: a board of three members appointed by the President to fulfill, with staff assistance, provisions of the National Environmental Policy Act.
Clearcutting	The subject of considerable controversy and varying definition, it generally refers to the harvest of all forest cover on a designated tract of land in order to meet requirements for regeneration of a desired tree species. Clearcutting is associated with even-aged management (in contrast to all-aged).
EIS	Environmental Impact Statement: a document based on review, with public participation, required of federal agencies in consideration of actions with significant impact on the environment. An environmental assessment involves a lesser study that may or may not lead to a full EIS.
FLPMA	Federal Land Policy Management Act of 1976: the organic act, or charter, of BLM, with bearing on national forests as well.
FORPLAN	Forest Planning Model: instituted in 1980 as a primary tool for analysis in preparation of national forest plans as required by NFMA (see below).
K-V (or K-V money)	A reference to the Knutson-Vandenberg Act of 1930: It requires purchasers of national forest timber to deposit money needed for reforestation, timber stand improvement, or other measures to improve the future productivity on timber sale areas.
Multiple Use	The Multiple-Use–Sustained-Yield Act of 1960 defines it as utilization of resources in combination to meet needs, with economic return not always the deciding factor. Another way of viewing multiple use: Search for balance in meeting demands for goods and services on a fixed land base.
NEPA	National Environmental Policy Act: the first piece of legislation signed by President Richard M. Nixon in January 1970 (although it was passed by Congress in 1969). It opened a new era of accountability in resource management.

NFMA   National Forest Management Act of 1976: It added new factors to the RPA process (see below), including interdisciplinary planning for the national forests.

NFPA   National Forest Products Association: organized in 1902 as the National Lumber Manufacturers Association, a major spokesman for the industry viewpoint.

NRDC   Natural Resources Defense Council: a watchdog organization born during the environment-conscious 1960s, exercising critical attention to costs versus benefits of logging and grazing on public lands.

OMB   Office of Management and Budget: an executive agency attached to the White House with strong influence in determining policy and budget.

RARE I   Roadless Area Review and Evaluation: begun in 1972 to determine the best use of roadless areas in the national forests not covered by the Wilderness Act. In 1973 RARE I identified 56 million roadless acres. The Forest Service selected 274 roadless areas, containing 12.3 million acres, for further wilderness study.

RARE II   Initiated by the Carter Administration in 1977 to accelerate allocation of use of roadless areas in compliance with NEPA requirements. RARE II identified 62 million roadless acres in 1,920 separate units, recommending 15 million acres for Congressional designation as wilderness and allocation of 11 million acres for further study and 36 million acres for nonwilderness use.

RPA   Forest and Range Renewable Resources Planning Act of 1974: It directs the Forest Service to assess all the nation's forest and rangeland resource needs and capabilities, to define alternatives, and to recommend a Forest Service program of management as a basis for budget requests—with updated program every five years, and assessment every ten years.

SAF   Society of American Foresters: the professional society that Gifford Pinchot was instrumental in establishing in 1900.

S&PF   State and Private Forestry: the wing of the Forest Service extending beyond the national forests with a cooperative program, developed as the alternative to federal regulation.

GENERAL BIBLIOGRAPHY

(For additional sources on particular subjects, see the notes for relevant chapters.)

Clawson, Marion. *Forests for Whom and for What?* Baltimore: Johns Hopkins Press, 1975.

Dana, Samuel Trask and Sally K. Fairfax. *Forest and Range Policy: Its Development in the United States.* Second edition. New York: McGraw Hill, 1980.

Douglas, William O. *My Wilderness.* Garden City, N.Y.: Doubleday, 1960. 2 vols. Vol. 1: *The Pacific West;* vol 2: *East to Katahdin.*

Freeman, Orville L. and Michael Frome. *The National Forests of America.* New York: Putnam's, 1968.

Frome, Michael. *Whose Woods These Are: The Story of the National Forests.* Garden City, N.Y.: Doubleday, 1962.

Marshall, Robert. *The People's Forests.* New York: Smith and Haas, 1933.

Nash, Roderick. *Wilderness and the American Mind.* New Haven, Conn.: Yale University Press, 1967.

Pinchot, Gifford. *Breaking New Ground.* Seattle: University of Washington Press, 1972. (First published by Harcourt Brace, New York, in 1947.)

—————. *The Fight for Conservation.* Seattle: University of Washington Press, 1967. (First published by Doubleday, Page, New York, in 1910.)

Shands, William F. and Robert C. Healy. *The Lands Nobody Wanted.* Washington, D.C., Conservation Foundation, 1977.

"Agriculture and the Environment." First Annual Report, Office of Environmental Quality. U.S. Department of Agriculture, 1980.

"One-Third of the Nation's Land." A Report to the President and Congress by the Public Land Law Review Commission. Government Printing Office, 1970.

"The Principal Laws Relating to Forest Service Activities." Agriculture Handbook No. 453. Government Printing Office, Revised 1978.

INDEX

AAAS. *See* American Association
for the Advancement of Science
Abourezk, James, 57
Acid Precipitation Act (1980), 229
Adirondack Forest Preserve, 22
Admiralty Island National
Monument, 74, 87, 144
ADVENT, 52
Afognak Island, 75
Agent Orange, 29
Agricultural Stabilization and
Conservation Service, 272
Agriculture, Department of, 36–37,
45, 129
Alaska, 163, 164
Alaska National Interest Lands
Conservation Act (1980), 74,
211
Alaska Region, 114
Allegheny National Forest, 206–207
Allott, Gordon, 287
Alpine Lakes Wilderness, 98
Al Sarena case, 202–203
AMAX, 211
AMC. *See* Appalachian Mountain
Club
American Association for the
Advancement of Science
(AAAS), 13
American Forestry Association, 13,
80, 308
American Mining Congress, 100,
199
American Paper Institute, 214, 215

American Recreation Coalition, 143,
144
Anderson, Arthur, 59, 60
Andrus, Cecil D., 209, 282
Angeles National Forest, 11
Apalachicola National Forest, 96
Appalachian Mountain Club
(AMC), 52, 142
Appalachian Trail, 35, 140, 142
Arapaho National Forest, 116, 151,
169
Arbor Day, 13
Archaeological Resources Protection
Act (1979), 85
Armstrong, Bob, 164
Ash, Roy, 279
Ashley National Forest, 11
Aspinall, Wayne, 204, 286, 287
Asset Management Program, 83

Back Country Horsemen, 115
Bacon, E. M., 186
Baird, Spencer Fullerton, 18
Ballinger, Richard A., 23, 75, 307
Ballinger Canyon, 58, 148
Bankhead-Jones Farm Tenant Act
(1937), 133–134
Barkley, Alben, 278
Barnes, Irston, 258
Barrett, Frank, 308
Bartram Trail, 142
Bears
black, 167
grizzly, 155, 160

349

habitat, 160
See also Wildlife; Wildlife projects
Behan, Richard W., 290, 292,
  294–295, 302
Benoit, Clifford, 118–119, 120
Berg, Norman A., 61
Bergland, Bob, 37, 59, 269
Berntsen, Carl M., 81
Bighorn National Forest, 80
Bighorn-Weitas Roadless Area, 93
Biltmore, 19
Biltmore Forestry School, 20, 298.
  See also Schenck, Carl A.
Biological Natural Areas, 58. See
  also Research Natural Areas
Bitterroot National Forest, 5, 29,
  35, 93, 112, 119, 139
Black Hills Forest Reserve, 102
Blacks, 64, 65, 67, 68–70
BLM. See Bureau of Land
  Management
Block, John R., 61, 84
Bob Marshall Wilderness Area, 43,
  78, 198–200, photo section
Boise Cascade Corporation, 296
Boise National Forest, 107, 139
Bolle, Arnold W., 5
Boone and Crockett Club, 12, 75
BOR. See Bureau of Outdoor
  Recreation
Borden, Tom, 239
Border Grizzly Project, 160
Bormann, F. Herbert, 225
Bostick, Vernon, 108, 300
Bottrell, Dale R., 248–249
Boundary Water Canoe Area
  Wilderness (BWCA), 140, 149,
  154, 178, 183, 197, 205, 224,
  291, 295
Bowaters (company), 234
Boyer, Don, 120
Brady, Bobby, 77
Brandborg, Guy M., 35–36, 44, 300
Brandeis, Louis D., 23
Brandy Peak, 36
Brett, Richard M., 215
Bridger-Teton National Forest, 19,
  80–81, 116, 137–138, 208–211,
  275
Brower, David, 181
Brown, George, 89

Brownlow Report, 277
Budget, Forest Service, 37, 46–50,
  115, 121
  appropriation of funds to major
    branches, 46
  subsidy system for project
    funding, 112–113, 116
Bureau of Indian Affairs, 106
Bureau of Land Management
  (BLM), 106, 129, 130, 183, 195,
  280, 282, 297
Bureau of Outdoor Recreation
  (BOR), 273, 274, 275, 276, 283
Burk, Dale, 36, 112
Burlington Friends of Safe Energy,
  215
Butz, Earl L., 30, 37
BWCA. See Boundary Waters Canoe
  Area Wilderness

Cache Creek, Wyo., 209–210
California Citizens Committee for
  the San Rafael Wilderness, 185
California Forest Practices Act
  (1973), 238
California Redwood Company, 17
Campbell, Robert, 258
Camp Roosevelt, 27
Carhart, Arthur, 24, 178, 197
Caribbean National Forest, 266
Carson National Forest, 74
Carter, James E., 7, 60, 96, 146,
  175, 188, 195, 249, 282
  administration of, 47, 175
Cartwright, Charles, 67
CCC. See Civilian Conservation
  Corps
Cedar-Bassett Action Group, 189
CEQ. See Council on
  Environmental Quality
CGG. See Consolidated Georex
  Geophysics
Chadwick, Douglas and Elizabeth,
  161
Challis National Forest, 139
Cheek, Roland, 115
Cheney, Dick, 206
China, People's Republic of, 303
Chittenden County Committee on
  Energy, 215
Christiansen, Neils B., 301

Chugach National Forest, 74
Church, Frank, 30
Church Committee report on
    clearcutting, 37
Cimarron National Grassland, 134
Citizens for Alternatives to Toxic
    Herbicides, 296
Civilian Conservation Corps (CCC),
    64, 65. See also Employment
Civil Service
    and education, 299–300
    See also Civil Service Reform Act
    (1978); Education
Civil Service Reform Act (1978),
    57, 59, 61–62
Clapp, Earle H., 24, 27, 221, 278
Clarke-McNary Act (1924), 25, 232,
    233
Classification and Multiple Use Act
    (1964), 183
Clawson, Marion, 114–115
Clean Air Act, 29
Clearcutting. See Forest
    management
Clearwater National Forest, 93
Clepper, Henry, 299
Cleveland, Grover, 19
Cleveland National Forest, 11
Cliff, Edward P., 5, 7, 28, 52, 80,
    81, 109, 111, 117, 130, 187,
    274, photo section
Coastal Zone Management Act, 29
Coconino National Forest, 21, 22,
    85, 203
Cohutta Wilderness, 188
Colorado wilderness study, 193
Committee for Rational Use of Our
    Forests, 214–215
Committee of Scientists, 93–95
Common Varieties Act (1955). See
    Multiple Use Mining Act
    (1955)
Computers, use by Forest Service,
    52–53, 169
    Forest Planning Model
    (FORPLAN), 52, 53, 243, 294
    Recreation Information
    Management (RIM), 52
Condor project, 158
Connaughton, Charles, 186

Conservation and Rehabilitation
    Program on Military and
    Public Lands Act (1974). See
    Sikes Act
Conservation of natural resources,
    1, 19, 20, 28, 29, 134, 170–171,
    177, 212–218, 243
Consolidated Georex Geophysics
    (CGG), 199, 200
Continental Divide Trail, 142
Cooper, Arthur, 93–94
Cooperative Farm Forestry Act
    (1937), 233
Cooperative Forest Management
    Act (1950), 233
Cooperative Forestry Assistance Act
    (1978), 221, 238–239
Cooperative Forestry Extension Act
    (1978), 238
Copeland Report, 26
Copper Mountain, Colo., 151–152
Coronado National Forest, 51
Cost of Living Council, 88
Coston, Charles T., 32, 96, 200
Council on Environmental Quality
    (CEQ), 30, 47, 50, 95, 146,
    147, 223, 248–249, 266, 281
Cradle of Forestry, 36
Crafts, Edward C., 5, 28, 111, 273
Crawford, Kenneth, 25
Crested Butte, Colo., 211
Crowell, John B., Jr., 8, 9, 47, 48,
    105, 165, 195, 270
Crystal Mountain, 150
Cushwa, Charles T., 224
Custer National Forest, 137–138
Cutler, M. Rupert, 8, 37, 147,
    167–168, 269, 271, 293

Daly, Marcus, 16
Dana, Samuel T., 182
Daniel Boone Trail, 142
Davis, Clinton L., 26, 36, photo
    section
Davis, George D., 200
DDT, 225, 250, 257, 260, 281
DeBell, Dean S., 227
Decentralization and organization
    of Forest Service, 34
"De facto" wilderness, 183, 186,
    187, 291. See also Roadless

Area Review and Evaluation (RARE I); Roadless Area Review and Evaluation (RARE II); Wilderness
Delaware State Forest, 240–241
Desert Land Act (1877), 17
DeVoto, Bernard, 35
Dickerman, M. B., 222
Dierks Company, 235
Dochinger, Leon S., 228–229
Domestic Council, 50
Douglas, William O., 5, 290, photo section
Dow Chemical Company, 29, 250, 251
Duerr, William A., 301
Duscha, Julius, 54

Eastern Region, 40, 69, 114
Eastern Wilderness Act (1975), 187. See also Wilderness; Wilderness Act (1964)
East Meadow Creek, 186
Economic influences on natural resources, 283, 284, 292. See also Forestry, economics of
Education, 20, 54, 234
  accreditation of universities, 304
  of biologists, 168
  curriculum needs, 300–304
  early history, 298–304
  of minorities, 68, 305–306
  opinions of educators, 301–306
  of recreation managers, 152, 153, 302–303
  relevance of, 300–306
  requirements, 54
  role of Forest Service in, 299–300
  universities, 298–306
  of women, 68, 305
Ehrenreich, John H., 302–303
Ehrlich, Anne and Paul, 170, 264
Eisenhower, Dwight D., 33
El Capitan Wilderness, 297
Emergency Jobs and Unemployment Assistance Act (1974), 65
Emerson, Ralph Waldo, 13, 175
Employment, related to natural resources, 63–69, 116
  equal opportunity employment, 65

Job Corps, 64
Senior Community Service Employment Program, 65
Youth Conservation Corps, 64, 65
See also Personnel management; Roosevelt, Franklin D.
Energy, 136, 296
  conservation of, 215–218
  development of resources for, 135
  exploration and conflicts with wilderness, 198–212
  "National Energy Program for Forestry," 213–214
  from wood, 212–215
Energy Security Act (1980), 201
Environmental Defense Fund, 258
Environmental Protection Agency (EPA), 47, 50, 279–282
EPA. See Environmental Protection Agency
Evans, Brock, 190
Evans, John, 84
Even-aged management, 173. See also Forest management; Clearcutting
Exxon, 208

FCCC. See Fort Collins Computer Center
Federal Insecticide, Fungicide, and Rodenticide Act, 29
Federal Land Policy and Management Act (1976) (FLPMA), 90–92, 126, 128, 129, 130, 132, 183, 200, 211, 290, 294, 296, 297
  application of, 200
Federal Timber Purchases Act (1799), 15
Federal Water Pollution Control Act, 29, 119
Ferguson, A. Claude, 57–58, 84
Fernow, Bernhard E., 14, 19, 20, 119, 221, 298–299
Feuchter, Roy, 152–153, 154
Finney, Don, 212
Fire
  significance of, 26
  "10 a.m. policy," 85

Fish and Wildlife Conservation Act
(1980), 162
Fish and Wildlife Service (FWS),
129, 158, 159
Fisheries, 47, 48, 107, 118, 123–124,
156, 157, 158, 164
anadromous systems, 164, 165
management techniques, 164,
165, 166
Fitzgerald, A. Ernest, 57
Flaming Gorge National Recreation
Area, 275
Flamm, Barry, 269
Flathead National Forest, 160, 200
Fleishman, William, 251–252
Florida Trail, 142
FLPMA. See Federal Land Policy
and Management Act
Food and Agriculture Act (1977),
221–222
Food and Agriculture Organization,
265
Forest and Rangeland Renewable
Resources Planning Act (1974).
See Resources Planning Act
(1974)
Forest Environment Research
Group, 222–223
Forest management, 4
all-aged forest management, 108,
117, 169, 171
clearcutting, 5, 8, 30, 104, 106,
108–112, 117, 163, 173
economics of, 4, 7, 103, 104, 112,
113–115, 117, 120
multiple-use management, 5, 8,
28, 30, 75–82, 117, 118, 170,
171, 177, 180, 182, 189
social aspects of, 8, 50
sustained-yield management, 111,
178, 189
See also Forestry
Forest Pest Control Act (1947), 233,
249
Forest planning, 100–101
Forest Practices Act (1973) (Calif.),
29
Forest Products Laboratory, 22, 45,
180, 216–218, 220–221, 265
Forest Reserve Act (1891), 19
Forest Resource Plan (Pa.), 243–244

Forestry, 4, 5, 6
agronomy approach, 109
cost-efficiency of, 116–117
economics of, 3, 5, 46, 112, 113
environmental responsibilities, 4,
5
future of, 307–312
institution of, 309–310
international, 266–268
natural approach, 109
role of, 310–312
tropical, 1, 2, 6, 159, 266–268
urban, 244–247
See also Forest management;
History of forestry; Industrial
Forestry; Private forestry; State
forestry
Forestry, Division of, 14, 19. See
also Forest Reserve Act (1891)
Fort Collins Computer Center
(FCCC), 52
FORPLAN. See Computers, use by
Forest Service, Forest Planning
Model
Fort Valley, Ariz., experiment
station, 22
Foster, Charles H. W., 301
Francis Marion National Forest, 159
Franklin, Jerry F., 172, 227
Freeman, Orville L., 29, 37, 183,
204, 278
Fremont, John C., 220
Friedman, Janet, 85
Fuelwood, 244. See also Energy
FWS. See Fish and Wildlife Service

Gallatin National Forest, 137–138
Galston, Arthur, 251
GAO. See General Accounting
Office
Garfield, James A., 20, 23
General Accounting Office (GAO),
104, 106, 136, 228, 272
General Land Office, 16, 203
Geological Society of America, 146
George Washington National Forest,
27, 97
Gibbs, Wolcott, 13
Gifford Pinchot National Forest, 74
Gila National Forest, 24, 51, 127,
179

wilderness, 190
Giltmier, James W., 294–295
Gisborne, Harry T., 85
Glass, Norman R., 229
Golden Trout Wilderness, 158
Graham, Frank, 258
Grande Ronde River system report,
124–125
Grand Mesa National Forest, 43
Grand Teton National Park,
137–138
Graves, Henry Solon, 12, 20, 23,
24, 37, 85, 299
Gray, John L., 301–302
Grazing, 13, 21, 22, 123, 127–128,
179
regulation of, 126–129
suitability of lands, 130–131
See also Overgrazing; Rangelands;
Range management
Great Bear Wilderness, 199, 200
Greeley, Arthur W., 184
Greeley, William B., 25, 179, 277
Green Mountain National Forest,
97–98
Grinnell, George Bird, 75
Grizzly bears, 155. See also Bears
Gunnison National Forest, 43, 169,
170, 211
Guttenberg, Sam, 301
Gypsy moth, 257–260

Habitat. See Wildlife
Hagenstein, William D., 82
Hahn, T. Marshall, 214
Haigh, John A., 90
Hair, Jay D., 230
Hamil, David A., 270
Hamilton, Thomas E., 100
Hammatt, R. F., 34, 35
Harris, Robert W., 227
Hartgraves, Charles R., 95
Hayakawa, S. I., 10
Height, Dorothy, 69
Hells Canyon, 290
Helms, Jesse, 10
Henry, Joseph, 18
Herbicides, 29, 119. See also Pest
management
Heritage Conservation and
Recreation Service, 276

Hetch Hetchy Valley controversy,
22–23
Hibbert, A. R., 119
Hill, Charles, 167
Hispanics, 64, 65, 67
History of forestry, 12, 13, 14, 15,
16, 17
Arbor Day (1872), 13
cattle syndicates, 13
Division of Forestry, 14
"land-grabbers," 16, 17
railroads, 13
trust busters, 12
See also American Forestry
Association; Graves, Henry
Solon; Pinchot, Gifford;
Roosevelt, Franklin D.;
Roosevelt, Theodore
Homestake Mining Company, 16,
102
Homestead Act (1863), 16, 17
Hoosier National Forest, 57
Hossack, John, 67
Hoover, Herbert, 83
Hough, Franklin B., 13, 14, 18, 22,
119, 219
Hronek, Bruce, 132–133
Hubbard Brook Experimental
Center, 225–227
Hughes, Jay M., 305
Humphrey, Hubert, 88, 174
Hurst, William D., 131
Hutchison, S. Blair, 222, 227
Hyde, William F., 3
Hydrology, 118, 119. See also
Watershed

Ickes, Harold L., 27, 36, 276, 277,
278
Idaho Batholith, 107
Idaho Environmental Council,
92–93, 210
Idaho Wildlife Federation, 107–108
Indiana Forest Resource Plan, 239
Industrial forestry, 232, 233
land ownership, 234, 235
mechanization, 233
political influence, 234, 235
stewardship of resources, 234
See also Logging industry;
Lumber industry; Timber

Industry, related to forest resources, 102–103. *See also* Industrial forestry; Logging industry; Lumber industry; Mining; Timber
Inland Waterways Commission, 22
Insect infestations, 112, 160, 248–249, 257–263
  control, 172, 248–263
Institute of Tropical Forestry, 45, 266
Interagency relationships, 269–282
  conflict, 271, 276–279
  with Department of Interior, 272
  Wild and Scenic Rivers, 275
  See also Land and Water Conservation Fund; Outdoor Recreation Resources Review Commission
Intergovernmental Personnel Act (1970) (IPA), 51
Intermountain Region, 40, 114
International Paper, 234
International Union for the Conservation of Nature, 223
International Union of Forest Research Organizations, 265
IPA. *See* Intergovernmental Personnel Act (1970)
IPM. *See* Pest management, integrated
Irby, Charles H., 67
Iverson, Floyd, 40
Izaak Walton League, 57–58, 80, 137, 138, 205, 290–291, 308

Jack Creek timber sale, 116
Jackson Hole report, 137
Jaquish, E. Delmar, 68
Jayne, Benjamin A., 303
Jefferson, Thomas, 244
Job Corps, 64
John Muir Wilderness, 194
Johnson, Evert W., 182
Johnson, Huey, 6, 189, 195
Johnson, John, 221
Johnson, Lyndon B., 29, 174
Jones, Dale, 156
Jones, Otis, 68
Journal of Forestry, 85–86, 114–115, 118, 290, 294–295

Jurisdiction, 1, 2
  mission of Forest Service, 5, 6

Kalm, Peter, 219
Kansas-Nebraska Natural Gas Co., 296
Kaufman, Herbert, 36
Kay, Jane, 251–252
Kelley, Evan W., 125
Kellogg, Anita, 51
Kennedy, John F., 53, 284
Kirkness, Walter, 107
Klamath National Forest, 158
Kneipp, L. F., 179
Knutson-Vandenberg Act (K-V Funds) (1930), 103, 112, 162
Koford, Carl, 158
Kootenai National Forest, 160
Kotok, E. I., 34, 35
Krug, J. A., 210
Krutilla, John F., 90
Kuron, 250

Land and Water Conservation Fund (LWCF), 47, 273, 275, 286
Land sales, 10
Lappé, Frances Moore, 136–137
Larson, Geri B., 67, 68
Larson, L. Keville, 6, 236–237
Lawson, Francis, 257
Lawsuits concerning natural resources, 290, 291, 293, 294, 295
  appeals, 291, 292
  See also Monongahela
Lea Act (1940), 233
Leahy, Patrick, 56
Leisz, Douglas R., 53
Leopold, Aldo, 24, 85, 157, 178–180. *See also* Gila National Forest; Wilderness; Wildlife
Lewis, Chuck, 152
Lewis and Clark National Forest, 200
Lewis and Clark Trail, 142
Likens, Gene E., 225–226, 228
Lincoln National Forest, 296–297
Little Missouri National Grasslands, 135
Livestock industry, 126, 129, 132, 135, 137. *See also* Grazing; Overgrazing

Logging industry, 157, 172, 178,
244
history, 232, 233, 242–243
impacts of, 164, 165
Lolo National Forest, 59, 92, 200,
294
Los Padres National Forest, 11, 58,
74, 201
Lowman, George, 159, 162
Lumber industry, 7, 17, 21. See also
Timber
LWCF. See Land and Water
Conservation Fund

McArdle, Richard A., 28, 33, 39,
76, 307, 311, photo section
McCarran, Pat, 130, 308
MacCleery, Douglas, 8, 9, 94
MacDonald, Charles, 44
McGee, Gale, 81
McGuire, John R., 2, 3, 5, 6, 31,
110, 309
McHenry, Douglas Bruce, 60
Machlis, Gary, 303
McIntire-Stennis Act (1962),
221–222, 300. See also
Education; Research
MacKaye, Benton, 35, 175
McLaughlin, William, 303
McSweeney-McNary Act (1928), 24,
221
Management by objectives, 121
Man and the Biosphere program,
265
Manti-LaSal National Forest, 209
Marsh, George Perkins, 18, 119
Marshall, Robert, 25, 180–181, 198,
277. See also History of
forestry; Wilderness
Mather, Stephen T., 45
Matheson, Scott, 83
Mattoon, M. A., 233
Mealey, Stephen P., 169
Menard, William, 282
Merit Systems Protection Board, 57
Merriam, Harry R., 163
Merriam, Lawrence, 308
Metcalf, Lee, 5
Miller, Henry, 16
Milner, Wendy L., 66
Minard Run Oil Company, 206–207

Minckler, Leon, 117, 236
Mineral King case. See Sierra Club
v. Morton
Mineral King Valley, 150, 271
Mineral Leasing Act (1922),
202–203. See also Mining
Mining, 16, 135
conflicts in wilderness areas,
198–212
leasing, 296, 297
mineral exploration, 135, 195,
196
See also Mining Law (1872)
Mining Law (1872), 201–202, 287
Misty Fjords National Monument,
74, 144, 211
Mitchell, W, 211
Monongahela National Forest, 78,
79–80, 110, 291
controversy, 79–82, 88, 292
Montagne, Monty, 58, 59. See also
Off-road vehicles
Moorman, James W., 291–292
Morgan, Rufus, 195
Mountain goats, 161
Mountain pine beetle, 262–263
Mountain States Legal Foundation,
200–201, 210
Mount Hood National Forest, 99
Mount Jefferson Primitive Area, 182
Mueller, Ernst W., 107
Muir, John, 22, 23, 175, 276
Multiple use management. See
Forest management
Multiple Use Mining Act (1955),
203. See also Mining
Multiple-Use–Sustained-Yield Act
(1960), 28, 30, 31, 77, 87, 88,
93–95, 119, 141, 157, 182, 293.
See also Forest management

NADP. See National Acid Rain
Precipitation Assessment
Program
Nantahala National Forest,
143–144, 207–208, 274
National Academy of Sciences, 19,
229
National Acid Rain Precipitation
Assessment Program (NADP),
229

National Association of
    Manufacturers, 100
National Association of State
    Foresters, 234
National Cooperative Refinery
    Association (NCRA), 209–210
National Council of Negro Women
    (NCNW), 69
National Environmental Policy Act
    (NEPA) (1969), 7, 50, 68–69,
    95–96, 119, 187, 281, 293
National Federation of Federal
    Employees, 53
National Forest Commission, 19, 22
National Forest Development
    Program, 284
National Forest Management Act
    (NFMA) (1976), 1, 7, 8, 9, 31,
    89–90, 96, 103, 115, 119, 141,
    165, 168, 169, 171, 284, 292,
    294. See also Resources
    Planning Act (1974)
National forests, 3, 7, 21, 24, 30,
    74, 126, 127, 133
    forestry practices, 233
    ranger districts, 43
    recreation opportunities in,
    143–144, 151, 154
    See also individual forest names;
    National Forest System
National Forest System, 19, 126,
    175, 231
    effects of timber cuts on, 111,
    120–121
    geographical distribution, 71,
    72–73(map)
    role of, 74–75
    wildlife responsibility, 157
    See also individual forest names;
    National Forests
National grasslands, 28, 126, 133
    historical uses, 134
    land ownership, 135
National Historic Preservation Act
    (1966), 85
National Oceanic and Atmospheric
    Administration (NOAA),
    279–280
National Parks, 33, 150, 156. See
    also individual park names
National Parks Association, 45

National Program of Research for
    Forests and Associated
    Rangelands, 222
National Rural Electric Cooperative
    Association (NRECA), 270
National Timber Supply Act (1969),
    30, 308
    hearings on, 82
National Trails System, 140
    National Recreation Trails, 142
    National Trails System Act
    (1968), 142
    trail clubs, 142
    See also Recreation, outdoor
National Urban and Community
    Forestry Leaders Council, 246
National Wilderness Preservation
    System, 174, 175, 181, 182,
    183, 184, 185
    allocation of wilderness,
    176(chart)
    public role in, 185, 186, 196–197
    See also Roadless Area Review
    and Evaluation (RARE I);
    Roadless Area Review and
    Evaluation (RARE II);
    Wilderness; Wilderness Act
    (1964)
National Wildlife Federation, 80,
    146, 229
Native Americans, 64, 65, 67
Natural Resources Council of Maine
    (NRCM), 260–262
Natural Resources Defense Council
    (NRDC), 88, 113–114
NCHMA. See North Central High
    Mountain Area
NCNW. See National Council of
    Negro Women
NCRA. See National Cooperative
    Refinery Association
NEPA. See National Environmental
    Policy Act (1969)
New England Pilot Fuelwood
    Project, 213
Nezpercé National Forest, 139, 223
NFMA. See National Forest
    Management Act (1976)
Nixon, Richard M., 31, 60, 279–282
NOAA. See National Oceanic and
    Atmospheric Administration

Noonan, Patrick, 274
Norris, George, 278
North Cascades National Park, 181, 275, 286
North Central High Mountain Area (NCHMA) (Pa.), 242
North Country Trail, 142
Northern Region, 117
Northwest Coalition for Alternatives to Pesticides, 262
NRCM. *See* Natural Resources Council of Maine
NRDC. *See* Natural Resources Defense Council
NRECA. *See* National Rural Electric Cooperative Association
Nutrient cycling, 225–230. *See also* Research

Ocala National Forest, 96–97
Odell, Susan, 67
Office of Environmental Coordination, 50
Office of Environmental Quality, 269
Office of Management and Budget (OMB), 37, 40, 88
Off-road vehicles, 57–58, 59, 146–149
O'Hara Creek Natural Area, 223
Older Americans Act (1965), 65
Olmsted, Frederick Law, 299
Olson, George A., 208, 274
Olympic National Park, 27
OMB. *See* Office of Management and Budget
Oregon Forest Practices Act (1971), 238
Oregon Range Evaluation Program (ORVP), 132
Orell, Bernard L., 234
Organic Act (1862), 45
Organic Act (1897), 19, 78–79, 88–89, 103, 118, 204–205
Organization of Forest Service, 32, 33, 38(chart), 41(chart)
  administrative control, 37, 47
  branches of, 37–38, 46
  chief and staff, 37–38
  corruption, 56
  decentralization, 34

headquarters in Washington, 39
  leadership, 33
  major programs, 46
  personnel, 34
  regions, 40
  tradition, 32, 33
ORRRC. *See* Outdoor Recreation Resources Review Commission
Osceola National Forest, 96–97
Otter Creek Wilderness, 188
Outdoor Recreation Resources Review Commission (ORRRC), 141, 273. *See also* Recreation, outdoor
Overgrazing, 21, 44, 123–125, 127, 128, 130
  effect on vegetation, 132–133
  effect on wildlife, 157

Pacific Crest Trail, 140, 142, 275
Palisades, Idaho, 210
Pasayeten Wilderness, 181
Pawnee National Grassland, 134
Payette National Forest, 107, 139
Peace Corps, 265
Penn, William, 15
Pennzoil Company, 74, 206
Personnel management, Forest Service, 34, 40, 48
  biologists, 167–168
  chief and staff, 37, 38
  inspectors, 34
  minorities, 51, 65, 67–68
  opportunity for advancement, 33, 51, 54
  performance standards, 49
  rangers, 11, 34, 43–45, 103, 104, 127
  range technicians, 127
  recreation managers, 152–153
  regional foresters, 40, 43–44
  salaries, 40
  supervisors, 34, 43
  transfers, 45
  women, 51, 65–67
Peshtigo, Wisc., fire, 13, 16
Pest management, 248–263
  biological control, 262–263
  chemical controls, 248–253
  forest losses to insects, 250–251
  integrated (IPM), 249–263

and public health, 251–252
pesticides use on national forest,
    253–257
    See also Insect infestations
Peterson, Ervin L., 271
Peterson, R. Max, 31, 51, 57, 94,
    107, 113–114, 147, 153, 156,
    200, photo section
Petroleum exploration, 135,
    192–212, 296
Petroleum Exploration, Inc., 296
Pettigrew Amendment. See Organic
    Act (1897)
Picture Rocks National Lakeshore,
    275
Pierce, Robert S., 226
Pinchot, Gifford, 12, 19, 21, 22, 23,
    33, 34, 37, 39, 45, 52, 76, 89,
    108, 119, 128, 175–176, 177,
    232, 241, 242, 265, 276–277,
    298–299, 310, photo section
Pinchot Institute for Conservation
    Studies, 36
Pint, William E., Jr., 132–133
Pisgah National Forest, 24, 167,
    207–208
Pocono Mountains, 240–241, 260
Politics, 60–62, 283–297
    budgets, 283, 284, 289
    committees, congressional,
        287–288
    and the executive branch, 289
    "Iron Triangle," 289
    legislation, 286
    proposals, 289
    See also Carter, James E.; Nixon,
        Richard M.; Reagan, Ronald
Pooler, Frank C. W., 179
Powell, Major John Wesley, 18
Power, Thomas W., 113
Predator control, 287
Preservation, 21, 22, 58, 155, 177,
    183, 189, 243
President's Council on Executive
    Organization, 279
Presley, Jerry J., 239, 240
Price, Overton, 12
Primitive areas, 183
    See also Wilderness
Private forestry, 24, 25, 114, 232,
    235–237

landowners' classification,
    235(chart), 236(chart), 237
practices, 233, 234
timber production, 237
wildlife considerations, 236
    See also State and Private
        Forestry Branch; State forestry
Property Review Board, 10
Public involvement, 35, 36, 96,
    98–99, 289, 290, 291
Public Land Law Review
    Commission (PLLRC), 204–205,
    272–273, 279, 286
Public Lands Commission, 128
Public participation. See Public
    involvement
Puuri, Carl R., 118

Quaker State, 206

Rahm, Neil, 5, 112
Railroads, influence on natural
    resources, 13, 16
Randolph, Jennings, 79, 88–90
Rangelands, 123, 137–138
    consolidation of ranches, 131
    history of use, 123, 126, 128, 134
    improvement program for, 129,
        132
    inappropriate use, 136
    See also Bureau of Land
        Management; Grazing; Taylor
        Grazing Act (1934)
Range management, 124, 125, 126,
    127
    funding for programs, 132
    goals of, 127
    grazing fees, 126, 129, 130
    overstocking, 123, 124, 127
    permit systems, 124, 126,
        128–129
    restoration of overgrazed lands,
        129, 130
    stocking methods, 131
    vegetative management, 132
    See also Grazing; National
        grasslands; Overgrazing;
        Rangelands
Rate, Hank, 55–56, 189–190
Raven, Peter H., 132

REA. *See* Rural Electrification
     Administration
Reagan, Ronald, 60, 83, 86, 283
     administration of, 7, 8, 9, 10, 47,
          60, 205
     personal ranch, 74
Recreation, outdoor, 28, 47, 116,
          117, 140, 242
     American Recreation Coalition,
          144, 145
     benefits of, 143, 144, 152, 153,
          154
     conflicts among recreationists,
          149–154, 314
     crowding, 145
     facilities development, 150, 151
     funding of, 143, 145, 148–149
     history of, 141
     impacts of, 145, 147, 150
     "Operation Outdoors," 28
     participation in, 140
     planners, 151
     regulation of, 145, 152, 154
     skiing, downhill, 149
     trails, 140, 142
     user fees, 143, 145
     See also National Forests;
          National Trails System; Off-
          road vehicles; Outdoor
          Recreation Resources Review
          Commission
Redwood National Park, 17
Reforestation, 9, 112, 115, 238. *See
     also* Forest management
Regions, 40
     United States Department of
          Agriculture Forest Service
          regions, 42(map)
Regulations of forestry practices,
          25, 26, 238. *See also* Forest
          management; *Use Book*
Reich, Charles A., 284
Reidel, Carl, 234, 303–304
Renewable Resources Extension Act
          (1978), 221
Renewable Resources Research Act
          (1978), 221
Research, 45
     budget and scope, 220
     at forestry schools, 221–223
     "Information and Education," 45

regional experiment stations, 45
     research branch of Forest Service,
          2
     use of, 225–230
Research natural areas (RNA),
          223–224
Resettlement Administration, 134
Resler, Rexford A., 61
Resources for the Future, 115
Resources Planning Act (1974)
          (RPA), 7, 31, 88–93, 96–101,
          119, 171, 194, 238, 284, 290,
          294, 295. *See also* National
          Forest Management Act (1976)
Reuss Amendment, 270–271
Reynolds, Richard T., 172
RIM. *See* Computers, use by Forest
          Service, Recreation Information
          Management
River of No Return Wilderness,
          139. *See also* Wilderness
Roadless Area Review and
          Evaluation (RARE I), 10, 187.
          See also Roadless Area Review
          and Evaluation II
Roadless Area Review and
          Evaluation II (RARE II), 10,
          194, 195
Roads, 5, 106, 112, 172
     costs of construction, 103
     design, 104, 106
     impacts of, 120, 161, 162
     subsidies for construction costs,
          112–113
Roberts, Patricia L., 199
Robertson, F. Dale, 120–121
Robinson, Gordon, 293
Rocky Mountain National Park, 277
Rocky Mountain Region, 43, 114,
          155–156, 168–171, 209
     budget, 168
     wildlife management, 168
     See also Regions
Rogue River National Forest,
          202–203
Roosevelt, Franklin D., 26, 63, 134
     Civilian Conservation Corps, 27,
          64
     New Deal, 27
     public work programs, 63
Roosevelt, Nicholas, 82

Roosevelt, Theodore, 12, 21, 22,
    45–46, 119, 265, photo section
Roosevelt National Forest, 116, 169
RPA. *See* Resources Planning Act
    (1974)
Ruckelshaus, William, 235
Rupp, Craig, 6, 86, 167–168
Rural Electrification Administration
    (REA), 270, 271

Sadler, Russell, 48
SAF. *See* Society of American
    Foresters
Sagebrush Rebellion, 9, 10, 82–84,
    130
Salmon National Forest, 139
Sample, V. Alaric, 101
San Bernardino National Forest, 11,
    194
Sandvig, Earl, 39, 125, 126, 127,
    131
San Rafael Wilderness, 185, 197
Sawmill Improvement Program,
    Forest Service, 216
Sax, Joseph L., 293
Saylor, John P., 174, 285–287
Scapegoat Wilderness, 199–200
Schenck, Carl A., 20, 298–299. *See
    also* Biltmore; Biltmore Forestry
    School; Education
Schlapfer, Theodore, 79
Schlesinger, James, 282
Schurz, Carl, 13
SCS. *See* Soil Conservation Service
Seaton, Fred R., 31
Seliga, Thomas A., 229
Selway-Bitterroot Wilderness, 93
Senior Community Service
    Employment Program, 65
Senior Executive Service, 40, 61
Sequoia–Kings Canyon National
    Park, 150, 271
Sequoia National Forest, 5, 150,
    158, 271
Sespe Sanctuary, 74
Seubert, Patricia, 66
Sevin, 258
SFRP. *See* State Forestry, State
    Forest Resources Planning
    Program
Sheridan, David, 147

Shoecraft, Billee, 251
Shoshone National Forest, 19, 80,
    137–138
Shuler, Jay, 159
Sierra Club, 80, 87–88, 100–101,
    150, 185, 186, 210, 296
Sierra Club v. *Morton*, 5
Sikes Act, 162, 168
Silcox, Ferdinand A., 25, 26, 34–35,
    39, 76, 198, 307
Sipsey Wilderness, 187–188
Sisquoc Sanctuary, 74
Siuslaw policy statement, 121
Skok, Richard A., 68, 300,
    304–305, 306, 309–310
Slinn, Ronald J., 214
Smail, John, 172–173
Smiley, Daniel, 259–260, 262
Smith, Dick, 196–197
Smith, Herbert A., 12
Smith, Lee, 3
Smithsonian Institution, 18
Smoky Bear fire prevention
    campaign, 36
Snoqualmie National Forest, 150
Socially Responsive Management
    (SRM), 50
Society of American Foresters
    (SAF), 5, 20, 25, 213, 237, 304
Soil and Water Resources
    Conservation Act (1977), 90,
    136
Soil Conservation Service (SCS),
    37, 90, 134, 136, 271
Soils, 2, 5, 63, 108, 111–112, 118,
    119, 146
    compaction, 118, 133, 136
    erosion, 2, 5, 118, 120–121, 133,
    134
Sorenson, Pete, 58
Southern Appalachian Multiple Use
    Council, 295
Southern Region, 159, 258
Speculation, timber, 105. *See also*
    Timber
Spotted Owl Management Plan,
    172
Spruce budworm, 260–261, 296
SRM. *See* Socially Responsive
    Management

State and Private Forestry Branch,
    45, 231
  cooperative programs, 231, 235
  funding, 232, 239
  history of, 232
  standards, 232
  See also Private forestry; State
    forestry
State forestry, 232, 234, 237–240,
    241
  cooperative programs, 237, 238
  land management assistance,
    238–240
  management, 234
  Pennsylvania State forests, 240,
    241, 242
  planning, 243–244
  regulations, 238
  State Forest Resources Planning
    Program (SFRP), 239
Stebbins, Robert C., 59, 147
Stewart, James L., 256
Stoddard, Charles, 293
Strauss, Joseph, 207
Stuart, Robert Y., 25, 63
Stull, Robert J., 147
Sullivan, Carl R., 164
Superior National Forest, 178
Swan, K. D., photo section
Swigart, Ray, 127
Synfuels Bill. See Energy Security
    Act (1980)

Taft, William Howard, 23, 241
Targhee National Forest, 137–138
Taylor Grazing Act (1934), 129,
    277. See also Bureau of Land
    Management; Grazing;
    Rangelands
Teeguarden, Dennis E., 301
Thoreau, Henry David, 13, 175,
    244
Thorsen, Del W., 274
Threatened and endangered species,
    155–156, 158, 159, 171
Threatened and Endangered Species
    Act (1973), 157
Three Sisters Primitive Area, 181
Three Sisters Project, 158
Thunder Basin National Grassland,
    135

Timber
  annual allowable cuts, 5, 6, 30,
    111
  appraisal, 103–104
  bidding on, 105
  exports, 111
  harvest, 2, 8, 27, 48, 170
  logging industry, 2, 15, 16, 25,
    27, 30, 39, 106, 107, 157, 178
  management plans, 29, 102
  market, 8, 9, 111
  production, 67
  sales, 47, 102–106, 113–114, 116,
    160, 285
  stand improvement, 103
  supply, 3, 28, 29, 287
  surveys, 104
  sustained yield of, 5, 30
  See also Speculation, timber
Timber and Stone Act (1878), 17
Timber Culture Act (1873), 13, 17
Timber Production War Project, 27
Timber Resources Review, 28
Timber Trespass Act (1831), 15
Tippeconnic, Bob, 67
Tongass National Forest, 74, 87,
    144, 211–212, 291
Tonto National Forest, 29, 123, 128,
    132–133
Towell, William E., 4, 5
Train, Russell E., 81, 281
Transfer Act (1905), 21
Tropical forestry. See Forestry,
    tropical
Tugwell, Rexford Guy, 63
Turner, John, 116
Tuskegee Forest Resources Council,
    68–69
Twelve Mile Strip, 167

Udall, Stewart L., 29, 278
Udall-Conte Resolution, 196
Uncompahgre National Forest, 43
Uniform Land Tenancy Act, 130
Union Carbide, 258
Unions, Forest Service employee,
    53–54. See also Personnel
    management
United Nations Environmental
    Program, 265

U.S. Borax and Chemical
    Corporation, 211–212
Use Book (Forest Service manual),
    21, 52
U.S. Plywood-Champion, 112

Vanderbilt, George Washington, 19,
    24
Vermejo Ranch, 74
Vessey, James K., photo section
Vietnamese, 69
Voigt, Garth K., 228
Volunteers, 51, 52
Volunteers in the National Forests
    Act (1972), 51

Wadsworth, Frank, 159, 266
Wallace, Henry A., 34, 63
Wallop, Malcolm, 210
Walsh, Richard, 193
Wasatch National Forest, 144
Washakie Wilderness, 206
Waters, William E., 250
Watershed, 2, 3, 5, 18, 26, 107, 130
    management of, 118
    water quality, 118
    See also Wild and Scenic Rivers
        Act (1968)
Watershed Protection and Flood
    Prevention Act (1954), 272
Watt, James G., 10, 83, 195, 200,
    205, 276, 296–297
Watts, Lyle, 27, 125
Weaver, James D., 95, 113–114
Weeks Act (1911), 24, 26, 79, 232,
    233
Weinmann, Raymond G., 118
Weitzman, Sidney, 110
Weitzman Report, 49, 50, 110, 309
Weyerhaeuser Company, 234–235
Weyers, Ken, 108
Whiskeytown-Shasta-Trinity
    National Recreation Area, 275
"Whistleblowing," 56–59
White Mountain National Forest,
    52, 145
White River National Forest, 19, 24,
    170, 178, 292
White River Plateau Timberland
    Reserve, 19
Wikstrom, J. H., 227

Wild and Scenic Rivers Act (1968),
    141, 286, 290
Wilde, Jetie B., Jr., 67
Wilderness, 10, 139, 158, 295, 296
    classification of, 185
    conflicts with logging, 186
    conflicts with mining, 198–212
    criteria for, 182–183, 184
    "grandfathered uses," 179, 184
    history of wilderness philosophy,
        175, 177–181, 191
    inventory, 187
    management, 182, 184, 190–191
    in Pacific Northwest, 181
    policy of Forest Service, 191, 192
    public role in, 185–186, 196–197
    "U" regulations, 179, 180
    values of, 192–195
    wild areas, 179, 187
    zoning, 181
Wilderness Act (1964), 10, 21, 28,
    29, 174, 177, 180, 183, 191,
    205, 286
    implementation by Forest Service,
        184–185
    and minerals, 200, 204–212
    See also Wilderness
Wilderness Protection Act (1982),
    195
Wilderness Society, 77, 100–101,
    185, 211, 308
Wildlife, 10, 12, 47, 48, 58, 106,
    107, 129, 134–135, 155, 236,
    241
    cover, 162, 163
    ecosystems, 156, 161–162, 169,
        170, 171
    endangered species, 155
    forage, 162–163
    habitat of, 156, 157, 159, 162,
        169, 171
    indicator species, 169
    management, 165, 166, 168
    nongame, 162, 172, 240
    policy, 166
Wildlife projects, 161, 162
    Condor project, 158
    eastern timber wolf, 158–159
    funding, 162
    golden trout project, 158
    Puerto Rican parrot, 159

rare and endangered birds, 159
    spotted owl, 172
Wildlife Society, Alaska Chapter,
    163–164
Williams, Paula J., 65
Wilshire, Howard G., 148
Wilson, E. O., 311
Wilson, James, 21, 232, photo
    section
Wind River Mountains, 112
Windsor, John, 116
Wirth, Conrad L., 273
Women, in Forest Service, 51,
    65–69
Wood products, 2, 3, 106
    demand for, 2
    foreign trade, 3
    recycling, 4, 217–218
    using residual wood material,
        106–107
Wood Residue Utilization Act
    (1980), 106, 216
Woods, C. N., 125–126

Worf, William A., 190, 191–192
Worm grunting, 77
Wright, Steve E., 190

YACC. See Young Adult
    Conservation Corps
Yard, Robert Stirling, 45
YCC. See Youth Conservation
    Corps
Yellowstone National Park, 18,
    137–138, 175, 206
Yellowstone Timber Reserve, 19
Yosemite National Park, 18, 22, 23
Young Adult Conservation Corps
    (YACC), 65
Youngs, R. L., 216, 228
Youth Conservation Corps (YCC),
    64, 65. See also Civilian
    Conservation Corps
Yurich, Steve, 256

Zahniser, Howard, 177, 181
Zeroth, Elaine, 170
Zon, Raphael, 12, 14

About the Book and Author

The Forest Service
Michael Frome

The Forest Service directly administers 8 percent of the nation's land mass, and its authority touches all aspects of forestry, conservation, and rangeland use. The largest agency in the Department of Agriculture, it is responsible for the delicate task of balancing the tremendous requirements for wood products, minerals, and other natural resources against the equally strong need for recreation areas and natural wildlife habitats.

In this astute, sympathetic, yet constructively critical assessment of the Forest Service, Michael Frome sketches its growth from its inception under pioneering conservationist Gifford Pinchot to its current position as an agency struggling to meet the often conflicting demands of environmentalists, recreation groups, private industry, and the general public. He also traces the history of the U.S. attitude toward its forests, examines global forestry concerns, and looks at the programs designed to deal with human, as well as natural, resources in the future.

This is a fully revised and expanded version of Mr. Frome's 1971 book on the Forest Service. It includes updated chapters on mining laws, timber harvesting, wilderness issues, and environmental forestry, adds new chapters on deforestation, the impact of government energy programs on national forests, and endangered wildlife and plant life, and offers a constructive assessment of national resources education.

Michael Frome, author, conservationist, and critic, currently serves as visiting professor at the College of Forestry of the University of Idaho. His books include *Whose Woods These Are, Strangers in High Places,* and *Battle for the Wilderness.* Much of this book was written while he was author in residence at the Pinchot Institute for Conservation Studies.

Westview Library of Federal Departments, Agencies, and Systems

Ernest S. Griffith and Hugh L. Elsbree, General Editors

†*The Library of Congress*, Second Edition, Charles A. Goodrum and Helen W. Dalrymple

†*The National Park Service*, Second Edition, William C. Everhart, with a Foreword by Russell E. Dickenson

†*The Smithsonian Institution*, Second Edition, Paul H. Oehser; Louise Heskett, Research Associate; with a Foreword by S. Dillon Ripley

The Bureau of Indian Affairs, Theodore W. Taylor, with a Foreword by Phillip Martin

The Forest Service, Second Edition, Michael Frome, with a Foreword by Carl Reidel and an Afterword by R. Max Peterson

†Available in hardcover and paperback.